P9-DKE-845

# THE LOBOTOMIST

## A MAVERICK MEDICAL GENIUS AND HIS TRAGIC QUEST TO RID THE WORLD OF MENTAL ILLNESS

JACK EL-HAI

JOHN WILEY & SONS, INC.

In memory of Sam El-Hai



Published by John Wiley & Sons, Inc., Hoboken, New Jersey
Published simultaneously in Canada

Photo credits: page iv, courtesy of George Washington University Archives; page vi, courtesy of Franklin Freeman.



**Library of Congress Cataloging-in-Publication Data:**

El-Hai, Jack.
The lobotomist : a maverick medical genius and his tragic quest to rid the world of mental illness / Jack El-Hai.
p. cm.
Includes bibliographical references and index.
ISBN 0-471-23292-0 (cloth)
1. Freeman, Walter, 1895–1972. 2. Neurosurgeons—United States—Biography. 3. Frontal lobotomy—History. 4. Depression, Mental—Surgery—History. I. Title.

RD594.F74E4 2005
617.4'8'092—dc22 2004014946
(Freeman)

Printed in the United States of America
10 9 8 7 6 5 4 3 2 1

Canst thou not minister to a mind diseas'd,
Pluck from the memory a rooted sorrow,
Raze out the written troubles of the brain,
And with some sweet oblivious antidote
Cleanse the stuff'd bosom of that perilous stuff
Which weighs upon the heart?

—William Shakespeare, *Macbeth*

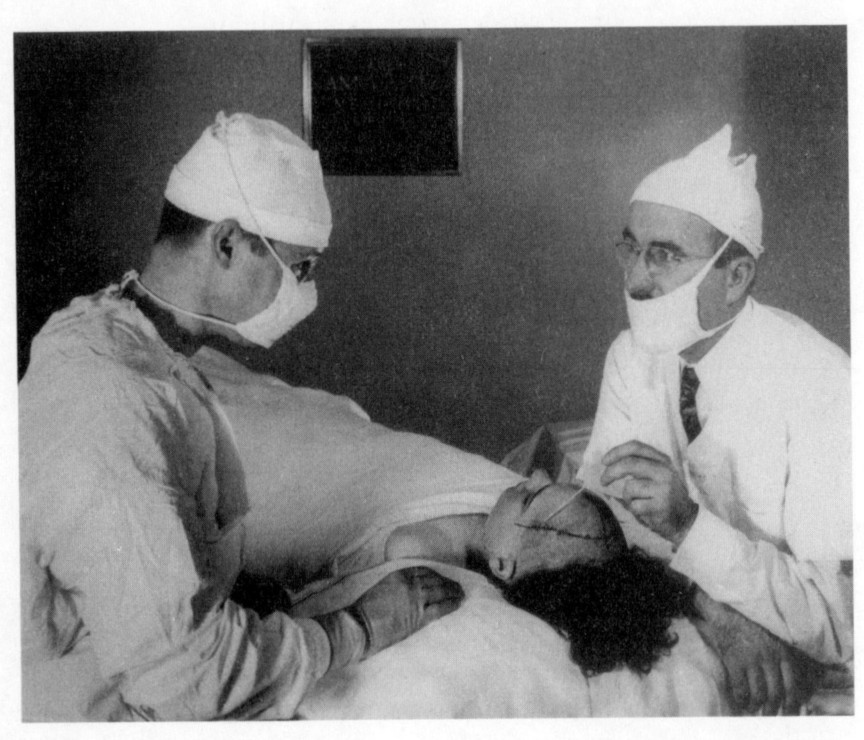

Walter Freeman (at right) with the neurosurgeon James Watts and a lobotomy patient, 1942.

# Contents

Walter Freeman after deep sea fishing in Florida, 1940.

# Prologue

Aside from the Nazi doctor Josef Mengele, Walter Freeman ranks as the most scorned physician of the twentieth century. The operation Freeman refined and promoted, lobotomy, still maintains a uniquely infamous position in the public mind nearly seventy years after its introduction and a quarter-century past its disappearance. The name of the surgery itself—a term Freeman and his partner, James Watts, coined to describe the cutting of the frontal lobes of the brain to relieve psychiatric disorders—produces a discomfort even stronger than other antiquated medical terms such as vivisection and bloodletting. When I tell people that I have been working on a book about a man who performed nearly thirty-five hundred lobotomies, including the first such surgery in the United States, I often see distress in their faces. I can almost see the images flashing in their minds: the filthy back wards in psychiatric hospitals of decades past, sick people in restraints, sharp instruments violating the brain, the vacant eyes and gibbering mouths of permanently damaged patients. I know they involuntarily summon these images promulgated by the movies and popular literature because I did the same in the early months of my research.

Perhaps what best evidences the intensity of the discomfort that the word *lobotomy* holds for many people is the alacrity with which rebellious youth and humorists have appropriated the term. In today's popular culture, lobotomy usually pops up not as the term for an obsolete treatment for the mentally ill but as a verbal fillip in song titles and the names of rock bands or as a mock explanation, hoisted up for laughs, for someone's stupidity. "I'd rather see a bottle in front of me than a frontal lobotomy," said Tom Waits during a magazine interview many years back—a joke that is now the first phrase that comes to mind for many people when they hear of psychosurgery.

When we turn lobotomy into a word designed to shock or use it as a quick generator of laughs, we make it easier to handle our discomfort with the procedure and its complex history. Trying to grasp the evolution and popularity of lobotomy can numb the thinking as much as can imbibing the mixed drink that bears its name—equal parts of amaretto, Chambord, and pineapple juice, sometimes topped with champagne. From the late 1930s through the mid-1950s, lobotomy thrived in the mainstream of psychiatric practice in North America, South America, Europe, Oceana, and parts of Asia, and it remained an occasionally used treatment into the 1970s. What accounts for the resilience of this apparently barbaric practice? How can we make sense of the attraction of so many physicians and patients toward a procedure that today seems so obviously wrong? One tempting way to approach these daunting questions is to reduce the controversy over psychosurgery to a conflict between good and evil. The opponents of lobotomy represent common sense, compassion, and the advocacy of medical ethics. Lobotomy's proponents stand for a kind of scientific recklessness and madness that overtook the psychiatric profession for years.

Some demonize the procedure and its practitioners, and Walter Freeman, the man most closely associated with lobotomy, has borne the heaviest burden of our condemnation. Today Freeman is widely remembered as a loose cannon who worked beyond the boundaries of accepted medical practice—a man intent on puncturing brains to appease his own personal demons. Many people erroneously believe that Walter Freeman carried a set of gold-plated ice picks and that he lost his license to practice medicine for performing lobotomies. They believe he was possibly insane.

Although I did not realize it at the time, my journey into the life of Walter Freeman began in 1996, when I entered a house in suburban Minneapolis, took a seat in the living room, and faced an elderly man who had been persuaded by a younger relative to tell me about his brother, Richard. The man was upset. Richard suffered from severe epilepsy, a condition that in the 1930s landed him in state institutions alongside people with psychiatric illnesses, mental retardation, organic brain diseases, and other incurable maladies. Richard, the man said, "did not accept his condition. He fought it all the time. That's why he went into the institution. . . . He wanted to do all the things normal people did, like swim and drive, but he wasn't capable." Richard spent nearly all of his adult life in psychiatric hospitals.

Before his death at age seventy-two, Richard experienced a steady mental decline. "The last time I saw him, he didn't remember anything or recognize anybody," the man recalled of his brother. "He was out of the picture, and I assumed he had just degenerated. With my sister's coaching, he recognized me. It broke my heart to see him [like that] all the time." As shocking as Richard's condition had become, his brother told me that he received a bigger jolt after Richard's death. "I didn't know about the lobotomy until recently," he said. That word, *lobotomy*, dropped from his lips awkwardly; it sounded ugly, repulsive. As an institutionalized ward of the state, Richard had undergone psychosurgery, and his brother was convinced that the purpose of the operation was to diminish Richard's complaining and lack of cooperation with the hospital staff.

Years earlier I had watched the scene in the movie *Frances* in which Jessica Lange, playing the actress Frances Farmer, is restrained on a table as a doctor slides a sharp metal tool inside her eyelid and pushes it into her brain. Until I spoke with Richard's brother, however, I never gave much thought to lobotomy. I believed that lobotomies turned people into human vegetables; they were heartless and savage surgeries promoted by hospital administrators like the ones in *One Flew Over the Cuckoo's Nest*, eager to crush the souls and spirits of uncooperative patients.

What I heard of Richard's experience supported my old notions of lobotomy, and they stayed with me for a long time. I held them when I wrote an article about the practice of psychosurgery in the Upper Midwest for a regional medical journal, a project that introduced me to the career of Walter Freeman. I harbored them when I later wrote a short account of Freeman's life for the *Washington Post Magazine*. And I still maintained them when I drew up my proposal for this book. Then I faced the mountain of documents left by Freeman in the wake of his half-century-long career as a neurologist, a psychiatrist, and a lobotomy promoter.

Freeman is a biographer's dream: an engaging writer with a substantial ego who recorded his thoughts in countless books, articles, letters, journals, and memoirs. Although sometimes guarded about his personal feelings, he never feared setting down his professional speculations, no matter how outrageous or controversial. In addition, many of the people who accompanied Freeman throughout his life—his closest medical colleague, James Watts, as well as his family and, most importantly, his patients and their families—often displayed keen insight into the tensions, conflicts, and dilemmas that accompanied the introduction and advancement of psychosurgery. Gradually, as I became better acquainted with the circumstances surrounding the development of lobotomy in the 1930s and the psychiatric environment in which it thrived in later years, I formed a pair of central questions: What accounted for Freeman's attraction to this drastic and damaging form of psychiatric treatment? Why did he stay with it for so long a time, even after most other physicians had abandoned it?

When biographers raise puzzling questions like these, it is their responsibility to find answers. The answers to my questions surprised me, given the images of lobotomy that I carried into the initiation of my research. I soon had to admit that answers have two faces. For one, an answer is a solution, the erasure of the question mark. But an answer can also be a response or a reply in a dialogue that begins once a question is posed. The voices that poured out of the Freeman documents when I asked my questions were overwhelming and deeply moving. I was prepared to hear the responses of Freeman's opponents, the psychiatrists and others who raised their voices in outrage during Freeman's career to declare that lobotomy was mutilating, ineffective in treating mental illness, and possibly criminal. These included some patients and their families, and their objections resounded with familiarity.

Other voices in the documents, though, sang a strangely unfamiliar tune in reply to my queries. Many of the era's most important medical figures—neurosurgeons, neurologists, psychiatrists, physiologists, and others—lent their support to Freeman's work. Medical practitioners of lesser reputation, doctors in private practice and on the staffs of psychiatric institutions, eagerly adopted his techniques. Patients, some of them writing and speaking with astonishing clarity, observed how their lobotomies had changed them. Their spouses, children, siblings, and parents often expressed gratitude for the lobotomies and considered Freeman a member of their extended family.

In short, the documents that occupied me in countless hours of reading and interpreting did not present a unanimous opinion of the medical soundness, ethics, or effectiveness of Walter Freeman's practices as a lobotomist. In the discordant streams of impressions I received, I heard Freeman's colleagues, family members, and patients arguing among themselves. At first prepared to condemn Freeman as a cruel, devious, and unprincipled man, I had to recognize the persuasive evidence that at times he acted in the best interests of his lobotomy patients, given the limitations of the medical environment in which he worked and the perilous nature of scientific innovation. Realizations like these—discoveries that a life holds more gradations of complexity than previously imagined—account for the addictive nature of writing biography. What drove me forward in my Freeman research was not my desire to vindicate the doctor but to understand him. And if I could fashion a narrative that gave me that understanding, could I succeed in answering the main questions that pulled me into Freeman's life at the outset?

In Freeman's last published book, a poorly received volume titled *The Psychiatrist: Personalities and Patterns*, he opens with the question, "What manner of man is the psychiatrist?" He goes on to explain the range of his interest in attempting to delineate the shared characteristics of people who specialize in psychiatry: "And so, in this study, the psychiatrist is examined as a member of the human race, Homo sapiens. He is considered not only in regard to his medical training and experience, but also as a member of the community, with interests in family, in education, in community service, in research. He is considered as a counselor, as a trustee, as a soldier, as a banker, as an administrator, as well as in other roles. Being human, he is subject to the ills of the flesh, to accidents, to emotional disorders, even, too often, to suicide."

Walter Freeman fell prey to some of the ills he listed, and he lived not only as a lobotomist but also as a *Homo sapiens*. He deserves, at the very least, the kind of all-inclusive scrutiny he hoped to give to others. The reader can judge whether this book gives Freeman his due.

# Chapter 1

# September 1936

The unconscious patient was wheeled from an operating room at the George Washington University Hospital in Washington, D.C., to her bed. Nurses elevated her head to reduce bleeding in her brain. They monitored her temperature, blood pressure, and pulse; they noted that the patient had vomited after surgery and could not control her bladder.

Walter Jackson Freeman, a neurologist and a psychiatrist, stood by her at a critically important moment in his career. He leaned over the patient and began a neurological examination. The patient's pupils were responsive and of equal size. Her face showed no asymmetry or other signs of paralysis. Her knee reflex was normal, and she curled her toes inward when Freeman stroked the soles of her feet. So far, the patient appeared to be doing well.

Four hours later, her anesthesia wore off. Alice Hood Hammatt, sixty-three years old, opened her eyes and focused on Freeman. He was then forty, slightly over six feet tall, solidly built, with a receding hairline and a trim goatee that some of his psychiatric patients enjoyed pulling—and Freeman liked having it pulled. His eyeglasses were scratched from frequent contact with the eyepieces of microscopes. Hammatt offered her hand to him in greeting.

Freeman, speaking in a resonant and slightly nasal voice, asked Hammatt how she felt. Her expression placid, she replied that she felt much better. She appeared better, Freeman thought. That evening, Hammatt was able to name her husband, describe his line of work, recite her address, and identify objects in the room.

Freeman returned to Hammatt's bedside the next day and found her alert and sitting up. Again, she greeted him by offering her hand. Curious about her emotional state, he asked if she still felt her old fears. No, she replied. Was she sad or happy? Happy, she declared. Did she remember what exactly had caused her such anxiety in the past? "I seem to have forgotten," she said. "It doesn't seem important now." As she spoke, she "continually rolled a paper handkerchief about in her hands, rubbing it about her face and arms as though drying herself," Freeman observed.

Over several days, Freeman watched Hammatt steadily grow more alert and active. She read magazines and could discuss what she had read. Her appetite was good, and she slept well. Most strikingly, she suffered from little anxiety. This, Freeman believed, represented a substantial improvement over her previous condition. Weeks earlier, Hammatt had come to Freeman's examining room to seek help for insomnia, anxiety, and debilitating depression. Emotional problems and nervous breakdowns had plagued her for most of her life. Freeman diagnosed her with agitated depression and checked her into George Washington University Hospital for observation. Her condition deteriorated. She "showed uncontrollable apprehension, was unable to sleep, laughed and wept hysterically," Freeman noted. Under other circumstances, permanent institutionalization in a mental hospital would have been Alice Hammatt's fate. As a patient of Walter Freeman in September 1936, however, she could choose one other course of treatment.

Now the treatment was over. During his visits to Hammatt's bedside, Freeman closely watched her recovery. She did not care about the shaved areas of her scalp and even showed off her bare patches, when previously the thought of having her hair shorn for surgery drove her to distress. The appearance of her anesthetist, which had made her cry out in fear a few days earlier, produced no apprehension now. When her husband entered her hospital room, Hammatt welcomed him with calmness; when he left, she did not panic.

These responses were exactly what her doctors had hoped to see. Freeman and his partner in Hammatt's treatment, neurosurgeon James Winston Watts, "were congratulating ourselves upon a brilliant result," Freeman declared. What they had just accomplished was the first attempt in American medical history to treat a psychiatric disorder by means of surgery on the frontal lobes of the brain. At their hands, Alice Hammatt had received the first lobotomy in the United States.

In Freeman's eyes, Hammatt's case history had made her an ideal candidate for the experimental procedure. A native of Emporia, Kansas, she was the youngest daughter of pioneer settlers of that state. Her parents spoiled her, and she learned to make her needs known by crying and throwing tantrums. Anxiety and depression left her incapacitated with stomach ulcers and emotional breakdowns.

In her mid-twenties, she married Theodore Dudley Hammatt, an even-keeled and patient man who worked in the Kansas State Department of Agriculture. Her first pregnancy was a nightmare of depression, agitation, and suicidal feelings. The child died at age two, but her next two pregnancies were more bearable, and the children survived. In time, the Hammatts moved to Washington, where her husband accepted a position in the Grain Futures Administration of the U.S. Department of Agriculture. Hammatt's emotional troubles grew worse. A perfectionist as a housekeeper, she distressed her husband and children with her fussiness. The murder-suicide of her sister- and brother-in-law in 1933 added to her stress. She developed a crush on another man, told her husband about it, and became miserable from revealing the infatuation to him. Sometimes she would grimace at herself before a mirror, urinate on the floor, and, standing nude at the windows, expose herself to the neighborhood. In addition, she was a "master at bitching and really led her husband a dog's life," Freeman noted. "She worried if he was a few minutes late in coming from the office and raised the roof when things did not suit her. She was a typical insecure, rigid, emotional, claustrophrenic individual throughout her mature existence." Freeman also found her to be vain, afraid of growing old, and overly concerned about her thinning hair. Alice Hammatt would soon show him how protective of her hair she could be.

For ten years, Hammatt had regularly used strong bromide sedatives to help her sleep. But her insomnia had worsened in 1935 after she and her husband lost their way while driving around New York City. Immediately afterward, "she was highly keyed up and unable to sleep and she had been troubled with sleeping ever since," Freeman noted. Hammatt continued to entertain thoughts of suicide.

In his initial examination of Hammatt, Freeman judged her "rather thin and flabby" but appearing younger than her actual age. She seemed scared and anxious; she wrung her hands and thrashed her limbs during the exam. In a loud and high-pitched voice, she demanded that Freeman explain the purpose of his questions about her general health. After registering her in the hospital for observation, Freeman reexamined a medical journal article written several months before by a physician in

Portugal. The Portuguese doctor, Egas Moniz, described a new treatment he had used on psychiatric patients—and many of those patients had shown symptoms of agitated depression similar to Hammatt's.

The previous year, Moniz and Almeida Lima, a neurosurgeon, had started a series of operations on mentally ill patients in Lisbon. The procedure, which Moniz called leucotomy, involved cutting the neural connections in the prefrontal regions of the brain. Moniz hypothesized that in severing these connections, he was disrupting detrimental emotional responses that had grown habitual in the patients during the course of their disease. Leucotomy, he speculated, forced their brains to develop new neural pathways and more beneficial emotional responses. Freeman, who had read Moniz's descriptions of the procedure in medical journals and corresponded with him, believed that the validity of Moniz's hypothesis "hardly seems to matter since this procedure is reported by Moniz to have cured five out of six patients with agitated depression and to have relieved the sixth to some degree." The results, not the theory, were what counted. Freeman knew that patients with agitated depression—those with the same diagnosis as Alice Hammatt—had in Moniz's account responded better to leucotomy than had patients with other diagnoses.

Since the spring of 1936, when Moniz began publishing his results, Freeman and Watts had been preparing themselves to perform the first leucotomy in the United States. From Paris they had ordered several leucotomes, the surgical tool that Moniz and Lima had used to perform their operations. They had acquired preserved brains from the bodies of cadavers and used the leucotomes to practice cutting the nerve fibers of the prefrontal lobes. Finally, with the appearance of Alice Hammatt, they had a suitable candidate for the surgery. They asked Hammatt and her husband for their consent.

Before making a decision, the Hammatts sought an opinion from their son-in-law, Archibald J. Brier, a physician in Topeka. He, in turn, asked for advice from Karl A. Menninger, soon to become the most famous psychiatrist in the United States. Freeman had known Menninger—as well as his father and brother, psychiatrists who both practiced at the Menninger Clinic in Topeka—for years. "Of course," Menninger later wrote to Freeman, "I was glad to tell him how well we knew you and how highly we regarded you."

With this stamp of approval, the Hammatts agreed to proceed. On the evening before surgery, Alice Hammatt received a preparatory enema. At the last minute, however, she abruptly withdrew her consent to be operated on. It was her hair. She wanted to keep it and balked at having her scalp shaved. "We got around her objection by promising to spare the curls if we could," Freeman wrote. He knew that sparing her hair would be impossible. The next morning, September 14, an anesthetist arrived to give her a dose of Avertin, a common anesthetic that produced unconsciousness. Her anxiety spiked once again. "Who is that man?" she cried. "What does he want here? What's he going to do to me? Tell him to go away." She struggled in bed so wildly, Freeman reported, that the staff "was scarcely able to control her sufficiently to administer the Avertin by rectum." The anesthetist bolstered the Avertin with a dosage of nitrous oxide gas. Then her head was shaved back to the ears.

Meanwhile, Freeman and Watts grappled with the significance of the step they were about to take. "I realized when I did the first operation that I was taking a big risk," Watts recalled. "In other words, . . . I'd been considered by people to be conservative. I knew as soon as I operated on a mental patient and cut into a physically normal brain, I'd be considered radical by some people." But a radical image appealed to his partner. The grandson of a famous surgical innovator, Freeman rushed toward controversy with the enthusiasm of a man determined to make his name as the slayer of a millennia-old scourge to humanity.

Freeman and Watts went to work on Hammatt. Behind their surgical masks, they made an unusual pair: Freeman, whose intent movements and unflagging energy revealed his determination to break new ground in medicine, and Watts, a baby-faced and laconic surgeon whose conservative orientation made him cautious of undertaking a radical operation. Taking turns performing the surgical duties, they closely followed the procedure that Moniz had described. They first cleaned the scalp. Using gentian violet, they daubed the scalp to show the location of the two holes they would cut in the skull. They made incisions 3 centimeters in length into the violet markings, stopped the bleeding with mastoid self-retaining retractors, and, using an auger, made holes in the skull over the left and right frontal lobes. They then inserted a leucotome 4 centimeters straight down through the hole on the left side into the exposed surface of the brain. The tissue had the consistency of warm butter. Pressing a stylet at the top of the instrument caused a wire loop to protrude from the other end. They rotated the loop a full circle and cut a round core of white neural fibers. Next they withdrew the loop back into the leucotome, pulled out the instrument a single centimeter, and cut another core. A third core was scooped at a depth of 2 centimeters.

After removing the leucotome from the brain, they reinserted it into the same hole but at a different angle and cut three more cores. They withdrew the leucotome and inserted the instrument into the hole over the right hemisphere, cutting three more cores, followed by another three at a different angle. The last cut, however, sliced a blood vessel that had been caught in the loop. A hose sucked away the surging blood as they untangled the vessel from the leucotome. After a total of twelve cores had been made, the operation was over. The patient's vital signs showed no indication of distress. The doctors washed the incisions with saline solution, used black silk sutures to seal them up, and bandaged the site. The operation had lasted about an hour.

Freeman felt confident in his diagnosis and selection of Hammatt as the first leucotomy patient. He had faith in their mastery of the technical aspects of the surgery and, most of all, in the outcome. As she lay in bed postsurgery, Alice Hammatt was calm, not obviously impaired mentally, and recovering rapidly. Word got back to Menninger in Topeka that "visitors have found her much improved immediately following the operation." He asked Freeman to fill him in on the details of this new treatment.

Six days after the operation, however, Freeman received an unpleasant jolt when Hammatt suddenly became disoriented and excited, and began stuttering. Although she seemed to understand what others said, words came to her with difficulty. Freeman's examinations showed "symptoms indicative of frontal lobe deficit but no paralysis or disturbance of sensibility." A few days later she was able to page through magazines and draw pictures—she misspelled words in her labels of her drawings—"but she cannot yet write legibly or carry on a conversation." Freeman was concerned that Hammatt seemed "almost too placid," and she resumed the odd "rubbing and rolling movements" she had shown soon after the surgery.

Freeman speculated that swelling or hemorrhage in the areas of the severed fibers were responsible for this setback, and with relief he noted that Hammatt gradually returned to normal with the passage of several more days. He believed it was too soon to determine whether she would show any permanent changes in her personality or brain function, "but the agitation and depression that the patient evinced previous to her operation are relieved," he declared. A quarter-century later, Alice Hammatt's operation still shone brightly in his memory. "The result was spectacular," he wrote.

She soon could speak normally and walk without difficulty. She said she looked forward to leaving the hospital. When Hammatt at last did go home, she was able to sleep without medication and live without a nurse's care. Freeman found that Hammatt could direct the operation of her household, "although her husband and her maid did most of the work. She was rather shrewish and demanding with her husband, outspoken with her friends, and unselfconscious." But Hammatt noticed a distinct change in her level of anxiety. "I can go to the theatre now and not think whether my shoes pinch or what my back hair looks like, but can really concentrate on the show and enjoy it," she said. She worried little, could think without distraction from anxiety, gained enough patience to spend time with a friend whose energy formerly exasperated her, and "was content to grow old gracefully," Freeman observed in her examinations after she left the hospital. "She was well dressed, talked in a low natural tone, volunteered relatively little, but upon questioning showed excellent appreciation of her changed condition." The changes were also noticeable to her husband, who thought she behaved more normally than ever before. He called the next five years the happiest of his wife's life, and they may have been the best of his own, as well.

Freeman and Watts lost little time in reporting Hammatt's treatment to their colleagues. Just a week after her discharge from the hospital, seventeen days after the surgery, they recounted the details of her case to members of the District of Columbia Medical Society. The meeting was one of the most raucous in the organization's history. When Freeman implied that the operation had relieved Hammatt of her symptoms, cries of protest arose from the audience. "Walter, you can't say that!" exclaimed Dexter Bullard, the director of a nearby psychiatric hospital. Others, roaring their agreement with Bullard, thought that the trauma of surgery, not the cutting of neural fibers, might have shocked Hammatt into a temporary remission. But Freeman and Watts were certain of their conclusions. Within a few weeks, their paper on this use of prefrontal lobotomy appeared in the *Medical Annals of the District of Columbia*.

Freeman, speaking later to a meeting of his colleagues in the medical school at George Washington University, explained the effectiveness of the surgery in terms that excluded the possibility of surgical shock and in terms different from those Moniz had used. Relief came to Hammatt, he thought, because the disruption of neural activity left her less distracted by her anxiety and the pressure of her disturbing thoughts. The anxiety was still there, he speculated, but Hammatt simply noticed it less and was thus allowed to direct her thinking along more useful lines. He acknowledged certain side effects resulting from the surgery, most noticeably the disappearance of the patient's spontaneity and an absence of initiative in starting conversations and taking physical action. But Freeman declared that the benefits of the new operation far outweighed any detrimental consequences.

In the seven decades since Alice Hammatt's surgery, research on the brain has shown that Freeman's explanation of the role of the frontal lobes in amplifying anxiety was not far off the mark. Today's neuroscientists see the frontal lobes as a gatekeeper of sensation and a regulator of emotion, the hub of decision making and planning. Humans have the most fully developed frontal lobes in the animal kingdom, and recent studies have demonstrated that people with frontal lobe damage suffer harm to their insight and their recognition of their own defects. By disrupting the links between Hammatt's frontal lobes and other regions of her brain, Freeman and Watts may very well have succeeded in reducing her ability to feel and act upon her anxieties. Unable to develop her anxious feelings into a conscious sensation that demanded socially unacceptable responses, Hammatt may have simply let them drift and fade.

Some months after surgery, Hammatt suffered a convulsion likely related to her surgery, fell, and broke her wrist. Her injury darkened her contentment. For a time, "she became more indolent and sometimes abusive," Freeman noted. But she continued to live with reduced anxiety, and she stayed out of mental hospitals. Five years after her lobotomy—the term Freeman and Watts eventually applied to their psychosurgeries—she contracted pneumonia and died at age sixty-eight on September 28, 1941.

By the time of her death, hundreds of other patients around the world had undergone lobotomies and similar operations. Alice Hammatt's lobotomy was the opening shot of a battle that would convulse the world of psychiatric medicine in the years to come. For a time, Walter Freeman and his allies—a prestigious coalition of psychiatrists, neurologists, and neurosurgeons—would appear victorious in pulling lobotomy into the mainstream of medical practice. In the United States alone, the number of lobotomized patients would soar to about forty thousand over the next four decades, and Freeman would take part in nearly thirty-five hundred of these surgeries. Many patients would receive discharges from psychiatric hospitals, and on their return home would either elate or annoy their families with their changed behavior. Some operations failed and left patients to grapple with profound inertia, debilitatingly childish behavior, convulsions, and incontinence, all on top of their previous psychiatric disorders. In other cases, patients felt relieved of their symptoms and returned to places of responsibility in their families and careers. Eventually, however, the development of new forms of treatment would send lobotomy into a precipitous decline. Throughout it all, Walter Freeman remained a forceful proponent of the procedure. A lover of battle and controversy, he advanced the cause in fights against hospital administrators, Freudian psychoanalysts, and even his closest partner. Only Freeman's own death silenced his advocacy of lobotomy, long after the operation had acquired an overpowering array of opponents.

In September 1936 and the weeks that immediately followed, Freeman merely knew that lobotomy held great promise. Others agreed, sensing that the ground had just shifted in the treatment of mental illness. "I felt somehow that we were in the presence of one of the milestones of modern medicine; I have seldom been more stirred," John Farquhar Fulton, a renowned Yale physiologist who knew both Freeman and Watts, noted in his diary after hearing an account of the operation. He hoped that the two men would "keep their feet on the ground and stay away as far as possible from the publicity the procedure is almost certain to bring. It will throw the psychiatrists into a convulsion and I am sure there are very few open-minded enough to accept the procedure in the spirit in which it is being proposed."

If Fulton ever directly expressed his cautions to Freeman, the neurologist ignored them. By disposition, Freeman felt compelled to work in solitude and to disregard the warnings of others. His upbringing and childhood experiences had taught him to act boldly with little concern for the consequences.

## Chapter 2

# Rittenhouse Square

In the summer of 1890, a small, balding, and energetic American physician named William Williams Keen arrived in Berlin to join dozens of his colleagues from around the world at the Tenth International Medical Congress. The other attendees, some of the brightest minds in late-nineteenth-century medicine, included such pioneers in the study and treatment of the brain as Emil Kraepelin, Valentin Magnan, and E. Victor Horsley. Keen was no underachiever in this illustrious assembly. Then fifty-three years old and a Philadelphian nearly all of his life, he was a renowned surgeon who would later become the president of the American Medical Association. He was a principal speaker at the Berlin conference, and his topic was his technique of tapping the lateral ventricles of the brain to reduce intracranial pressure. The conference had been planned during an earthshaking time in the history of neurological medicine. For several decades, European investigators such as Paul Broca and Carl Wernicke had been working to localize the functions of the many regions of the brain. Which parts of the brain controlled motor functions and which received and translated sensory information? Where were the seats of speech and intelligence? Slowly, through animal experimentation and through research into the experiences of people who had suffered head injuries and tumors, the brain was being mapped.

Keen shared a spot on the Congress program with Gottlieb Burckhardt, a bespectacled and distinguished-looking psychiatrist. Burckhardt directed the Prefargier Asylum, a mental hospital founded a half-century earlier by a snuff merchant in the French-speaking town of Marin, Switzerland. Burckhardt subscribed to the recently developed medical theory that localized regions of the brain controlled specific sensory and physical functions. At the end of 1888, he had begun performing surgery on the brains of patients in order to reduce the severity of their visual and auditory hallucinations, dementia, emotional excitability, and impulsive behavior. He speculated that the brains of these patients processed sensory information abnormally and relayed aberrant signals to the brain's motor centers. It might be possible, Burckhardt thought, to relieve their symptoms "by creating an obstacle" between the sensory and motor regions of the brain. He operated in secrecy on his first such patient, a woman prone to hallucinations and verbal outbursts. Burckhardt sliced out an inch-wide strip of brain tissue weighing 5 grams that extended along the parietal lobe, the section of the brain that lies behind the frontal lobe. Burckhardt later operated three more times on this patient's brain. The procedures calmed her behavior, he said, but her dementia remained unchanged.

Thus encouraged, Burckhardt went on to operate on five more mentally ill patients. One of them, a thirty-one-year-old lithographer identified in Burckhardt's reports as Friedrich August N., was a special case. Like Burckhardt's first patient, Friedrich N. experienced hallucinations and excitability but with the addition of violent behavior, unpredictable movements of the body, and delusions of grandeur. Burckhardt, whose knowledge of brain anatomy was better than that of most of his contemporaries, believed—for reasons unknown today—that Friedrich N.'s psychiatric problems could be traced to abnormalities in his forebrain. So for the first and only time in his brief career as a brain surgeon, Burckhardt targeted a patient's frontal lobes. On April 17, 1889, he cut out sections from the first and second frontal convolutions of Friedrich N.'s brain. Although the patient's behavior calmed, he began suffering seizures.

The highly experimental nature of Burckhardt's use of surgery to treat psychiatric illness made his colleagues uneasy. His report at the Berlin conference startled many who heard it and generated no strong interest. How could Burckhardt be sure he was treating the correct sections of the brain, if indeed injuring any specific part of the brain could relieve psychiatric symptoms? After the conference he returned to Switzerland, where he wrote an eighty-eight-page paper describing his surgeries and their outcomes. When his report met a storm of criticism regarding his destruction of seemingly healthy brain tissue, Burckhardt defended himself by explaining the dire condition of the patients he operated on and their slim chances of ever recovering under traditional treatments. Burckhardt's professional ostracism continued, however, and he never again cut into the brains of his patients. He retired five years after the publication of his paper without further writing on or researching the topic of psychiatric surgery and died, unheralded, in 1907.

Satisfied with the gratifying reception of his own talk, William Williams Keen left the Berlin conference and went home to Philadelphia. This city, then America's third largest, was unusually fertile ground for the study and practice of medicine. It was the home of excellent medical schools, a strong community of physicians, and a great deal of wealth—and therefore the means to support innovative treatments and procedures. Partly for those reasons, Philadelphia has often been considered the birthplace of American neurology and psychiatry, a city in which the treatment of mental disorders and the neurological problems that sometimes cause them had always received more attention than elsewhere. Neurology and to a lesser degree psychiatry were specialties that interested Keen. And in the years to come they would greatly absorb his most prominent descendant, Walter Jackson Freeman.

In 1752 the Pennsylvania Hospital opened in Philadelphia, the first modern medical center in the American colonies. For many years, Benjamin Rush (1746–1813), perhaps the first doctor in the New World to study mental illnesses, directed its psychiatric ward (at the same time that he played a role in the drafting of the U.S. Constitution). Rush believed that mental disorders grew from overactive blood circulation in the brain, and he designed treatments to inhibit and stimulate the circulatory system. He wrote of treating his patients with invigorating exposure to sunlight, fresh air, and physical activity, although visitors to his patients' quarters at the Pennsylvania Hospital sometimes reported squalor and confinement. Rush did not shrink from developing unorthodox devices: one, the "tranquilizer," attempted to reduce circulation of blood in the brain by confining the patient in a chair and setting a box over the head to dull the senses; another, the "gyrator," forced blood into the head by spinning the patient like a propeller. Rush gave his patients alcoholic drinks, medicinal herbs, coffee, tar, and opium in other attempts to modulate circulatory flow. (Rush also gave Meriwether Lewis a crash course in medicine before Lewis and William Clark set off on their famous transcontinental expedition.) His book, *Medical Inquiries and Observations upon the Diseases of the Mind* (1812), was the primary textbook on psychiatric disease for much of the nineteenth century, and in 1965 the American Psychiatric Association recognized Rush as the father of American psychiatry.

In 1885 the nation's first outpatient facility for mental patients opened at the Pennsylvania Hospital. By that time, another Philadelphian named S. Weir Mitchell had made his mark as the leading American neurologist. Mitchell, a Renaissance man who distinguished himself as a speaker, a novelist, and an essayist, devised a "rest cure" therapy for his predominantly upper-class patients with neurasthenia, the nineteenth-century term for debilitating physical exhaustion that lacked an identifiable physical cause. Mitchell believed that neurasthenia had organic, not psychological, roots. Bed rest, electrical stimulation of the nervous system, massage, and a diet high in milk constituted his treatment, which was widely imitated and remained popular through the turn of the twentieth century.

Mitchell was known around the world, although evidently not by sight. During his travels in Europe, he once felt run-down and decided anonymously to consult with the famous French neurologist Jean-Martin Charcot. Charcot examined Mitchell and referred him to an American doctor, none other than S. Weir Mitchell. (The French physician was dumbfounded when Mitchell finally revealed his identity.) His reputation allowed him to critique the practice of medicine, as well as the behavior of its American practitioners, without engendering hostility. As the keynote speaker at the fiftieth annual meeting of the Association of Medical Superintendents of American Institutions for the Insane in 1884, for example, Mitchell chastised the embryonic psychiatric profession for its isolated practice in rural hospitals, its lack of scientific rigor and critical examination, and its preoccupation with institutional red tape. Speaking for all neurologists, Mitchell challenged asylum superintendents and doctors to practice medicine more scientifically. "We think your hospitals are never to be used save as the last resource," he told them.

It was in this city, the Philadelphia of neurological and psychiatric renown, that W. W. Keen, on his return from Berlin, practiced his surgery and raised his family of four girls. He reported on Burckhardt's startling experiments to some of his colleagues, and soon afterward another physican, Emory Lamphear of St. Louis, published a summary of Burckhardt's experiments. It is possible, however, that as the years passed and medical breakthroughs increased in frequency, Keen forgot about Burckhardt's efforts to surgically treat mental illness. On November 14, 1895—within months of the publication of Sigmund Freud's and Josef Breuer's landmark book on hysteria and Wilhelm Roentgen's discovery of X-rays—Corinne

Keen Freeman gave birth to Keen's first grandchild. This baby, Walter Jackson Freeman, would grow up to be a favorite of Keen's. He would also grow up to know a great deal about Gottlieb Burckhardt's experimental work. "It is remarkable," Walter Freeman would write decades later, "that Burckhardt's work in 1888 was not eagerly followed up by the enterprising surgeons of his day."

Walter Jackson Freeman was also the name of the baby's father. A moderately successful otolaryngologist, he had married Corinne Keen soon after completing medical school. His ancestors were seventeenth-century emigrants from England who scattered themselves among the American seaboard colonies. Two prominent physicians, Stephen Camp (a founder of the New Jersey Medical Society, the oldest in the country) and Lewis Morgan, formed part of the New Jersey contingent. In his middle years, the younger Walter Freeman treasured a book of Lewis Morgan's scribbled school notes, which had been passed down through his family.

The senior Freeman was born in 1860 and grew up in Beverly, New Jersey. His father was a railroad executive and his mother was a member of a well-off mercantile family. He initially inclined toward a career as a naturalist, but his parents' disapproval of that profession led him to medicine. As a medical school student in New York City, he met Corinne Keen, grew infatuated, and began taking additional classes in Philadelphia so that he could court her. Commuting by rail between New York and Philadelphia, he completed a double load of course work and received two M.D. degrees. At one point, stress from this heavy workload produced in him a painful three-day-long suppression of his urine.

Perhaps that overload of medical education fatigued his spirit as well as his bladder, for despite his talents as a medical examiner and technical skills in probing the sinuses, the elder Walter Freeman did not love his profession. Unlike his father-in-law Keen, he avoided medical meetings and published few papers. Consequently, his professional reputation was unknown outside of Philadelphia. His relationship with Keen seemed respectful with an occasional pinch of understated teasing. He once admonished Keen after the surgeon removed a 111-pound ovarian tumor from a teenager who weighed 68 pounds after the operation: "Now Father Keen," the son-in-law said, "you must be very careful or some day you might throw away the wrong piece." Another time he listened to a description of his father-in-law's diverticulosis, pondered the situation, and pronounced it "a blowout of the innertube."

On Corinne's side, the family roots extended back to Jurgen Schneeweiss, a native of Saxony whose migrations as a soldier during the Thirty Years War of the seventeenth century and as a settler in America's New Sweden colony gradually transformed his name into Jöran Kyn. Kyn, or Kien, is Scottish for "cows." Kyn's descendants settled in Philadelphia, sold produce to Washington's Continental Army at Valley Forge, and established themselves as a family of merchants. (In taking up medicine, William Williams Keen had bucked the family's commercial tradition.) Born in 1868, Corinne was the oldest of four girls. Her mother, a distant cousin of Lizzie Borden—the infamous Falls River, Massachusetts, woman accused but acquitted of murdering her father and stepmother with hatchet blows to their heads—died in 1886 when Corinne was still a child, and Keen raised the girls. One of Corinne's sisters was Dora Keen, who would grow up to become a noted mountain climber and a teller of her outdoor adventures in the pages of *National Geographic* and other magazines.

Corinne and Walter produced a boy, Walter Jackson Freeman, the first of seven children, who was often unwell. When young Walter was fourteen months old, Grandfather Keen was called upon to excise thirty enlarged lymph nodes from one side of the baby's neck: a procedure that resulted in a permanent paralysis of the trapezius and sternomastoid muscles and a lifelong droop of his shoulder and head. The boy later had a tonsillectomy and endured a bout of diphtheria to become possibly the first person in Philadelphia who was treated with a new diphtheria antitoxin developed in Germany. He also was laid low by the then-common childhood diseases of measles, scarlet fever, whooping cough, mumps, and pinkeye.

Walter thrived despite these infections. His earliest memory was a dramatic and disturbing image: the point of a pickaxe breaking through the wall of his nursery when the neighboring residence was being demolished. Otherwise, Walter recalled little of the house in which he was born, except for long hallways and a towering staircase. For reasons he never knew, he was always afraid of horses.

A few years after Walter's birth, his family moved into a home at 1832 Spruce Street, which, like their previous residence, was just a half block from Philadelphia's Rittenhouse Square. He later remembered Rittenhouse Square as "a rather dingy place where nurse maids wheeled baby carriages and gossiped," but the enclosed park had once been the focus of Philadelphia's premier neighborhood. Named in the memory of David

Rittenhouse, an eighteenth-century Philadelphia engineer and astronomer who, like Grandfather Keen, once served as the president of the American Philosophical Society, the square measures 540 feet on each side. Through the 1860s, when grazing cows—joined by errant pigs and chickens—still sometimes nibbled its grasses, Rittenhouse Square was a bucolic retreat much quieter than the city center. For years, an iron fence surrounded the square and a square-keeper daily patrolled the grounds and clanged a bell to announce its nightly closing. Wealthy families began building homes on its perimeter after the Civil War. By the time Walter was born, the neighborhood's zenith had passed, and Rittenhouse Square was beginning a gradual decline that would go unchecked until Walter was in his twenties.

Rittenhouse Square served as the backdrop to many of Walter's indelible childhood memories. It was where Walter's father—with an old-fashioned doctor's circular mirror strapped to his forehead—would banish him when he was playing too loudly in the backyard of their home. Red brick and white–shuttered houses surrounded the square, where fat dogs chased fat squirrels in the shade of its trees. As a roller-skating track, Rittenhouse Square was without rival, and Walter once smashed his two front teeth on its pavement while trying out a new and unfamiliar pair of skates. Most importantly, though, living a half block from the square placed the Freemans in privileged company. In the 1890s the neighborhood was still full of doctors and other professionals whose ancestry commonly included colonial Americans and *Mayflower* passengers—just as did Walter's family. (He knew that somewhere in his lineage was John Howland, a *Mayflower* arrival who fell overboard during the voyage and was rescued with a boat hook.) Like their neighbors, the Freemans employed maids, cooks, laundresses, and governesses.

Walter was a good-looking boy with piles of dark curly hair. He later characterized himself as a city child of average interests and attainments. "On the whole I think I was a sensitive, imaginative boy, docile, shut in a bit, and full of questions," he wrote. The questions earned him the family nickname "Little Walter Why Why." He grew up attentive to the practice of medicine, which surrounded him, and soaked up the words and phrases. Once in school he penned a tale about two steamships that he named *Cathartic* and *Emetic* without knowing the meaning of those medical terms. Finding out that the first term signified a medication to produce an evacuation of the bowels and the second a compound intended to induce vomiting, "I blushed furiously when informed," he remembered.

As an adult, Walter recalled many dramatic events from his boyhood, such as waking up around midnight in Grandfather Keen's house to watch the illumination of the statue of William Penn atop Philadelphia City Hall—an event that marked the community's entry into the twentieth century. Yet he could remember few friendships. His best pal came from within his own family. With Morris De Camp Freeman, a slightly younger cousin more outgoing than Walter, he played with building blocks and then moved up to collecting baseball cards, cigar bands, adventure books, and postage stamps. The boys joined together to patrol the streets on bicycle, suffer bullying from the neighborhood toughs, swim, and watch trains. One year they got the idea of mounting stilts to get a good view of the Mummers Parade on New Year's Day. Walter and Morris nursed a mutual contempt for girls and made grand plans for the Society for the Prevention of Useless Girls—SPUGs for short. Disdaining the company of other children, they set up another exclusive secret society, just two members strong, which they called the Wal-Ris Club. They became closer than brothers, Walter believed. But after they grew up—even though Morris was best man at Walter's wedding—Walter generally saw little of his childhood companion, who, he noted, "became a neurotic dilettante and eventually an alcoholic."

The first school Walter attended was located just three doors down from the Freeman home on Spruce Street. For her clientele of children from wealthy families, the schoolmaster, Mrs. Farnham, emphasized the three Rs. When he was eight, Walter transferred to the local Episcopal Academy, where all the teachers were men and where the sound of Latin and Greek drills echoed in the hallways. Freeman later gave good marks to the quality of education he received there. "The drills in writing, spelling and arithmetic have given me a certain intolerance for those less trained in accuracy," he wrote. Walter also made "machine-like progress" through Latin and Greek grammar and acquired the ability to concentrate on his studies amidst the turmoil of music and play in the Freeman home. When he completed his studies at the academy, Walter earned the graduation prize in Greek.

As a schoolboy, Walter was shy and unathletic. He found girls "bothersome" and never dated until he was in medical school. "I think I actively disliked girls until I went to college," he remembered. His family offered little additional intimacy or friendship. As the years passed, six siblings eventually joined Walter: brothers Richard, Norman, Jack, and William, and sisters Virginia and Dot. Although all the siblings shared the Freeman name, a series of German governesses, and the same upbringing and background, they never grew close. Walter contentedly assumed the Victorian role of the distant but dominant oldest brother, and he developed the only relationship that could be called warm with Norman, who eventually followed in Walter's footsteps to become a renowned vascular surgeon. Years later, paging through an edition of *Who's Who*, Walter observed that he "was gratified to find the name of a distantly related brother," Richard, an art professor.

Walter's father practiced medicine from a home office. When Walter was in college, he recorded a lurid description of that examining room:

> The chair was in a far corner partitioned off from the rest of the room by a glass-topped instrument table with a row of glass-stoppered bottles about the periphery, a high screen at the chair's back hiding unknown terrors behind its imitation-leather exterior, and two walls. . . . A cylindrical boiler beneath which a ring of gas jets was burning suggested untold horrors that were ably supplemented by the bleary eye of the working light, all ready from behind the patient's ear to throw a glare down his own throat when the doctor should choose by his mirror to throw the beam in that direction. A hydraulic fan made gurgling noises not unlike strangulation and half a dozen sprays pointed their nozzles at him like a firing squad. Nearby was an oven just large enough to bake a baby. But the instrument table—horrors. There were the uniform medicine bottles sitting like a jury along the back and sides; the likeness the more striking because each had its own little stall. There they sat overlooking the evidence—nickel-plated forceps, probes and all the scientific paraphernalia of laryngology.

During vacations, the elder Freeman took his sons on hiking, fishing, and camping trips—his great solace and escape from his medical practice. Walter's own lifelong love of the outdoors certainly originated in these trips to Maine, New Brunswick, and his father's farm in Beverly, New Jersey. The elder Freeman sometimes took advantage of these times to fish, hike, and swim with his sons; more often he relaxed with a book to read or a letter to write while the boys explored the area, occasionally with a paid guide. Thus the trips never bonded the boys with their father. The elder Freeman never drew close to his sons or shed his formality with them. Walter remembered him as a shy, socially awkward, and humorless father whose example taught his son to regard emotional expression as something strange and frightening.

The father, for example, had trouble expressing anger and disappointment. Once when a truant officer captured Walter out of school, the elder Freeman horrified his son by producing a whip—and flailing himself. When it came time for Walter to hear from his father on the facts of life, "it was a painful experience to both of us and I doubt whether he repeated it for his other sons," Walter recalled. "I had been showing interest in the external anatomy of my young girl cousins. With the aid of his ancient textbooks on anatomy and gynecology illustrated with woodcuts, he dilated upon internal anatomy, reproduction, and especially venereal disease, threatening to have me followed or even tempted by operatives who would report to him. I was thoroughly uncomfortable—but remained a virgin. He never alluded to it again." (Later, in his own medical practice, Walter advocated an approach opposite to his father's in explaining the details of sex: he approved of the use of the "dirty story" to reduce everyone's anxiety.) Emotional distance even tarnished the father's moments of generosity, such as when he gave Walter a gold coin as a present for his church confirmation and immediately ordered the boy to give up the coin to the collection plate.

He was a distant husband, as well, and compared poorly in professional esteem and achievement with the other physician in Corinne's life, her father. Her husband's idea of heaven was a place where people had no noses and he would no longer have to bother with otolaryngology. He suffered his unhappiness in solitude. As he allowed her little involvement in his life, Corinne would energetically involve herself with her children. She served tea to their friends on Sundays, encouraged the youngsters to dance and work with wood, bought them subscription tickets to orchestra concerts, and sent them to church and Sunday school. On her initiative, the Freeman boys toured nearly every cultural and industrial attraction in Philadelphia, including steel mills, locomotive factories, the Fleer's chewing gum plant, and the U.S. Mint. She organized vacation trips to Jamestown, Gloucester, Cape May, and other inexpensive resorts. An enthusiastic singer, pianist, and organist, she set her children before all kinds of instruments. Corinne regarded attending church, doing household chores, meeting with friends, and practicing music as moral duties—the fulfillment of social obligations that would advance her children in the world. She established her bedroom sofa as a safe haven for the kids when they needed to talk with their mother about their troubles and fears.

In Walter's case, all of Corinne's talk of duty failed. He declined to paint or get involved in woodworking. He fumbled with the cello and "never acquired more than a *largo* speed, mostly *maestoso*," he remembered. He eventually gave up the instrument and began a lifelong avoidance of art museums. He refused to unravel his worries to Corinne on the bedroom sofa. Walter kept to himself and remained determinedly self-contained. Self-reliance grew into a habit. "Actually, I suppose my worst handicap is this tendency to go it alone, but I get satisfaction from it," he wrote as an adult. "I like my own company."

Around the age of thirteen, Walter took up the perfect hobby for a boy more interested in observing others than revealing himself—photography. Shooting landscapes and capturing unusual street scenes attracted him more than taking portraits of people. "I was a watcher rather than a performer, and my camera was as much a shield as it was a recorder," he acknowledged. For most of the rest of his life, he would regularly set himself behind the camera and use its lens not to examine personality and behavior but to record physical details of the places and things he admired, including the bodies of his patients.

Walter's aloofness exasperated Corinne. Rudyard Kipling had just published his *Just So Stories* collection in 1902, and she told her son that he reminded her of the tale of the unapproachable feline, of whom Kipling wrote, "For I am the cat that walks by himself and all places are alike to me." Walter was not moved. Long after Corinne's death, he confessed that he had never loved his mother.

The Freeman residence on Spruce Street contained many of the activities that Corinne planned for her children. The house, which still stands, rose four stories, was heated by three coal furnaces, and was shaped like an L. The top floor housed servants and offered a spot for seclusion and meditation, probably a frequent destination for Walter. In the large backyard, the Freeman kids swung from the branches of a wisteria tree, played noisy games of football and baseball, and dug holes. Inside, the parents liked order and enforced it. Walter was rewarded when he arrived downstairs early for the seven o'clock breakfast, and he was fined at a rate of one cent a minute for lateness.

The afternoon dinners on Sundays often climaxed the week's activities. These were held across Rittenhouse Square at the home of Walter's grandfather Keen. "Here was a noble feast at which we met cousins and guests," Walter remembered. "The table would accommodate eighteen on special occasions and the overflow ate in the library next door." After the meal, the children would migrate to the library and leave the sitting room to the adults. At three o'clock sharp, Grandfather Keen withdrew for a nap, and the gathering was over.

W. W. Keen dominated more than his family's weekly schedule. He was a towering figure not only in the life of his grandson but also in the medical life of Philadelphia. Born in 1837 in that city, he earned his bachelor's and master's degrees from Brown University, stayed at Brown for supplementary study in the sciences, and enrolled at Jefferson Medical College in Philadelphia in 1860. Typical of medical schools at the time, Jefferson offered little more than the basics. "Seven professors, one demonstrator and that is all," Keen remembered. "No laboratory, no library, no hospital, no specialties . . . no patients for students to examine, no ward classes, no microscopes."

But there were classmates with talent and promise. One day while Keen was poring over *Gray's Anatomy* (a medical textbook he would later edit), another Jefferson student caught his attention. "The afternoon sun was hot, and I had the Venetian blinds slanted so as to exclude the direct sunlight," Keen wrote years later. "Suddenly the slats were changed to a horizontal plane, and, as I turned my head to see who was there, I saw a pair of eyes looking at me and heard a voice outside say, 'Doctor, don't you want to help me with some experiments on snakes?'" The speaker introduced himself as Silas Weir Mitchell. Mitchell's life would intertwine with Keen's for decades as the friends built distinguished careers in medicine, until Mitchell's death in 1914.

Service in the Civil War interrupted Keen's medical education. In the summer of 1860—still with no more than a year of medical school behind him—he became a surgeon attached to the Fifth Massachusetts Regiment, which saw action at the first battle of Bull Run. Imprinted with memories of the chaos of battle and appalled by the ignorance of many of his medical colleagues in the military, he left the field of battle. With his M.D. in hand in 1862, Keen completed additional medical studies in Paris and Berlin.

After Keen's return from Europe, Mitchell asked him to go to work at the U.S. Army Hospital for Diseases of the Nervous System, which Mitchell had played a role in creating. The hospital, located in Philadelphia, treated soldiers for epilepsy, all kinds of nervous diseases, palsies, and a frightening variety of wounds sustained in battle. Mitchell and Keen formed two-thirds of the administrative staff. After the Battle of Gettysburg in 1864, so many casualties filled the hospital that it had to relocate to a new four-hundred-bed facility. A potent reminder of the brutality of war, the hospital also served as a uniquely fertile laboratory in which Keen and his colleagues could study nerve injuries. The research team of Keen, Mitchell, and neurologist George Morehouse produced important and influential papers on the treatment of gunshot wounds and nervous disorders caused by explosions.

After the war, Keen taught at several Philadelphia institutions, including the Philadelphia School of Anatomy, where the artist Thomas Eakins assisted him in anatomy classes. As a medical school professor, he contributed to the education of some ten thousand doctors. His openness to new ideas, however, destined him for distinction outside the classroom. In 1876 James Lister, the English proponent of antiseptic surgery, came to lecture in Philadelphia. Nine years earlier, Lister had begun applying carbolic acid to wounds, fractures, and surgical incisions to kill the germs that often caused life-threatening infections. The British medical journal *The Lancet* had trumpeted Lister's achievements. Among his many early successful cases were those of a seven-year-old boy who was run over and seriously injured in both legs by the wheels of an omnibus but saved by Lister from amputation, and of Queen Victoria, from whom Lister had removed an abscess in the left armpit without occasioning infection, though not without accidentally spraying the monarch in the face with the antiseptic acid. For the first time, patients who underwent surgery could feel some confidence that infection would not kill them if their operations were successful.

In Philadelphia, Lister demonstrated his methods of avoiding surgical contamination by liberally applying antiseptic agents to the patient's body and by cleaning his hands and tools before surgery. Keen—who during the Civil War had seen surgeons whet their knives on their boots before operating—became an instant convert. "I was amazed at the remarkable results," Keen wrote. "From that day on, I always followed Lister's principles (founded chiefly on Pasteur's researches), although I modified the methods to the aseptic as time went on. . . . Thus was antiseptic surgery first introduced into Philadelphia."

Keen's reputation as a surgeon grew. He became known as a doctor who was unafraid to aggressively intervene and was at his best in difficult situations. (One of his favorite expressions when taking control of a medical emergency was, "Now we have the whiphand of it," a phrase that one could imagine coming from the mouth of his grandson decades later.) In 1893 Keen twice assisted in surgery on President Grover Cleveland; the operations, performed in New York harbor aboard the presidential yacht *Oneida*, were kept secret from the public. Cleveland suffered from a malignant growth on his jaw. Using a metal mouth retractor that Keen had purchased in Paris a quarter of a century earlier, Keen and the other members of the president's medical team removed a sizable portion of Cleveland's bone and replaced it with a rubber prosthesis. (The retractor and part of the president's jawbone are still sometimes exhibited at the Mütter Museum in Philadelphia.) Keen felt proud that after the surgery Cleveland was able to maintain an image of robust health and energetically orate on behalf of maintaining the gold standard in a time of financial panic. In subsequent years, Keen became something of a specialist in the maladies of presidents and their families, with the McKinleys, Wilsons, and Roosevelts among his patients. He was part of the medical team that diagnosed Franklin Delano Roosevelt's polio in 1921. Keen's fame made him wealthy, and in the years after 1890, his annual income frequently exceeded $30,000 (about $570,000 in today's dollars).

It was not a president's ailments but the suffering of a twenty-six-year-old carriage maker from Lancaster, Pennsylvania, that led Keen to perhaps his career's most renowned achievement. Although always a general surgeon whose knife traversed all regions of the human body, Keen had emerged from the Civil War with a special interest in treating diseases and disorders of the nervous system. In 1887 Theodore Daveler, a man suffering from headaches, seizures, paralysis, and other symptoms of a brain tumor, repeatedly pleaded with Keen to attempt surgical removal of the growth. Keen could take Daveler as his patient only at great risk. Rickman Godlee in London had performed the world's first such operation only three years earlier, but an infection had killed the patient within a month.

Keen, still reeling from the death of his wife, Emma Corinna Borden Keen, from dysentery the previous year, agreed to operate—without the yet-developed assistance of local anesthesia, X-ray photography, reliable artificial lighting, and blood transfusions. After supervising the scrubbing of an operating room at St. Mary's Hospital in Philadelphia with carbolic acid, the boiling of instruments for two hours, and the cleaning of his hands and those of his assistants, Keen bored into Daveler's skull, probed the opening with his ungloved finger, manually pulled out the tumor, tied off the bleeding vessels, stitched the incision with catgut, and in two hours saved his patient, who lived for more than thirty years after the surgery. When Daveler died in 1918, Keen returned his attention to his old patient in order to review the postmortem study. No tumor had returned. Keen's accomplishment as the first American surgeon to successfully remove a primary brain tumor, along with other achievements such as performing America's first colostomy and developing a new technique of cardiac massage, earned him lasting fame.

Fame only partially motivated Keen. There was a spiritual element to this gifted surgeon's innovations. Keen, a deeply religious Baptist whose childhood was marked by the end-of-the-world predictions of the evangelist William Miller in 1843, believed that the precise boundaries of the localized areas of the brain that controlled specific muscles of the body furnished proof of God's work. He operated "with a missionary zeal," wrote his literary editor and descendant, W. W. Keen James. He abhorred alcohol, revered and financially supported Baptist missionaries in Asia and Africa, and never considered remarrying during the forty years that remained of his life after his wife's death. Living as a widower with his unmarried daughter, Florence, Keen suffered a grief so intense that it made it difficult for him to utter his wife's name, especially her pet name, Tinnie. "Every year, on the anniversary of her death, my grandfather would shut himself away from the family and mourn her loss," Walter remembered. "He could never speak of her without tears flowing."

Keen was short, just five feet two—he had to stand atop a box while operating—but had a stoutness, florid complexion, and strong gaze that made him appear larger. In childhood, he had accidentally chopped the tip off his left forefinger. He impressed nearly everyone as a paragon of energy, healthy competitiveness, and integrity. His output of writing and editing over the years was prodigious, including at least nine books on medicine and religion and scores of contributions to professional journals. Although he retired in 1901, he continued to travel, attend medical conferences, stay abreast of advances in medicine, and write without respite until he reached the age of ninety. By the end of his illustrious life, Keen had earned eleven honorary degrees and had been honored as an officer of the Belgian Order of the Crown, an officer of the French Legion of Honor, and the recipient of the Henry Jacob Bigelow Medal of the Boston Surgical Society.

Just as Walter did decades later, Keen harbored conflicting feelings about the popular press. His appreciation of the power of the press is undeniable, and Keen was not afraid to harness it to advance his own agendas. He wrote hundreds of articles in professional journals and such popular magazines as *Ladies' Home Journal* and the *Saturday Evening Post* on everything from brain surgery to defenses of animal experimentation, and he penned such a large number of letters to newspaper editors that he earned the label "gadfly." But Keen also distrusted journalists for their frequent inaccuracy. Once, after Keen had unsuccessfully used his technique of cardiac massage in an attempt to revive a patient, a reporter from the *New York Evening Journal* called to ask whether in this case Keen had made a great discovery that succeeded in resurrecting the dead. Keen's deadpan reply was "that they were quite right, except that I had made no great discovery and since my patient had not recovered, I had not brought the dead back to life." The irony was wasted on the newspaper, which soon afterward published an article about Keen's miraculous heart therapy and ability to treat the dead. A fictitious interview with the patient was included.

Similarly, in 1912, a journalist from the *Philadelphia Press* called Keen for a comment on a report that his daughter, Dora, had been abandoned and doomed to starvation by fellow climbers on an expedition to reach the summit of Mount Blackburn in Alaska. Keen replied that Dora had provisions to last many weeks on the mountain and that she could certainly return to civilization as easily as the other climbers if she chose to. He then asked the *Press* to suppress the report of her abandonment. The story ran the next day and was soon shown to be false. The newspaper "made no excuse for printing false news, nor any apology for the anxiety and pain which its publication caused me," Keen later wrote.

Walter drew close to his grandfather in the spring of 1911, when he accompanied Keen and his aunt Florence on a month-long Caribbean cruise. Passengers on the S.S. *America*, they all found the tropical weather and airborne streaks of flying fish invigorating after a dreary Philadelphia winter. The first port of call was Havana, followed by Santiago de Cuba. It was the third place of embarkation, however, in Colon, Panama, that made Walter see the respect that his grandfather commanded in others. At the Panama Canal, still under construction, Dr. William Gorgas, one of Keen's former students, served as chief medical officer. Gorgas, who would gain fame through his success in taming the region's rampant yellow fever, cleared his schedule to treat the threesome to a guided tour of the locks and gorge as well as a trip to old Panama City on horseback, which, much to his surprise, Walter enjoyed. The cruise concluded with stops in Curaçao, Trinidad, Martinique, Haiti, St. Thomas, and Puerto Rico. Walter, busy with his No. 2 Brownie camera, transformed his cabin into a darkroom.

By the month's end, Walter knew his grandfather much better. What developed between them was not exactly warmth but mutual admiration. Keen regarded Walter as a bright boy and a loyal grandchild. In turn, although Walter was not yet interested in medicine as a career, he felt something of the veneration due a doctor who worked hard, wove himself into the inner circles of medical societies, was open to new techniques and ideas, and treated his patients with humor and directness. In his grandfather, Walter saw a physician very different from his father—one who engaged the world, followed speculation with action, and enjoyed his profession. Medicine could provide a stimulating intellectual journey that flitted between the realms of life and death. It could also offer distraction from grief and anger, as Keen had discovered in his retreat into his profession after his wife's death. The surgeon's power, which derived from a mixture of technical skill, improvisational virtuosity, and knowledge, was obvious even to a youngster.

Once, when Keen mentioned that he was shrinking in his old age and attributed his loss of height to the gradual compression of his spine, young Walter spiritedly disagreed. "No," the grandson asserted, "it is due to the weight of the ideas in your head." Big ideas, the boy recognized, were a bit of a burden, but they could also lead a curious and decisive person to the heights of achievement. Through the example of his grandfather, as well as through his own education and developing self-reliance, Walter Freeman was preparing himself for an astonishing career.

## Chapter 3

# The Education of a Lobotomist

Although Freeman lamented that his father distrusted blue-chip investing and "had a ready ear for salesmen in goldmines and Florida real estate," the young man's family was able to gather sufficient resources to send him to Yale in the fall of 1912. In his first year in New Haven, Freeman found no academic interests to focus upon—he moved without inspiration from his classes in English, German, and history—and felt unprepared for the demands of his professors. "In prep school a daily stint of 30 lines [of Greek translation] was standard; in college it was 30 pages," he remembered. "The jump from geometry and algebra to calculus was equally baffling. I barely squeaked through my first year studies." This is a common experience for first-year students at academically tough schools, especially those, like Freeman, who matriculate at sixteen years old.

Freeman channeled his awareness of his immaturity into pain and despair. He avoided girls, received no invitations to join a fraternity, and remained mortified years later by an incident in which, as a lowly reporter for the *Yale Daily News*, he spilled an alphabetically arranged file of subscriber cards before the eyes of the newspaper's editor. He was soon dismissed from the paper's staff. In the annual pushball competition between first- and second-year students, he jockeyed his way to the front of the freshman group straining against the six-foot ball, stumbled to the ground, and was flattened. "With an aching leg I limped to the rear," he noted. His efforts to participate in sports were just as ungratifying, mainly due to his young age. He tried wrestling "but not long enough to develop my arms to the point where I could lift my opponent at the appropriate moment," he recalled. When he joined a swim team, his lack of confidence made him practice only when nobody else was around to watch, and his coach never let him compete. His classmates sensed Freeman's social discomfort, and he and his equally awkward roommate were nicknamed "Minnie and Lizzie."

Dissatisfied with himself, he returned to Philadelphia for Christmas and Easter vacations and spent his time there moping and listening absently to his mother's unwanted expressions of concern. He avoided his family, went to late-night parties, and "was sulky around the house the next afternoon in prospect of another party that night." The parties failed to lift him. "Being a Yale man gave me a certain prestige but I was not man enough to make the most of it," he noted. Everyone at home, including his siblings and his old friend, Morris De Camp Freeman, seemed childish or uninteresting to him. When he returned to Yale, he once again faced his loneliness and intimidating course work. "The first year of college was misery," he concluded.

Even if it caused him misery, young Freeman's early enrollment in college may have saved his life. Nobody in 1912, of course, knew that World War I loomed just a few years off. Had Freeman taken a more leisurely path through prep school and entered college at the usual age of eighteen instead of sixteen, he would have faced military conscription by the time he was ready to receive his bachelor's degree. Many of his Yale classmates died in battle, including a close friend. Freeman was not old enough to serve in the military until he was in medical school, by which time his career choice shielded him from duty on the front lines.

Freeman's sophomore year was nearly as bad as the first. He bailed out of the study of classical languages, abandoned mathematics, and tried introductory classes in the physical and social sciences. "Still there was no spark. I drifted," he declared. He found some solace in his longtime hobby of photography. He enjoyed taking pictures of sporting events, and he took pride in one photo he captured by climbing a flagpole to obtain an aerial view of people waiting in line for game tickets. Later he scored a coup by devising a clever plan to photograph members of the Scroll and Key, a Yale secret society, as they sang their anthem outdoors at midnight. He mounted the camera in a window across the street. "With a confederate to work the shutter, I walked along the street in front of the Keys with a magnesium flash-gun until I came opposite and set it off," he remembered. The society members angrily protested, and the *New York Times* published the picture.

Freeman began to find his way at Yale after the summer of 1914, between his sophomore and junior years, during which he worked as an apprentice machinist at a General Electric factory in Lynn, Massachusetts. He made his share of expensive mistakes—including damaging a milling machine, "which probably cost more to repair than the total [pay] I received"—but the eleven-hour days, twelve-mile bicycle commute from his family's summer rental in Swampscott, and grimy manual labor seemed to raise his self-confidence. Before returning to Yale, he got himself fitted for a cutaway suit with top hat and pearl gloves.

Back in New Haven, he followed his summer as a machinist by entertaining his mother's notion that he should become an engineer. A class in English Romantic poetry grabbed his attention and sparked his interest in writing verse. When he became an upperclassman and boosted his social status, he borrowed his grandfather Keen's limousine and chauffeur to transport his growing group of friends. By the end of his third year, his interest in engineering had solidified, and he enrolled in summer school to improve his math skills. His failure in descriptive geometry convinced him that he should not become an engineer after all. While killing the engineering bug, however, the summer gave him another one: a case of typhoid fever, which he believed he contracted by eating raw clams. By summer's end, he was a patient at Polyclinic Hospital, just a few blocks from his family's house in Philadelphia.

Typhoid, the first of several illnesses that would strike him during the next twelve months, laid him up for all of the first semester of his senior year at Yale. In the hospital, he saw firsthand what it was like to be doctored and nursed. After his discharge, he had a long convalescence in bed at home. Listless, scrawny, and inflamed, he transferred his feelings to a cattleya orchid sent to him by his mother's sister, Florence Keen. "I looked at the dusky lips, the raw-red throat and the coated tongue and nursed my misery," he wrote. "That damned flower just wouldn't die, came back day after day to my bedside. I don't think I was seriously ill or delirious but that orchid haunted me."

Freeman's weakened condition made him unable to keep up with his studies or attempt anything strenuously academic. Initially he could do little more than read, play cards, and listen to phonograph records. Although his mother spent hours reading to him and keeping him company, Freeman preferred the attentions of "a large-eyed nurse I called my ox-eyed Hera." She tolerated the mild jokes and pranks he could muster the energy for. "At home I was weak as a kitten," he remembered. "The first time I took a hot bath I could hardly get out alone. My pulse shot up to 140."

He shared a room with his father, who inquired about his future plans and repeatedly recommended against going into the medical profession. In some other line of work like engineering or the law, his father said, you could retire early, possibly by age fifty. Then fifty-five, the senior Freeman was still putting long hours into work he disliked. The son had no particular interest in becoming a physician, but his father's words did not persuade him to rule it out.

When he regained some strength, Freeman ventured out of doors. Bundled up in blankets, he sat in Rittenhouse Square and closely observed the swirl of life around him. By Christmas 1915 Freeman was gaining strength and putting on weight. Before returning to Yale, he enjoyed a final period of rest in South Carolina at the homes of family friends. There he met a girl who became his first real crush, and for the first time in months he tried dancing. He also made his maiden voyage behind the wheel of an automobile; his erratic control of the vehicle would characterize his driving for the rest of his life. "I stepped on the throttle instead of the brake and recovered just in time to avoid a man," he observed. One evening he took part in an opossum hunt, "but my slow gait brought me to the tree after the quarry was bagged."

Typhoid had stolen half of Freeman's senior year, but in return it had given him an opportunity to ponder his future. Engineering had long vanished from his thinking. He liked literature, especially the study of poetry, but saw no future in it. Medicine kept crossing his mind. His father's admonitions against it rang in his ears, but Freeman found fault with the source of the advice, not with the profession. He viewed his father as a failure, a man unmotivated to involve himself in his work to a degree that would bring him real satisfaction and achievement. "I compared my father unfavorably to my grandfather [Keen] who, though retired, carried on a compelling interest not only in medicine but in related scientific fields," he noted.

Freeman jumped back into his Yale studies. He was assigned a room in Connecticut Hall, the oldest building on campus, where his six-foot-high frame filled much of the space in the low-ceilinged room and the floor groaned under his weight. Every day his mother filled his mailbox with long accounts of her activities and thoughts, but Freeman's interests lay elsewhere. After attending a chorus rehearsal, he kissed a girl for the first time. Meanwhile, the intensifying of World War I infused his course work in history, his major field, with significance. "Drill and drudgery were past and the courses had relevance and held my interest," he said. The onset of a streptococcal infection did not slow him at all, and by the end of the semester, he racked up a succession of A grades.

That spring, on a solitary Sunday walk, Freeman made his career choice. He had decided to take the train to Waterbury, Connecticut, and hike the twenty miles back to New Haven. Strolling along the road, he suddenly realized that he did not want to retire early, as his father advised. Like his grandfather Keen, Freeman hoped to remain intellectually active well into old age. "I returned after this walk, sat down and the same evening wrote that I was going to study medicine, and the various reasons for it," Freeman recalled. "This letter dropped like a bombshell into the family. [Brother] Jack, then 15, is reported to have said: 'Gee, for once he's serious.'" His whole family, father and grandfather included, congratulated him on his decision. "I was walking on air."

Fortunately, Freeman had a surplus of academic credits, and his semester in convalescence did not prevent him from graduating with his class in 1916, at age twenty. He left Yale with few distinctions. "I learned less at college than at prep school," he later observed. "I was a nonentity in the class from the standpoint of education, athletics, social position. I had few close friends and no girls. I never cultivated any of the teachers. I was an observer, a recorder, a photographer. I was glad when it was over." Freeman remained what his mother had labeled him years earlier: "the cat that walks by himself."

But medical school lay ahead. For someone who had paid scant attention to the sciences as an undergraduate, admission to medical school would be difficult. He enrolled in summer school at the University of Chicago, where he planned to catch up in science course work. There, to his own astonishment, he got by in biology and found his class in organic chemistry "a delight." He discovered that he had an aptitude for learning formulas, working in the laboratory, and tinkering with such instruments as the spectrometer and polarimeter. Freeman thought about redirecting his career into biochemistry. He rented a room thirty minutes from the campus and heightened the strenuousness of his daily walk by hauling with him "a box of bones." At the same time, he took shorthand and typing courses at a nearby business school, which aided him in transcribing his class notes.

Once again, however, Freeman's health threatened to derail his academic plans. He began to feel pain and swelling in his ankles and wrists. Unable to study, he spent days in bed. "I wrote home saying that I guessed God didn't want me to study medicine," he recalled. "In reply I received a stern admonition not to think that way, much less to mention it." Eventually, Freeman's joints lost their tenderness. His father believed the problem stemmed from infected tonsils, and he performed Freeman's second tonsillectomy (about fifteen years after the first) later that year. After that, Freeman had no problems with his joints.

He completed two summer quarters in Chicago and, convincing a dean of his commitment to medicine, enrolled that fall in medical school at the Philadelphia campus of the University of Pennsylvania. There, in one of the first lectures Freeman sat through, Professor Alonzo Englebert Taylor sternly predicted that many of the students would flunk out—a positive outcome, given that they would not make good doctors anyway. "This went on for an hour with embellishments, but also with promises of the good life for those who succeeded," Freeman said. "I came out with my head high, accepting the challenge."

Filling books with his course notes, written in an intensely scribbled, wavy shorthand, Freeman inclined toward biochemistry but took note of the unusual teaching methods of his first-year anatomy instructor. This man awed students with his ability to draw anatomical sketches on the chalkboard with both hands simultaneously. Freeman was impressed not only by the ambidextrous display but also by the professor's discussion of the nervous system. Using his new skills in shorthand, Freeman could keep up with the avalanche of new information in the lecture, even when his classmates were baffled.

He was less skilled in the laboratory, where his interest in precipitating and examining silver salts once led him to mix silver nitrate and gelatin, a highly explosive combination. Only by luck did he avoid producing a life-threatening blast. An even closer call came when he was working with a toxic solution of sodium cyanide. Accidentally he placed the wrong, tainted end of a pipette in his mouth. "In a matter of seconds I was panting for breath, hanging on to the work bench while the world began to swim, my knees turned to spaghetti and sweat broke out on my forehead," he wrote. Freeman knew exactly what had happened and the danger he faced. He was alone in the lab "with the rush of death's wings over my face." A minute passed, then another. Gradually Freeman recovered. He did, however, remove himself from the lab for the rest of the day.

As a boy, Freeman had sometimes entertained himself by counting as high as he could manage and had once set a goal of reaching one million. His powers of concentration came in handy as a medical student, and his predilection for solitary work was an asset. His interest in chemistry did not flag, and even when attending weekend symphony orchestra concerts he thought about his laboratory experiments. Freeman frequently attended meetings of student medical societies at the University of Pennsylvania. He enjoyed the collegial give-and-take of the case reports, presentations, and discussions. And as he did for much of the rest of his life, he found that he could function well on six hours of sleep a night. Freeman's grades were good, and he regained his confidence in his academic abilities.

That spring, Freeman made his debut as a presenter of a scientific paper. Showing his research on the crystallization of silver salts, he was the only first-year student to offer a paper to the undergraduate medical association. He dreaded the questions he anticipated receiving, but none came because "nobody else knew anything about the subject," he wrote. He believed his distinction as a presenter opened the door to his use of lab facilities when he wanted them.

During the first year of medical school, Freeman's fascination with the nervous system led him to develop a heightened interest in the brain. He decided to return to the University of Chicago for another summer of study, this time enrolling in a laboratory course on brain chemistry. Freeman's medical interests were evolving. He studied psychiatry with Professor Charles W. Burr, a fierce anti-Freudian who "was approaching retirement and who taught us that Freud had a dirty mind." Nevertheless, Freeman enjoyed Burr's lectures on mental illness enough to find them "morbidly appealing." (Freeman later visited Burr's home as part of "a delegation to talk over the possibility of making psychiatry more interesting to the students.") At his father's urging, he investigated gastroenterology. He spent time in the gastrointestinal service during a summer at the Mayo Clinic, but the specialty failed to absorb him. Neurology and neuropathology did hold his interest, however, and his comfort in the lab made him feel at home with the detailed work of the neuropathological examinations he witnessed.

In the background of Freeman's medical studies were the dark clouds of World War I. Too young to participate in the ROTC units forming while he was at Yale, Freeman took an active role in the military as a student doctor. In Britain, student physicians had been allowed to enter the infantry—a policy that produced a medical crisis when many were injured or died in battle. The U.S. government created the Medical Enlisted Reserve Corps (MERC), which permitted student M.D.s to remain in school until their medical services were needed. Freeman joined MERC at the end of his first year of medical school and proudly wore a uniform that he quickly damaged in the lab with acid burns and chemical stains. During his summer studies at the University of Chicago in 1917, he joined reservists and ROTC participants for target practice at the Great Lakes Training Station. At 200 yards, Freeman led his group in shooting accuracy.

One morning during the summer of 1918, Freeman awoke from his sleep with the clear memory of a dream in which George Goodwin, a roommate from Yale who was serving in the military, had died. A few days later came a letter from Goodwin's father giving the date and time that his plane was cut in two by another. "Checking this time and allowing for the longitude I believe it was at the exact moment [of the dream]. I have no explanation for it," he recalled. "I had no concerns about him previously." Although shocked, Freeman concluded that it was nothing more than a coincidence.

His own time of active duty had already arrived. Freeman had not wanted to move far from medical school, so he asked his grandfather Keen to write letters to the surgeons general of the army and navy, both former Keen students, suggesting limited duty for his grandson. In June 1918, after a few weeks spent drilling rivet holes in ship hulls at Hog Island, Freeman was given the rank of sergeant and assigned to an army hospital at Camp Dix, close to Wrightstown, New Jersey, and only a short distance from Philadelphia. For several weeks, he worked in the pathology laboratory under the direction of one of his University of Pennsylvania professors; he measured blood counts, conducted urinalyses, and assisted in autopsies and embalmings. Working with army cooks and bakers, he took stool samples and checked them for evidence of hookworms and other parasites. This duty earned Freeman the nickname "Honey-dipper." All in all, Freeman's assigned tasks at Camp Dix were educational and not too difficult. But at the end of June, when he returned to the army hospital from a three-day furlough to visit his family, he saw that it had been transformed by one of the worst medical epidemics in modern history.

The first cases of Spanish influenza had appeared. During the next five months, an overwhelming outbreak of this illness would kill 21 million people worldwide, including more U.S. soldiers than had died in action during the war. Within a week of Freeman's return from his furlough, sixty sick soldiers—many suffering from low blood pressure, terrible coughing, and profuse sweating—filled a ward of the hospital. On any given day, half of the influenza patients would die. "It was a frightening experience day after day," Freeman wrote. "We all wore masks, of course, but in order to make blood counts we technicians had to remove our masks to get blood into the pipettes. Coming within inches of the face of a man who the previous day was full of life and now was at death's door—the remembrance of it brings an unpleasant thrill." Freeman saw that autopsies of those killed by influenza revealed lungs engorged by fluids, weighing up to ten times more than normal. The patients absorbed insufficient oxygen, produced too little blood, and died of shock.

Freeman and his colleagues worked around the clock. Many of the medical officers caught the flu, but Freeman never did. He speculated that a head cold he had contracted just before the outbreak might have given him immunity. When he left Camp Dix to resume his medical studies in September 1918, he waited for his train along with sixty coffins at the rail station.

But Freeman learned that the influenza epidemic had spread from military installations to the cities, and the University of Pennsylvania closed its medical school. Along with other third-year students, Freeman went to work at the Medico-Chi Hospital in Philadelphia, which was filled with flu patients. "None of the frills like blood counts, X-rays and other laboratory tests were heeded. We did what we could to feed, bathe and give medicines to the patients," he recalled. Entire families were stricken, and three of Freeman's siblings came down with influenza.

When medical school classes reopened in October, Freeman lived in a dormitory along with other students serving as army medical officers. He was still a sergeant, and he commanded a platoon. Each night, his company commander and a subordinate entered the dormitory in the wee hours of the morning to make sure the medical students were all in bed. "Finally one night, awakened as usual by the heavy tramp and the wandering flashlight, I angrily told them to get the hell out of there and stop disturbing the men of my platoon," Freeman wrote. "I picked up one of my shoes but did not throw it." Freeman was demoted the next day, he said, for "insolence and insubordination and threatening an officer with a shoe." But the nighttime bed checks ended.

When the official announcement of an armistice arrived on November 11, 1918, Philadelphia erupted with happiness, the wildest celebration Freeman had ever seen. Trailing strands of snake dancers, crowds surged through the streets. "I remember clinging to the Statue of Liberty model at the corner of Broad and Chestnut Streets waving something and yelling until I was hoarse," Freeman wrote. He spent that bitterly cold winter trying to win the affections of Clover Todd, a girl who lived in New York City and whose father taught at Columbia University. "I made occasional weekend trips to take her dancing or to the theatre," Freeman wrote. "She was beautiful but too sophisticated, too spiritual, maybe, for me." She went on to marry Allen Dulles, who later served as the director of the Central Intelligence Agency.

By his senior year in medical school, Freeman was free to focus more specifically on his own medical interests. Over several months, neurology had begun attracting him more strongly. It was a relatively new specialty, with a great deal of uncharted territory in the knowledge of the brain, nervous system, and neurological disorders. He especially enjoyed the clinical application of anatomy and physiology to neurological problems, an approach Freeman absorbed in the classroom lectures and demonstrations of William G. Spiller. Hunched and gaunt, Spiller looked as though he did not get out in the sunlight very often, and he spoke in a soft monotone. "His greatest pleasure, he said, was sitting down with a fine long German monograph on some obscure disease of the nervous system," Freeman wrote. Another sign of the intensity of Spiller's neurological interests was revealed in an anecdote about the professor that Freeman thought had the ring of truth. Returning with his colleagues from a neurological meeting in Atlantic City, the story goes, Spiller listened to the other academics discuss their hobbies and leisure activities. They mentioned golfing, boating, and riding horses. Spiller brought the conversation to a halt with his admission that he enjoyed nothing better than retreating to his laboratory to cut sections of the spinal cord. He was "absorbed in his small world, reading, writing, teaching and practicing his specialty," Freeman observed.

While other students grew restless watching Spiller's detailed and undramatic presentations, "I received great inspiration from the minuteness and thoroughness of his examinations," Freeman remembered. The student sought out the professor as a mentor, and Freeman frequently visited Spiller's lab to examine specimens and slides. "I claimed him as my ideal, worked hard for him, sought his advice and when the time came did my first major work in his department." Freeman had watched Spiller examine a young woman who was unable to move her eyes to one side, a sign of the neurological disorder Foville's syndrome, which is usually caused by a brain tumor. Freeman took photographs of the patient and developed an interest in the disorder. When Spiller diagnosed a tumor of the pons, a prediction found to be true after the patient died and was autopsied, Freeman was impressed. He spent hours abstracting cases of the disorder and wrote a paper on Foville's syndrome, for which he won honors for the best research by a University of Pennsylvania medical student. Spiller acknowledged the young man's interests and lent his professional assistance, but he seemed incapable of responding with warmth, pride, or trust. He never let Freeman work independently in the lab, and he never invited the student to his home.

Despite the disappointment of his mentor's cold demeanor, Freeman embraced the study of neurology and the clinical work in which he could further explore the specialty. He chose to work in what he called "the great storehouse of neurological material" at the Philadelphia General Hospital, where he learned how to apply the neurologist's arsenal of tongue blade, reflex hammer, and ophthalmoscope. Freeman also spent two weeks in the home delivery service of another Philadelphia clinic. He did not like delivering babies in the slum neighborhoods of the city. "When things went rather slowly at one home, I incautiously lay down on the sofa to catch 40 winks and for the next week was scratching and rubbing on sulfur ointment: scabies," he wrote. Obstetrics was out. When he graduated from medical school at the close of the 1920 academic year, he was second in his class.

Freeman's high achievements had come at a cost. Medicine "held my interest to the point where I excluded many other things," he noted. "In fact I was barely aware of my family, do not recall what they were doing or where they were during this period." Freeman's father was dying of liver cancer, and Freeman spent little time with him. In Freeman's mind this was regrettable but unavoidable because "I think he had little desire for companionship, preferring in his quiet way to shut out the world and endure his suffering in solitude." Wounded animals may want to die alone, but it is doubtful that a man with a wife and seven children would prefer to die without company; he did so simply because he did not know how to reach out to his family.

Freeman's one regular activity with his father was to help the sick man shave with a thirty-year-old straight-edged razor that the elder Freeman had diligently stropped to a keen edge. Decades later, Freeman still remembered his father's grimaces as the son applied the razor. He did not allow enough time for his father's whiskers to soften in hot water, he observed, because "the task was distasteful and I finished it as quickly as possible. I'm sure my mother would have been more gentle, but she considered shaving a man's job, and I was the only one at home."

When his father died at age fifty-nine on December 20, 1920, the end came as a relief to Freeman. He was convinced that contracting cancer had horrified his father and that the sick man had disliked talking about it or showing his suffering. To Freeman, the most vivid moment of the events that immediately followed the death was the "impressive, almost horrifying sight [of] seeing his coffin wheeled slowly into the glowing furnace of the crematorium." Riding in a limousine with his four brothers after the funeral, Freeman declared, "Well, now that he's gone, I'm the head of the household." Freeman wasted little time in continuing his father's habits of financial investment. Within a week of his father's death, Freeman plunked down $250 (about $2,500 in today's dollars) to buy shares in a scheme by the Amalgamated Shale Corporation to extract oil from shale. The investment proved nearly worthless.

From his father's estate, Freeman inherited about $7,000 in securities. His mother was left with a small income—her husband did not support the idea of life insurance—and she became expert at finessing income tax laws and acquiring scholarships at boarding schools for her younger children. She took a secretarial job and later moved to Paris, where she lived inexpensively through the late 1920s.

Scheduled to begin his internship at University Hospital in Philadelphia in February 1921, Freeman had several months to fill after graduation from medical school. After taking a mishap-filled canoeing and camping trip along the Delaware River with his brother Jack—a trip in which they broke a hole in the canoe, patched it with chewing gum, and fell victim to the rain and hungry mosquitoes—Freeman began a temporary internship during the summer of 1920 at Pennsylvania Hospital, the two-hundred-year-old Philadelphia institution where Benjamin Rush had once directed the psychiatric ward. Freeman mainly worked in the hospital's emergency room but also spent some time on ambulance duty in which, he said, his job "was to clang the gong."

Though he gained valuable experience setting fractures, sewing up lacerations, and treating gunshot wounds, he went through his months at Pennsylvania Hospital only half interested in his work, as he knew it did not count toward the two years of internship required of him. He followed the temporary internship with several months of study in neuropathology at Philadelphia General Hospital, where the pathologist N. W. Winkelman sharpened Freeman's skills at examining neurological specimens, including sectioning, staining, and photographing brain samples. To Freeman also fell the task of sorting jars containing tissue specimens "that had been on the shelves of the old morgue since time immemorial." The hospital's antique ledgers holding descriptions of autopsies performed in the previous century especially intrigued Freeman, who quickly showed promise as a neuropathologist.

When his formal internship at University Hospital at last began at the beginning of 1921, Freeman served time in a variety of specialty services, and he initially found little to interest him. He had no patience for what he called "scut work," sometimes poured urine samples down the drain, and, he admitted, "was impatient and careless." Freeman eventually realized that as an intern he was supposed to be a member of the hospital team—and the role of team player was never one that Freeman filled well. His own interests superseded those of the team.

Freeman's attention heightened when he entered the neurology and neurosurgery service of the hospital. He fell under the influence of Charles H. Frazier, a well-known neurologist and neurosurgeon on the faculty of the University of Pennsylvania. Like many of his fellow interns, Freeman was awed by Frazier's precise and efficient work in the operating room. Yet the teacher's abrasive manner—it was said that Frazier only spared the hospital's custodial staff from his brutal and sarcastic remarks—seemed to attract Freeman as well. "Somehow I love that man. . . . Perhaps it is due to the moments when his sensitive spirit is off its guard and shows through a little from its rough exterior. . . . His harsh greeting and irritating laugh with apparent detestation of all sentiment have made him seem hard at times, and again his biting sarcasm has made enemies of those that did not understand him and look beneath the surface or try to penetrate the mask," Freeman wrote a couple of years later.

Like Spiller, Frazier treated worst those who tried to draw closest to him. Nevertheless, Freeman rearranged his rotation schedule and traded hours with other interns so that he could spend as much time as possible under Frazier's supervision. He marveled at the teacher's endurance and noted how Frazier often stood "at the operating table for four and five hours, undertaking the most painstaking technical procedures, then eating a few crackers and a glass of milk and making rounds with dressings." Freeman confessed great pride when Frazier at last warmed up enough to call him Walter.

At the same time, Freeman discovered "it was a great satisfaction to see the living brain after the skull and dura were opened." Unlike many of the others under Frazier's wing, however, Freeman did not decide to pay tribute to Frazier's operating room technique by specializing in neurosurgery. Surgery in general bored him. Freeman liked laboratory work and what he called "the preliminary neurological work-up" far too much to become a surgeon. Even after his neuropathology course at Philadelphia General Hospital had ended, Freeman continued returning to the lab there until his supervisors at University Hospital warned him he was neglecting his duties as an intern.

Freeman felt most comfortable and engaged as an intern when he was examining patients. His first professional publication resulted from his examination and diagnosis of one man's arm pain and pupil contraction, which Freeman believed were caused by a tumor at the first rib. When the patient died, an autopsy proved Freeman's supposition correct. This was the art of deductive reasoning at its headiest, and Freeman greatly enjoyed the game. Soon another patient commanded Freeman's curiosity: a young man who arrived at the hospital with his penis in dire shape. Inflamed and dark, the organ was encircled by a ring that the patient's girlfriend had thrust over it but was unable to remove. Freeman ended the patient's agony by filing through the ring and twisting it free with forceps. "The boy asked for the ring but I told him it was a specimen and that I would have to keep it," Freeman wrote. "I had the ring repaired and the Freeman crest engraved on it." For years afterward, Freeman wore the specimen on a gold chain. Later in his career, his keepsakes of patients would take the form of photographs and meticulously tracked follow-up records.

One of the first patients with whom Freeman attempted a psychiatric approach was a girl suffering from rectal pain. Given the inadequacy of his training in psychiatry, it is not surprising that Freeman botched the diagnosis. He traced the pain "to frustration in love and other domestic problems," he recalled. "I was chagrinned when another doctor examined her and found a painful fissure in the anus, which when properly treated, healed up nicely."

Far better professional results arose from Freeman's interest in a patient suffering from bad headaches without fever. Freeman found oddly textured, budlike cells in the patient's spinal fluid. He tried to culture the cells, which turned out to be yeast. After the patient's death, Freeman found cysts throughout the brain, and these cysts were filled with the yeast cells. "This was exciting," he wrote. "There was nothing about yeast infections of the brain in the text-books." He hit the library and found occasional references to infections by one organism, *Torula histolytica*, going back to 1866. Freeman's intense interest in this obscure malady and his thorough photographic documentation of the infecting agents made him an expert of sorts.

Nobody could stop Freeman from following his strong interest in laboratory work and neuropathology. When he heard that an assistant pathologist position was open in the lab at Philadelphia General Hospital, where he had worked informally the previous year, Freeman convinced his superiors at University Hospital to let him conclude his internship a few months early to take the job. On the first day of 1923, he went through an exceedingly busy day in the laboratory for a doctor of his limited professional experience. With the help of one assistant, Freeman performed nine autopsies. His recall and ability to reconstruct his actions were impressive: he was able to commit to memory the details of several autopsies before having the opportunity to set them down for the records. Freeman focused his time as a pathologist, however, on the practice of neuropathology. By this time he knew he wanted to become a neurologist, with a strong research and laboratory component to his work.

Freeman spent another year as a resident at University Hospital. He soon met Madeleine James from Doylestown, Pennsylvania, a tall young woman with a fondness for fine dresses and long ear pendants. She was more interested in art and poetry than in medicine or science, but Freeman increasingly spent his hours off duty with her during her trips into Philadelphia. Freeman also visited her home in Doylestown "and made good progress with her, particularly one Saturday night when a bat got into her home and I started after it with a tennis racquet." The bat escaped, but Madeleine's feelings grew. Freeman, who kept some of his family in the dark about the relationship, presented her with a ring, and she responded with the gift of a ring bearing his initials and the engraved words "*Dieu tu garde*" (May God protect you).

The first great excursion of the young doctor's life, a trip to Europe for the graduate training in neurology that would establish him as a specialist, inspired Madeleine's sentiment. Freeman had been selected as the recipient of an American Field Service Scholarship, which would fund a lengthy period of study in Europe, where war injuries had sparked the most ambitious research and training programs in clinical neurology. In May 1923 Madeleine accompanied Freeman to the dock where he boarded a liner for France, the first leg of a journey that he expected would last two years. As Freeman was saying his good-byes, Madeleine declared that she "was ready to stay on board, without luggage of any kind, saying the captain could marry us when we were past the three-mile limit. I declined, asking her to wait until I returned or asked her to join me."

Although Freeman refused to take along a bride, he did not turn away introductory letters bearing the signature of his grandfather Keen. He was first bound for Paris, where La Salpêtrière, a famous neurological institution founded as a hospital in the seventeenth century and renowned for the psychiatric investigations of Charcot during the 1870s, accepted foreign students as assistants. In decades past, the hospital had attracted Sigmund Freud and others interested in Charcot's studies of hypnosis, forms of paralysis that resulted from neurological disease, and postmortem staining methods. Although the hospital was no longer on the cutting edge of neurological study, the current head, Pierre Marie, had assembled a talented corps of medical students from all over the world.

La Salpêtrière initially struck Freeman as "a great old barn of a place" most notable for its out-of-date facilities. As he became better acquainted with the hospital, though, he gained an appreciation for the variety of neurological cases Pierre Marie had assembled there. While most of the patients were "old women worn out in the struggle of life and not yet ready to take the leap into eternity," Freeman felt himself drawn to the wards that Marie had set up as showcases of unusual ailments. Bed after bed contained patients suffering from diseases that Freeman found fascinating: pseudo-bulbar palsy, Huntington's chorea, rare hormonal disorders, and many others. Freeman helped staff the chilly wards at La Salpêtrière and also taught himself French—well enough to translate French medical journal articles in later years. He paid special attention to

Marie's examinations of patients. In one such assessment, a young woman complained of her fear of injuring herself with sharp objects and of sudden death. Marie's examination found nothing physically wrong, so he treated her anxiety by prescribing vigorous exercise. In the years to come, Freeman would similarly admonish anxious and neurotic patients to replace self-absorption with physical recreation.

Freeman's mother, as it turned out, was living nearby in a hotel on the left bank of the Seine. She had sold the house at Rittenhouse Square along with its furnishings and brought her two youngest children to France for the inexpensive schooling. Freeman rented quarters in a small hotel and became a habitué of the local cafes. Initially, his primitive French—it took him about three months to feel fluent—limited his friendships to other expatriates. There were other barriers to assimilation. "The seal ring from Madeleine kept me from getting too involved with local talent," he noted. Remembering an incident eleven years earlier when he had nibbled two snails "but nearly threw up," he avoided escargot. His inexperience with alcohol, an early manifestation of what he later called his "deep prejudice against alcoholics," also set him apart from the people he met in cafes. He had been bothered by the drunkenness of other students at Yale, "especially when I was trying to sleep in the dormitory while revelers were making the night hideous." As a result, not until his arrival in Paris had he even tasted wine, and he found that he liked it. After his return to America, he frequently took one or two glasses a day.

It was in Paris as well that Freeman grew his first beard. Consciously or not, he cultivated "a timid goatee" that imitated the facial hair of many of the great nineteenth-century neurologists. "Beards have always interested me," he later wrote. "Both my grandfathers carried them, and my father raised a generous one when he was a student in Munich. . . . So I come by my affection for them honestly." And he retained that affection loyally. He discovered that fingering his beard helped his concentration, and the growth provided other tactile benefits as well. "Those who have never grown beards cannot appreciate the delicious feeling of a breeze blowing through it on a warm summer day as the car covers the miles," he declared. "There is the softest titillation, like the caress of a beautiful woman. And when a beautiful woman reaches up with a little hesitation and strokes the beard, or even gives it a shy little tug—well, it has to be felt to be appreciated." Freeman kept the beard, or sometimes just a mustache, for nearly all of the rest of his life.

When he felt like visiting one of the culturally important attractions of Paris—a museum, a concert hall, or a church—he sometimes accompanied his mother. Near the end of his life, when he published his book *The Psychiatrist*, he included in a chapter on psychiatrist-poets some of his own verse inspired by one such visit to the Cathedral of Notre Dame: "Go and return my children! / Sinners, re-enter my womb! / Kneel in the fetal pose, / Bow, in rebirth, on the threshold."

At the time, however, Freeman did not experience his sojourns to the churches and cathedrals of Paris as a reawakening of his religious spirit. In fact, he dated his break from organized religion from the time of his year in Europe. "At this time I was having religious doubts," he wrote, "was rather shocked by what I considered the worship of the various images, idols, so prevalent." The full break was accomplished later in Rome, "where men used the pillars and steps [of churches] as toilet facilities," and where Freeman had his pocket picked while ascending the steps of the Scala Santa on his knees in emulation of the Italians.

Midway through his stay in Paris, Freeman took time off to make his first trip aboard an airplane to meet his grandfather Keen at the International Surgical Congress in London. "My grandfather had been president at the previous meeting three years before and was among the honored seniors this time," Freeman wrote. "I rode his coattails as much as possible, until one of the younger members called me rather disagreeably: 'Grandson!' My feelings were hurt and I stopped spreading myself."

Freeman returned to his studies at La Salpêtrière. During the fall of 1923 at the meeting of a national neurological society, Marie gave him the chance to make his first presentation of some cases before a French audience. Freeman's talk followed a controversial presentation that sparked an argument among the neurologists in attendance, and he began hesitatingly. His French did not fail him, and the patients demonstrated their symptoms on cue. Freeman felt relieved when his three minutes were up, and he considered his international debut a success.

His six months in Paris ended dismally, with a long stretch of rainy days that contributed to "the worst attack of pansinusitis I've ever had," he wrote. "It felt as though the whole center of my face was sticking out about three inches. . . . Convalescence was slow and a gloom settled over me. I hadn't seen the sun for thirty days." After his landlady and a doctor staying in his hotel helped him recover, Freeman decided it was time to move on. Pierre Marie had just retired from La Salpêtriére, and

Freeman found his successor unimpressive. Freeman arranged to continue his study of neurology with Giovanni Mingazzini, the founder of the laboratory of pathological anatomy at the Clinica Psichiatrica in Rome. Mingazzini also served as a professor of neurology and psychiatry at the University of Rome.

Freeman left Paris in early December on a night train that made for a startling arrival in the bright sunshine of Rome. He soon found a room in a pension that suited him. He had a roof garden with a view of the city, plenty of good food, and low rent. The fact that the elevator moved slowly and that his room was high up from the street did not discourage him. Once he descended the 209 steps, he was a short walk from his laboratory, which was situated in a modern and well-equipped psychiatric clinic.

Freeman was the only graduate student then working with Mingazzini, whom the American described as "a genial old man, [who] looks somewhat like a ham actor with his unshaven cheek, very long, long hair and white bow tie." Fluent in English, French, and German, as well as Italian, Mingazzini addressed Freeman in a rapid mixture of European languages that was more difficult for him to understand than plain unadulterated Italian, which Freeman did not know at all. Mingazzini kindly gave the younger man unfettered run of the lab and freedom to do as he pleased. Without taking time to visit any of Rome's great sights, Freeman began work. He spent time on a technique he had discovered for staining nerve fibers, and he began a study of the stages of maturation of the nervous system. Freeman quickly learned that he needed more experience as a lab worker. Once, while attempting to sever the vagus nerve of a live dog, the animal died, even before Freeman caught sight of the targeted nerve. "I might have continued, for animals are plentiful, but I saw no use in sacrificing dogs before I was able to do what I wanted to do to them and before they would be of some scientific value," Freeman wrote. "It hurt somehow to kill a dog when no good comes of it."

One odd highlight of his time at Mingazzini's Clinica Psichiatrica was the opportunity to watch the autopsy of an elephant that had died in a zoo. In an effort to gain access to the brain, the skull of the elephant was subjected to four hours of attack by crowbar, maul, saw, and pickaxe. Though the rough handling damaged the organ and the brain stem—the part that most interested Freeman—he made a mental note to somehow find a way to examine a better-preserved elephant brain in the future. A jackhammer might break the skull more cleanly, he speculated. Years later, Freeman would again grow preoccupied with alternative methods of accessing the brain.

Meanwhile Freeman worked on building his nonexistent knowledge of the Italian language. After three months of study with a native, journeys into Italian translations of Russian novels, and attempts to read Italian neurology textbooks, he pronounced himself fluent.

After his first six weeks in Rome, Freeman once again found himself in the company of his mother. The overcast skies of Paris had bored her. In preparation for her arrival, Freeman made his first visits to the main tourist destinations of the city, including the Forum, the Coliseum, and the Vatican, and, quickly considering himself an expert, he offered this same tour to his mother two days later. After that, admitting "my interests were elsewhere," he spent only Sundays with Mrs. Freeman. He devoted his remaining hours off to exploring Rome's modern highlights, including a political rally officiated by the new premier, Benito Mussolini. Although "I could not understand what he was shouting," Freeman noted, "my spine tingled as the crowd shouted back, gave the fascist salute and sang 'Giovinezza.' That was a rallying hymn, equaled only by 'The Battle Hymn of the Republic' or the 'Marseillaise.' It meant nothing to me politically, but I could see the enthusiasm that seemed to end the drifting of the Roman crowd." Like many other Americans in Italy at the time, Freeman felt fascinated, not repulsed, by fascism's rise.

In the spring of 1924, Freeman finished an article on the anatomy of the nervous system and submitted it to an obscure and short-lived medical journal in Estonia. By chance, the journal published Freeman's study in an issue honoring the career of Ludwig Puusepp, the Estonian surgeon who fourteen years earlier had experimentally severed the neural connections between the frontal and parietal lobes in the brains of patients suffering from manic depression. Freeman at the time seems to have taken no special notice of Puusepp's work. Because of its many typographical errors and photos reproduced upside-down, it is unlikely that many neurologists took notice of Freeman's work in the journal. Freeman, though, felt proud of its publication, and he later republished the article in the American *Journal of Nervous and Mental Diseases*.

Needing a vacation, he traveled a roundabout course through Naples, Capri, Salerno, and Venice. In the latter city, he grew depressed. "I would have given almost anything to have Madeleine with me," he wrote. "The haunting canals with their gondolas and the picturesque gondoliers were just too much for a single man. . . . I decided I had been selfish in not bringing Made along." He tormented himself with questions of whether she would wait for him to return, whether he could make her happy, and whether he was man enough for her. He decided to uproot himself once again for more neurological study in Vienna, at the Brain Institute directed by Otto Marburg. Mingazzini presented him with a letter of introduction. Freeman intended to stay for a while in Vienna before moving on for further study in the Netherlands or Germany.

Vienna, he found, still suffered from the aftermath of World War I and was enveloped in an economic depression. He quickly rented a room. "I had a pair of dancing slippers that were beyond repair and I put them in the waste-basket of my pension," he wrote. "The maid asked me almost tearfully whether she could have them for her husband." Austrian inflation also confounded him, and Freeman felt embarrassed when he belatedly discovered that his gratuity to a helpful servant—100 crowns—converted to about a tenth of a cent.

Soon after arriving in Austria, however, a letter he received from his grandfather Keen cut his stay short. Unknown to Freeman, the venerable old doctor was making use of his professional connections to find a good job for his neurologist grandson. Admiral Edward Rhodes Stitt, the surgeon general of the U.S. Navy, was seeking candidates to fill the post of senior medical officer in charge of laboratories at St. Elizabeths Hospital in Washington, D.C. Though concerned by the heavy administrative responsibilities of the job, Keen nominated Freeman and was seconded by a recommendation from Spiller. William Alanson White, the director of St. Elizabeths, was highly impressed with Freeman's laboratory experience in neuropathology and wanted to hire him. "I have never known a golden apple to fall into any man's lap as nicely as this has fallen into yours," Keen told his grandson.

Freeman, who had never heard of the hospital, found out from Marburg that St. Elizabeths was a well-regarded federal institution for the insane. "I was rather stupefied, for I had had no idea of doing such a thing," Freeman told a relative. "I thought it over for quite a while that day. For the right man it was quite a good place, [with] opportunities for research on various lines in insanity." But Freeman had doubts whether he was that man, and he debated the merits of an administrative job focused on the laboratory versus a teaching position, like Spiller's, which included both clinical and laboratory components. With some misgivings, in the end he resolved to accept the job at St. Elizabeths, and he did so because of his attachment to Madeleine James. The two had been corresponding throughout Freeman's year in Europe, and he had regularly sent her jewelry and reassurances of his fidelity. ("Only once have I been anywhere near drunk, and I was the soberest [of] the party except the chaperones," he had written from Paris. "If women put their arms around me, they do the same to everybody, and neither they nor I nor anyone else thinks anything about it. . . . As for women, there's only one, dear.") Although he cautioned Madeleine that she came in second to his medical career, Freeman decided that the $5,200 salary offered by St. Elizabeths would enable him to return home and propose marriage.

"You will be surprised to hear that I was in doubt at first as to the advisability of acceptance," he wrote back to Keen. "As I see it, the director of laboratories has about reached the pinnacle of that path, from which there is no graduation. To attain the pinnacle before the age of 30 is good for few men. What I have heretofore held steadfast towards as my goal was research, a teaching position, and then a professorship. Now it seems as if I were to be swerved aside. It hurts a little to have cherished plans swept aside, but there was never a real doubt as to what course should be pursued. This offer is too big a one to countenance any refusal, and the possibilities of it grow upon me as I think it over." To his mother, Freeman confessed, "It is not the kind of job I want." Nevertheless, he made plans to leave Vienna.

Before departing, Freeman did realize his dream of examining an elephant's brain stem. Marburg had brain sections on file in the institute's collections. Freeman wanted to see the mass of nerves that supported the functioning of the trunk, and he found that the elephant had highly developed motor and sensory divisions at the top of the medulla. Not destined, however, was an encounter between Freeman and psychiatry's most towering figure. Sigmund Freud was practicing psychiatry in Vienna during Freeman's brief sojourn there, but the creator of psychoanalysis and the future developer of psychosurgery never met. Freeman later learned, however, that one of his own areas of neurological research—a comparison of related structures of the brain stem and the spinal cord—had also interested Freud decades earlier, and the two men "had come to similar conclusions."

Freeman sailed third class from France to New York, finally reaching Philadelphia over the Fourth of July weekend. (Customs officials had raised concerns over some of the photographs he carried with him, including pictures of nude children he had taken at La Salpêtrière to document neurological reflexes.) He tried to reach Madeleine James on the phone, but no one was home. Then he decided to wait on the steps of the Yale Club for his brother Bill, a member, to show up. Bill's roommate, Roger Franklin, let Freeman inside. Freeman explained that he was bound for Washington and knew practically nobody there, so Franklin gave him a letter of introduction to his mother, who lived in the capital with two daughters employed at the Tariff Commission and the Library of Congress.

The following day, Freeman made the nerve-wracking trip to Doylestown to propose to Madeleine. Her impulsive feelings at the dock had long faded, and she promptly said no. Freeman was stunned, and the rest of their reunion "was a strained one. She, too, had been through a period of doubt, and she gave me back my ring, but refused to take back the one she had given me," Freeman wrote. "I think both of us were relieved when we said good-bye." Time healed their wounds over the next few weeks, and Freeman returned in August for "a very amicable gathering of the James family." He picked up some of his belongings and accomplished a less wrenching separation from Madeleine, in which they agreed to be friends. His goal in setting a final meeting with Madeleine, he told his mother, was to resolve the relationship peacefully and "if possible to avoid any ingrowing complex that might arise in either of us. You see I'm a psychiatrist now."

Before reporting for duty at St. Elizabeths, Freeman made one last try to obtain the kind of teaching position he dreamed of. He paid a call to Spiller in Philadelphia with every intention of withdrawing from the job in Washington if his old teacher would help him find an assistant professorship in his department at the University of Pennsylvania. Spiller "seemed shocked [and] turned me down, saying there were several others ahead of me." Saddened that Spiller would not extend himself on behalf of a former student who idolized him, Freeman prepared to move to Washington. He was alone and unsure of his future.

## Chapter 4

# In the Hospital Wards

Freeman arrived in Washington in July 1924. Twenty-eight years old, he was still tender from his rejections by Spiller and Madeleine James. During his nine years as a full-time staff member at St. Elizabeths Hospital, however, Freeman would become a rising star in neurology.

He had visited Washington two times previously. The first time, in 1908, he was a twelve-year-old traveling in the company of his mother and his sister Dot. After they hired a carriage driver to show them the sights of the city, an excursion Walter and his sister found intolerably dull, their mother, Corinne, began to doubt the sagacity of their driver. When asked about the many statues and monuments they passed, he claimed to know nothing. Finally, in exasperation, Corinne sarcastically observed of one statue, "I think that must be Nebuchadnessar." The driver replied blandly, "Ma'am, I think you're right. That's just who he is."

The second time, when Freeman came to Washington to begin his active duty with the army, he left without seeing any sights of the capital. In 1924 Washington was a much larger city. World War I had swelled the size of the federal government, and many agencies still occupied makeshift temporary quarters. Freeman wanted to see the sights this time—the sights that would interest a fledgling neuropathologist. He made repeated visits to the Army Medical Museum and the Surgeons General Library. At the museum, he viewed fractured bones from Civil War field hospitals, many of which—along with hundreds of other donations—had been acquired from his grandfather Keen. He treasured the library for its immense holdings of scholarly journals, in which Freeman felt he could research virtually any medical topic.

Freeman soon became acquainted with Washington's rich human resources as well. The city's governmental and institutional agencies employed some of the world's foremost medical and scientific experts. Once, when he wanted to learn more about the cranial capacity of Peruvian Incas, Freeman paid a visit to the National Museum and its large collection of human skulls. He found the curator Ales Hrdlicka endlessly willing to discuss skulls. But Hrdlicka, intrigued to find before him an example of an authentic Colonial American, insisted on payment of a curious kind. He made Freeman submit to measurements for Hrdlicka's anthropological research on skull types. At the U.S. Patent Office, Freeman found workers who—perhaps accustomed to citizens trying to circumvent Prohibition—quickly led him to information on how best to distill pure alcohol for staining microscope slides. When Freeman wanted to do some research with pentose, a sugar with one less carbon atom than glucose, he tracked down an employee of the Bureau of Standards who had synthesized it.

Freeman soon established himself as one of these valuable Washington experts. At the Navy Laboratory, the pathologist Winthrop Hall once showed him a slide of a strange neurological specimen that had arrived from California. Hall could not identify it. When Freeman held it up to the light, all of his earlier work on yeast infections of the brain rushed back. "I think it's torulosis," he told a surprised Hall without hesitation. After he found yeast cells in the specimen, Hall did not keep his astonishment to himself. Freeman quickly discovered he was distinctive in yet another way. Arriving in Washington, "I fell into a neurological vacuum," he wrote. For several years he remained the only specially trained neurologist in town, and he had to travel to Philadelphia to meet and exchange ideas with neurological colleagues.

Freeman rented a room near 18th and Columbia Road, a streetcar ride away from St. Elizabeths in Congress Heights. He ate most of his meals in restaurants. At the hospital, he first met with the assistant superintendent, A. P. Noyes, who led him through the paperwork of entering him on the payroll, then was directed to the hospital's superintendent, William Alanson White.

One of the most prominent figures in American psychiatry and an early proponent of psychoanalysis, White had been running St. Elizabeths for eighteen years. He was a smart and polished administrator, fiercely dedicated to defending his hospital while he successfully juggled careers as a superintendent, a university professor, and a book author. Frequently called as an expert witness in criminal cases, he aided the defense in the celebrated trial in 1924 of Nathan Leopold and Richard Loeb, accused of the murder of Bobby Franks. White's cool demeanor would later save the hospital when U.S. representative Tom Blanton of Texas, marshalling accusations from former patients and former staff members, held an inquiry on allegations of mistreatment at the hospital. After the lurid charges appeared in the newspapers, White faced the congressional committee and, as Freeman saw during a couple of visits to the hearings, genteelly deflected them. White even managed to turn the accusations to the hospital's advantage. Congress eventually acknowledged that St. Elizabeths' overcrowding and lack of facilities were due to a chronic shortage of funds, and White successfully pressed for increased appropriations. If he had learned anything as a hospital superintendent, it was the necessity of delegating work to underlings. He and Freeman would have a warm professional relationship for the next dozen years—until Freeman began advocating the use of psychosurgery in the hospital.

St. Elizabeths Hospital, whose fearsome brick towers and barred windows overlooked the Anacostia neighborhood of the District of Columbia, housed mental patients from the District as well as from the branches of the military and some departments of government service. Its use dated back to 1855, when the mental health activist Dorothea Dix fueled its construction, and it was known as the Government Hospital for the Insane. Under White's directorship, the hospital ended many of its most punishing practices, including the use of straitjackets and other harsh restraints. White believed that physically restraining mental patients was akin to admitting that his hospital could not devise medically sound methods of recognizing and overcoming their illnesses. Many, perhaps most, of the patients at St. Elizabeths suffered from disorders of the nervous system, including neurosyphilis, the final stage of the ancient venereal disease.

A pragmatic psychiatrist who could simultaneously advance both psychoanalysis and neurophysiologically based treatments, White had supervised perhaps the first attempt in an American hospital to treat neurosyphilis patients by introducing the malaria bacterium into their bloodstream. The technique, which used the high fever of malaria to destroy syphilis spirochetes within the body, had been pioneered by Julius Wagner Ritter von Jauregg, a Viennese psychiatrist. In 1917 von Jauregg—who had been studying the relationship between fever and psychosis for the previous thirty years—exposed three neurosyphilitic patients to malaria drawn from the blood of a wounded soldier, and their recovery spread the treatment across the globe. It was a rare instance of a reliable treatment for a psychosis emerging from the specialty of psychiatry instead of neurology.

Freeman had met the jowly von Jauregg, who wore a fierce brush mustache, at his clinic in Vienna several weeks before starting work at St. Elizabeths. The work of the elderly physician, whom Freeman described as "not a friendly person, rather aloof, reserved, even austere," began a period in which researchers experimented with many physical treatments for psychoses in which no injury to the nervous system was apparent. In 1927 von Jauregg received the Nobel Prize in medicine, the only psychiatrist ever so honored.

Though largely forgotten today, neurosyphilis was a terrible disease, a near epidemic that left its targets—mainly men thirty and over—in a wasted, twisted condition, riddled with bedsores and unable to speak coherently. Debilitated victims accounted for as many as one-fifth of the patients in mental hospitals. They frequently grew demented, psychotic, incontinent, and unable to control their muscles. Death, from pneumonia or infection, was a near certainty. From the U.S. Public Health Service in Puerto Rico, White had imported a shipment of mosquitoes carrying benign tertian malaria. A careless technician, however, had packed the insects in cages sealed with adhesive tape, and eleven of the twelve mosquitoes died or were damaged in transit when they stuck to the tape. As a result, a ten-ton truck—the only Public Health Service vehicle available to complete the delivery—was sent to St. Elizabeths to convey a single living mosquito. The insect was housed in a tiny cage designed to be strapped to the skin, and it succeeded in infecting a nonsyphilitic patient, whose blood supply was used as a malaria reservoir for those with neurosyphilis. "What happened to the mosquito is not recorded," Freeman wrote.

This fever therapy, introduced two years before Freeman's arrival, produced remarkable results at St. Elizabeths. Medicine, for once, contributed to the recovery of psychiatric patients. "No longer were the emaciated patients bent like pretzels, covered with sores and stinking to high heaven," Freeman observed. "It is true that many patients did not recover fully, but at least they were spared from the death that usually occurred within two years in spite of other treatment." In 1963 Freeman heard from the daughter of one such patient, who had died twenty-one years after receiving malaria fever treatment for syphilis. An autopsy revealed that her brain was scarred by syphilis, but there were no active spirochetes.

Freeman started his first day at St. Elizabeths with some words with White and an inspection of the Blackburn Laboratory, the new but as yet unfurnished facility of which he was in charge. White's plan was to combine facilities for X-ray, pathology, neuropathology, and bacteriology within a single laboratory. The lab had recently been enlarged from six rooms to fifty, with a neuropathologist and several technicians working under Freeman. In order to make it a functioning laboratory, Freeman had weeks of completing applications, requisitions, and budgets ahead of him. He amused himself by noting his position of authority. "Here I was, not yet 29," he wrote, "and three steps from the Presidency of the United States. My boss was the Superintendent, his was the Secretary of the Interior, and latter's the President."

Freeman joined the St. Elizabeths staff at a transforming point in the history of American psychiatric hospitals. For decades, such psychiatric investigators as R. Emil Kraepelin had classified and described scores of mental disorders; they had, for instance, broken schizophrenia down into four subcategories. Meanwhile, mental asylums had experienced substantial increases in admissions. Alcoholism and a rise in the incidence of syphilis—with the neurological devastation of its final stages—both contributed to the increases. Institutionalization of the mentally ill had become far more common than during most of the nineteenth century, when confinement in poorhouses and the care of families provided for most of the sick.

Between 1903 and 1933, American psychiatric hospitals more than doubled in size, and some institutions, like Central State Hospital in Milledgeville, Georgia, swelled into vast complexes of nearly ten thousand patients. Even into the 1920s, hospitals lacked treatments that healed, or even helped, many patients, and institutions like St. Elizabeths had grown into huge warehouses of the sick. Snake pits, hell holes, institutions of despair—they were called many things, but psychiatric hospitals of the 1920s really could do little for patients with mental disease except house and feed them until there was a spontaneous remission, which occurred with some frequency. It was a stretch to call these hospitals medical facilities, and the popular name "asylum" was perhaps the most apt of all. "One could cure nothing. There was little scientific understanding of mental illness. And one lived in rustication far from the medical centers with their state-of-the-art labs and great libraries," wrote the psychiatric historian Edward Shorter. "Younger, often idealistic psychiatrists bridled at this sterile incarceration and sought alternatives."

Freeman took a liking to other colleagues on the St. Elizabeths staff. One was Ben Karpman, whom Freeman had first met in 1916 in a chemistry class at the University of Chicago. Freeman admired Karpman's sense of humor and described him as "an aggressive Russian Jew who never overcame the gutturals of his native language." Karpman, who worked in the hospital's criminal division and was the founder and editor of the journal *Archives of Criminal Psychodynamics*, hoped to use psychoanalysis to treat psychotic criminals. He sent a letter to Sigmund Freud outlining his ambition and, instead of encouragement, received from the Viennese master a terse comment: "When you've got your results, send me a postcard."

Freeman initially regarded most of the forty-three hundred patients at St. Elizabeths as pitiful and disgusting. They made him experience "a weird mixture of fear, disgust and shame," he wrote. "The slouching figures, the vacant stare or averted eyes, the shabby clothing and footwear, the general untidiness—all aroused rejection rather than sympathy or interest." Rather than dwell on the patients, Freeman went to work setting up the lab, where his attention to patients "was not on the personal level, but rather in learning all I could about the brain of the psychotic." He was initially thrilled to have at his disposal a Sartorius microtome, a machine that could slice brain specimens into paper-thin sections. Then, however, "I got a fingertip between the carriage and the shaft and let out a yell. That finger was sore for weeks." When he realized that the device required a full-time technician at an annual cost of $5,000, he mothballed it. He was nearly drowning in paperwork when White pulled him aside and gave some advice. "Freeman, you have one pair of hands and one pair of eyes. You can do your own work but no more," White said. "Now you have twenty pairs of eyes and hands in the laboratory and if you use them right you can get twenty times as much accomplished." Years later Freeman noted that White's observations on the efficiency of a large staff had no merit whatsoever, but the director's interest in him gave Freeman much-needed humility and respect for his boss.

Freeman made rapid improvements in the efficiency of the lab. Colleagues were quick to praise how swiftly reports were turned around and how ably a large volume of autopsies were performed—more than two hundred during his first year. Freeman took great pleasure in studying neuroanatomy by examining the remains of animals from the Washington Zoo. In his first few months at St. Elizabeths, he conducted autopsies on a lion, a stork, a monkey, a deer, an alligator, a pelican, a leopard, and a boa constrictor. "I am also learning a lot about the mind and its reactions, and the study of psychiatry which was so neglected in the medical course in Philadelphia is becoming more and more interesting to me," he noted. He especially admired the staff members who tried to use the resources of the hospital to advance the knowledge of insanity and psychiatry. Even so, he initially doubted he would stay at St. Elizabeths for long. A few weeks after starting work, he wrote to a friend that "it is not quite the sort of thing I want and I'm not sure yet whether it will lead any place. . . . I want to teach and do some clinical work along with the laboratory—in other words to be another Spiller. . . . I'm not married, and have the ambition of life work still as the greatest thing ahead of me. When I do get married I shall have a truer perspective, I suppose, and then I shall probably relegate work to its proper place." He never fulfilled this last prophesy, however.

That first summer in Washington, Freeman quickly grew dissatisfied with the "uninteresting" people he was meeting. Tired of setting appointments to meet his father's friends, many of whom lived too far from Freeman's part of town to reach without a car, he dug out the note of introduction that his brother's Yale roommate, Roger Franklin, had written to his mother and sisters in Washington. Freeman appeared at their door one hot night wearing a nice straw hat. The Franklins received him with great friendliness. One of the daughters, Marjorie, made her appearance, and Freeman was impressed by her good figure and healthy head of hair. Marjorie's initial impression of this newcomer was less favorable; she later remembered predicting to herself, "Here is another Yale friend who would probably turn out to be a creep like the last two."

Marjorie, thirty-one years old and three years Freeman's senior, evidently pushed aside these doubts, because she invited him to accompany her to the U.S. Tariff Commission, where she worked as an analyst. She had forgotten her wristwatch at the office, she explained. They climbed into Marjorie's car, and in trying to open the windshield to let in some cooler air, the two somehow managed to smash the glass. Thus began the swift courtship of Walter Freeman and Marjorie Franklin.

Marjorie Franklin was an unusually accomplished young woman. A graduate of Barnard College, she remained there to earn a master's degree in economics, and at age twenty-five she became an assistant professor of economics at Bryn Mawr College. During her five years teaching at Bryn Mawr, she earned a Ph.D. in economics at Columbia University. Her thesis study on Philadelphia's system of taxation uncovered so much evidence of corruption that her faculty adviser urged her to withhold it from publication. She moved to Washington, and in 1923, with her knowledge of four languages, she secured a job in the foreign division of the Tariff Commission specializing in French colonial tariffs. "I was struck from the start by the energetic and vigorous intellect of the girl," Freeman wrote his sister Dot, although he added that "her mother is a clinging vine sort of person." He judged her sister, who had lost a fiancé during World War I and would never marry, "a slat like girl." These family dynamics had required Marjorie to run the household since the death of her father a few years earlier. Unlike Freeman's mother, Marjorie was not a social climber, and she was beginning to doubt that she would ever get married—in fact, she generally approached unfamiliar people with caution. Her lifelong dislike of being photographed neatly contrasted with Freeman's love of cameras and pictures.

Their trip to the Tariff Commission led to two important developments: a golf date the following weekend and Freeman's purchase of a green and tan Oldsmobile. Freeman lost the golfing match but won the girl. Soon Marjorie was bobbing her hair at Freeman's request. Just a few weeks earlier he had been convinced that he was an "untamed" man, unwilling to find time to spend with even a suitable woman, but Marjorie dashed those thoughts from his mind. At the same time, he acknowledged that his stinging rejection by Madeleine James made him more careful in appraising the women he met. "No longer would a pretty face and a charming manner be enough," he decided. Marjorie was not pretty in the traditional sense: she had a large nose and what Freeman considered an overly prominent mouth. No matter; Freeman felt drawn by Marjorie's independent spirit and fine intellect. She was willing to put her own professional goals on hold to support Freeman in his career pursuit. "Am I not lucky?" he observed.

"I really monopolized [Marjorie] from the start, since I had no other acquaintances in Washington," he remembered. "Crossing the Potomac by the 14th Street Bridge one evening a month later, I asked her to marry me." Marjorie must have been unprepared for this proposal, because she mentioned her responsibility to care for her mother, but she didn't say no.

It wasn't long before she thoroughly considered the matter and accepted Freeman's engagement ring. Freeman's brother Bill was getting married in October, and their mother would be home from Europe for only a short while. With that incentive, Marjorie and Freeman quickly planned their wedding. They were married in a small ceremony at Bryn Mawr College on November 3, 1924, the same day that marked the anniversary of Freeman's parents. They had known each other only three months.

The Franklins were Catholic—Marjorie's brother, Gerald, was later ordained as a priest—and Marjorie needed a dispensation from the cardinal of Philadelphia in order to wed Freeman. Their meeting with this clergyman gave Freeman "my final and immunizing dose of Catholicism," he wrote. The cardinal, his eyes closed and his fingertips pressed together, ponderously lectured the couple. Later, at the wedding, Freeman received "another prolonged lecture about my responsibilities in marrying a Catholic." The effect of this monologue was deep and lasting but not what the priest intended. On their wedding night, Freeman made Marjorie promise in writing not to bring up their children as Catholics. Twelve days after their wedding, Freeman wrote to his sister, "With a charming and devoted wife, a good job and a car, what else could a man want to make him happy—except children which no doubt will come in time." As it turned out, they came in record time. His wife was, as Freeman termed it, "fecund," and they conceived a child probably on the first night of their marriage.

They honeymooned for a week in a quiet old lodging house near the Delaware Water Gap, where they enjoyed ample solitude. Freeman found the late autumn weather bracing for his hiking and canoeing outings, but Marjorie was less tolerant of the cold. She shivered through the first night and spent the other nights in a flannel nightgown. They returned to Washington to set up house in a fourth-floor apartment at 3039 Q Street, N.W., where wedding presents, a scarcity of furnishings, and little food awaited them. "We ate our first meal sitting on packing cases," Freemen recalled. "It consisted of several boxes of wedding cake." Although they could have lived in staff housing at St. Elizabeths, Marjorie refused to do so.

They quickly developed a daily routine that barred the unexpected, except for Marjorie's bouts of morning sickness. They would leave together for work just after 8:00, and Freeman would drop Marjorie at the Tariff Commission before continuing to St. Elizabeths. While they were at work, their maid, a Howard University student hired for $25 a month, cooked and cleaned. At the end of the day, Freeman would return to pick up Marjorie. Reaching home by 5:30, they dined together and settled in for the evening. Marjorie's pregnancy produced in her a temporary dislike of Freeman's pipe, so he smoked cigars in her presence and puffed the pipe late at night when he worked or wrote letters to his friends in Philadelphia and Europe.

All that changed with the birth of their daughter, Lorne, on July 31, 1925. Freeman celebrated by bringing cigars and ice cream to the laboratory. For several months, Marjorie put up with the inconvenience of hauling the baby's carriage down four flights of stairs for daily walks in the neighborhood, often to Dumbarton Oaks or Dupont Circle. But after a year of marriage they relocated to an apartment on the second floor of the same building. They celebrated their anniversary with a dinner of oysters on the half shell, crown roast, and pumpkin pie. "As the oyster course ended a knock came at the door, and a boy from the jeweller's brought in a package containing six oyster forks," Freeman recalled. "I had ordered them that morning." The Freemans briefly considered moving to Albany, Syracuse, Rochester, or Buffalo, but their future was in Washington, they ultimately decided. At any rate, their time in their new apartment was unexpectedly brief, because Marjorie was pregnant again, and they bought a house at 4035 Connecticut Avenue. Marjorie's waistline swelled to sixty-six inches. Twin boys, Walter and Franklin, more than seven pounds each, were born on January 30 and 31, 1927. Marjorie nearly bled to death when she hemorrhaged after delivery and was unable to awaken her night nurse. She later needed a blood transfusion.

Freeman and Marjorie, who suffered from headaches and was weakened for weeks, suddenly found themselves heading a family of five. Freeman dashed off a telegram to his mother, now living at Rittenhouse Square near Grandfather Keen, which read: "Twin boys, healthy and heterologous. Marjorie doing well." Excited by the arrival of sons, he instantly declared Walter an extrovert and Franklin an introvert—an appraisal he never altered. The Freemans had hired domestic help after Lorne's birth, but the arrival of the twins demanded more serious assistance. A governess named Virgie, who had reared six of her own children, joined the family to manage the Freeman brood. Her help was greatly appreciated because Marjorie, as Freeman put it, "was not particularly adept as a mother, being too easily upset by commotion. Virgie took it all in her stride and then went home to her own hubbub." (Virgie later proved to be a carrier of typhoid. When she transmitted the disease to two of the Freeman children, she retired, and a couple of her daughters replaced Virgie in the household.) On weekends, the Freemans led their tribe through such local recreation spots as Rock Creek Park, the Washington Zoo, and Glen Echo, "and when Marjorie and I left for work Monday morning she would almost sigh with relief."

Freeman earned a scolding from their family physician when Marjorie again became pregnant, as the twins were only four months old. Baby Paul arrived on February 23, 1928. In their first forty months of marriage, Marjorie and Freeman had produced four children. Despite their written accord, the Freemans agreed that Walter and Franklin would be baptized as Episcopalians, Lorne and Paul as Catholics.

When lecturing, Freeman took the position that fathers understand their children better than mothers because men are less emotional in judging offspring. He believed that fathers should aggressively fill their important role as child rearers and not allow women in the household to shove them aside. "I like to get into the nursery—with cap and gown and mask, it is true—but I want my child, and I want him raw," he declared. "I know he's tough and won't be hurt by my big paws, and I can manage those little buttons and pins and things as well as the next man. Then I give him the works. I tap him on the forehead and watch him blink; I sit him up and then let him fall back and watch the way his arms spread out; I give him two fingers to hold and watch which one he puts in his mouth; I roll him on his stomach and watch how nearly he can hold up his head. . . . From these reactions I can tell what sort of a baby it is going to be, what temperament it will have and what to expect later."

Although she once confessed that she did not really like children, Marjorie took a gentler approach. "She really taught me how to love somebody," son Walter said. "This was important because my father didn't have that capacity for a close emotional tie—he was always somewhat distant. That's not to say he didn't have strong emotions and couldn't express them in his own way. When I had nightmares as a child, he would always come and comfort me, see me through the night. . . . But it was not quite the same kind of warmth that just came naturally to my mother." Added brother Franklin, "We all knew that he loved us."

The children grew up aware that their father "had some weird patients who were somehow a bit threatening," Walter remembered. Once, when the kids were at home with a cook, a man appeared at the door saying he wanted to kill Dr. Freeman. The cook, keeping her composure, replied, "Well, you'll have to go down to his office and kill him there, not here." She then telephoned the office to warn Freeman of the ex-patient's intentions. Freeman sometimes took his children along with him when he went to work at the laboratory at St. Elizabeths. Strolling through the campus among the old brick buildings and hearing the howls of the patients made for an eerie experience. "I remember being cautioned not to put my hands on the railings because the patients would spit on their hands and run them along them," Walter said. Such warnings were not always heeded. Freeman recalled with horror the time when a colleague shouted in alarm in the laboratory offices. Freeman turned around to see little Walter "running his tongue along the edge of the autopsy table."

The boy was simply indulging his curiosity—something St. Elizabeths gave Freeman the opportunity to do as well. One neurological question that puzzled him was the means by which fish maintain their equilibrium in water. Freeman had read that severing the lateral nerve of a fish made it swim on its side, and he hypothesized that fish remain perpendicular to the surface of water by keeping the water pressure equal on the lateral and vestibular nerves. To test his idea, he first put a goldfish in a two-gallon container, attached the container to a clothesline, and swung it in a colleague's backyard where there was plenty of room. When the centrifugal force of Freeman's swing made the water level vertical, the fish maintained its dorsal fin perpendicular to the water surface. Then Freeman asked a burly assistant to get the fishbowl going in a stronger swing that held the water in the container when it was upside-down. Again, as Freeman's photographs proved, the fish kept its position perpendicular to the water surface. A patient at St. Elizabeths who was working as the housekeeper of Freeman's colleague witnessed the experiment in the yard. When her employer returned home, the patient was asked whether anything interesting had transpired during the day. "No ma'am," the patient replied, "just Dr. Freeman was over here this afternoon swinging goldfish on the clothesline."

Freeman knew that schizophrenic patients were more likely than others to contract tuberculosis, while other investigators had suggested that paranoid patients were prone to cancer. In his research, Freeman seemed continually to find correlations between patients' mental diseases and their body measurements. Were certain kinds of mental patients physically different from others? "I went through the autopsy material at St. Elizabeths seeking a relationship between type of psychosis and various physical illnesses," he wrote. Freeman also sought to answer that question by tapping his boyhood skills in photography to document and measure the bodies of deceased patients before autopsy.

Photographing the front, back, and side of the bodies was best accomplished by clamping tongs into the ears of the corpses and having an assistant use a chain to hoist the bodies to a hanging position in front of a large grid marked in 10-centimeter squares. Freeman duly noted and measured the patient's weight, height, chest, and pelvis size, among other observations. He admitted that the procedure produced "a rather grisly scene," and indeed the photographs, many of which survive in the archival collections of George Washington University, show physical ravages and waste that are shocking to anyone unfamiliar with the deterioration of patients who have spent decades of their lives in custodial psychiatric care. Ultimately, Freeman reported, "nothing of note appeared." His paper summing up the results of twelve hundred autopsies at St. Elizabeths and subjecting the statistical outcome to biometrical evaluation was something less than influential. Decades later Freeman noted, "I have never seen this paper referred to by any other author and I don't recall being asked for a reprint."

"I came to St. Elizabeths imbued with the idea that some abnormality had to be found in schizophrenics," Freeman later recalled. But his physical measurements revealed none, and when he took samples of tissue from cadavers in the autopsy room, he found that microscopic examination unearthed little of importance. For a while he grew excited by his discovery that some schizophrenic patients displayed a shortage of catalytic iron in their nerve cells, "and for a time I deluded myself that nerve cells in schizophrenics were unable to metabolize oxygen adequately because of deficiency in cellular iron. . . . I had visions of restoring psychotic people to normal by increasing this metabolic iron. But my dreams failed to materialize," he wrote. The apparent normality of these patients' brains perplexed him, "and the question often occurred to me, 'How is it that a person with such a normal appearing brain can have been mentally sick for forty or fifty years?'" He came to harbor the perverse thought that the brains appearing the most normal would invariably have come from schizophrenic patients.

All of Freeman's work to scrutinize the brains and bodies of psychotics in search of differences led to nothing of consequence. "I approached the study from the standpoint of anatomy and chemistry; and when these gave no answers, I turned to constitution and disease susceptibility or resistance," he wrote. "Biometrical studies on the endocrines yielded no results of importance."

While Freeman's efforts to find physical differences in patients at St. Elizabeths came up dry, an excitement was in the air. In addition to von Juaregg's fever treatment for neurosyphilis, other biologically based therapies were producing changes in psychiatric patients. An assortment of lethal substances, including carbon dioxide, sodium cyanide, and sodium amytal, were under investigation in many parts of the world. Arthur Loevenhart, the chair of the neuropsychiatry department at the University of Wisconsin–Madison, discovered that catatonic patients sometimes broke out of their unconsciousness for brief periods when he gave them mildly poisonous doses of sodium cyanide. Then, in the late 1920s, Loevenhart and his collaborators, W. F. Lorenz and R. M. Waters, began experimenting with carbon dioxide by using the gas to produce what they called "cerebral stimulation" in patients severely afflicted with schizophrenia, depression, manic depression, and other illnesses. For up to twenty-five minutes, formerly mute and rigid patients became talkative and were able to move freely before returning to their original inaccessible, statue-like state, sometimes freezing in awkward postures in midsentence.

Another University of Wisconsin faculty member, the psychiatrist William Bleckwenn, injected catatonic patients with a solution of sodium amytal, a new anesthetic first commercially produced in 1923. He proposed that the aroused state that catatonics experienced should be used for psychotherapy sessions, although the patients might have had other activities in mind for their precious minutes of liberation. Other researchers experimenting with sodium amytal found that the drug gave patients with catatonia a sense of well-being and a wish to remain stimulated, but it did not in any way alter their delusions.

Freeman treated catatonic patients at St. Elizabeths with both carbon dioxide and sodium amytal but saw little improvement in the severely ill. He also devoted time to the study of the effects of variations in air pressure on his patients. He hoped that by manipulating the level of oxygen in their bloodstream, he could improve the functioning of patients' brains. Using the pressure tanks in the Washington navy yard, Freeman was not above experimenting on himself. "Low pressure gave me a headache and mental dulling," he reported. "Pressure of 45 pounds per square inch, equivalent to three atmospheres, brought about a hypomanic state in me

. . . while the only observable reaction in our catatonic patient was that he ate a sandwich instead of being tube-fed. We could not get him to speak. Increased oxygen, therefore, seemed a poor method of bringing about 'cerebral stimulation.'" A 1931 issue of *Time* magazine detailing his research ran Freeman's photograph above a caption reading, "He: Give the nitwit oxygen."

The air pressure tests were not the only instance of Freeman's use of aggressive treatments and procedures. He startled many of his colleagues with his preferred method of tapping the spinal fluid of patients. Instead of recruiting help to secure patients in a deep bend while sitting, then inserting the needle of a collection syringe between the vertebrae, Freeman employed what he was fond of calling the "jiffy spinal tap." Without assistance from other staff members, Freeman directed patients to sit backward on a chair and deeply bend their neck over the chair back. Carefully navigating the opening at the base of the skull, he then pushed a needle into a reservoir of spinal fluid located just inside but perilously close to the base of the brain. Even a slight error in the insertion of the needle could permanently injure the patient. Although Freeman did not invent this unorthodox procedure—he had seen it used often at La Salpêtrière as a graduate student—he employed it frequently, seemed unperturbed by its risk, and took pride in its speed and efficiency.

Two beliefs led Freeman into these explorations of unconventional treatments. The first was his conviction that many psychiatric illnesses were organic in origin and thus medical therapy for the brain would ultimately cure more mental diseases than any amount of psychoanalysis or talk therapy. As a doctor with one foot in the neurological theories and practices of the nineteenth century, he took little interest in the mental apparatus of psychiatric illnesses—the experiences, habits, and traumas that helped disorders take control of a patient's mind—but instead focused on their physical causes. While Freeman placed importance in the gathering of patients' personal histories and lamented the frequent lack of such information in the St. Elizabeths files, his view of psychoanalysis and its Freudian proponents was dim during his years at the hospital, and it grew dimmer as the years passed. In 1928, for instance, the editor-in-chief of the *Archives of Neurology and Psychiatry* asked Freeman to write a review of *The Book of the It*, a psychoanalytic volume by Georg Groddeck. Freeman ridiculed the book, particularly its author's use of such colorful phrases as "penis with a personality" and "woman has no penis." The journal's editor responded with, "Do you expect me to *publish* this?" and rejected Freeman's submission. Rereading *The Book of the It* in 1968, Freeman noted that Groddeck—who advocated psychoanalysis even as a treatment for organic disease—was "by all odds my favorite in the galaxy of the Freudian constellation," apparently because of the intoxicating blend of "astonishment, amusement, disgust, disbelief and what have you" that Freeman derived from Groddeck's work.

Also pressing on Freeman's mind, though, was the second urgent belief: that he was witnessing a catastrophe in progress at St. Elizabeths, a terrible squandering of human potential, and that there must be something, however untried and yet unthought of, that he could do as a neurologist to help halt it. For many years, Adolf Meyer, a goateed and balding neuropathologist and psychiatrist who taught at Johns Hopkins University School of Medicine, had articulated a new role for psychiatric practitioners—that of a fighter for the smooth operation of human society, whose harmony could be maintained only by ensuring that as many individuals as possible remained mentally healthy and contributed to the human effort. Societal vigor, as well as the fitness of individual patients, was the business of the modern psychiatrist. The post–World War I military, in particular, adopted psychiatry as a maintainer of social order. These ideas attracted Freeman. Over time, he came to believe that he had a responsibility to fortify the social contract. "I looked around me at the hundreds of patients and thought what a waste of manpower and womanpower," he remembered. The patients themselves were being deprived of happy lives, of course, but society was losing out on potentially productive members. Gradually, Freeman formulated a goal to benefit society by returning to usefulness some of its most seriously lost causes. Dysfunction was the enemy as much as disease.

By this time Freeman realized that moving to Philadelphia to become another Spiller, an academic researcher with some time to spend on clinical neurological investigation, was growing less likely every month. Philadelphia already had Spiller, and the old professor repeatedly discouraged his former student from returning home. For a time Freeman considered relocating to his home city to open a private practice, but gradually he recognized the opportunities open to him in Washington. He was the only neurologist practicing at St. Elizabeths and one of only two members of the American Neurological Society in the entire city. "Fame and fortune really beckon me to Washington," he reflected in a letter to his mother. "There is almost a virgin field. . . . Moreover a couple of outside consultations have brought me some outside recognition," including a patient he cured of encephalitis—the first person whose life he had actually saved. Freeman saw that his best chance of mixing research with clinical work would come from staying put. While remaining on the St. Elizabeths staff, he signed a three-year lease on an office for the private practice of neurology at 1801 I Street, N.W.

He shared the space with H. H. Schoenfeld, a surgeon with experience in treating patients with nervous disorders. Maintaining a private practice was common among doctors in the government service, and many of them treated other government employees who wanted appointments after normal office hours. Freeman's passage into private practice reflected a growing trend among physicians who concerned themselves with mental disease. After a century of being rooted within the walls of asylums, psychiatric treatment had slipped through the barred windows and out into the mainstream. In 1921, when the American Medico-Psychological Association changed its name to the American Psychiatric Association, the asylum doctors for the first time affiliated themselves with neurologists in private practice who were treating a wide range of psychiatric and neurological conditions. They all now marched under the banner of psychiatry.

Meanwhile, psychoanalysis, the lengthy talk-centered treatment based on the world of the unconscious and the experiences of childhood, had spread far beyond Vienna, the home of Sigmund Freud and his early followers. (Washington had seen the birth of its first psychoanalytic society in 1914.) Freud's theories focused on the causes of forms of neurosis, which were less disabling than the serious psychoses that typically resulted in hospital commitment. The early American practitioners of psychoanalysis, both neurologists and psychiatrists, served a largely middle-class group of patients who functioned well enough to stay out of the hospital. These doctors worked almost exclusively in private practice.

Freeman, the laboratory director, an examiner of brain tissue, and an eager student of all kinds of biology-based treatments for psychiatric illness, had little patience for psychoanalysis. His brand of medicine followed a completely different line: treating psychiatric patients by paying attention to disorders and diseases of the nervous system. But Freud's work and the growing American fascination with the role of psychology in literature, business, and behavior made it possible for Freeman and hundreds of others like him to draw private patients newly aware of the importance of mental health. The psychiatric historian Edward Shorter wrote that Freud's approach to psychiatry inspired this leap "from the asylum to Main Street as the venue for practice. The price of this advance was that psychoanalysis infiltrated American psychiatry far more deeply than elsewhere, causing scientific stagnation and increasing disengagement from the rest of medicine." Freeman would later suffer under the influence of Freudianism as much as anyone.

For several years, Freeman had been dividing his time in Washington between St. Elizabeths and other interests. Admiral Edward Rhodes Stitt, the friend of Grandfather Keen who had put up Freeman's name as a candidate for the laboratory job at St. Elizabeths, wanted to do more to help the young doctor's career. Freeman paid a call on Stitt during his earliest weeks in Washington. "As we sat there in the warm gloom of his old home with the furniture in its slip-covers, a stray cat came walking over the sill of one of the ground windows," Freeman wrote. "In his leisurely way Admiral Stitt welcomed the cat and gave it a saucer of milk from the kitchen. 'The family is away,' he explained, 'and took the dog with them. But not all the fleas. Now, fleas always prefer a cat to a human, so this cat is my providential flea trap.'"

One can only imagine the agony with which Freeman, the cat who always walked alone, endured such leisurely paced visits. He emerged from Stitt's house, however, with the admiral's help in securing an appointment as a lecturer in neuropathology at the U.S. Naval Medical School, which Freeman held for more than a decade. Freeman wanted to make deeper inroads into academe, though. Within a few months, he also paid a call on Eugene R. Whitmore, a professor of pathology at the medical school of Georgetown University. Freeman may have sensed that the future of psychiatric research was at academic institutions, whose departments of psychiatry and medical laboratories would vastly outdistance mental hospitals in new financial resources for research in the coming decades. Labs at schools like Yale, Cornell, and Harvard, not places like St. Elizabeths, would generate most of the discoveries in neuroscience.

Freeman was surprised to find Georgetown's medical school in disrepair and disarray, with papers and journals heaped in stacks, but he accepted Whitmore's offer of an associate professorship at no pay. For the next seven years, second-year medical students from Georgetown attended Freeman's Saturday morning autopsies at St. Elizabeths. Freeman was only a little older than many of his students.

The autopsy theater at Blackburn Laboratory was well lit and could contain the students, but it was not very comfortable. The students sat on concrete benches without cushions. The poor ventilation did little to minimize the odor, especially when the bodies of patients with festering bedsores were under examination. Once, when a patient with colon cancer was presented, Freeman began incising the bloated abdomen. "As soon as the knife cut through the peritoneum, there was a whoosh," Freeman wrote, "the abdomen collapsed and a cloud of nauseous gas escaped. . . . Fortunately, the sense of smell is easily fatigued, so that after a gasp or two I could continue the autopsy without going outside for a breath of fresh air."

How well the students held up that morning is not known, but the weekly program rarely failed to hold their attention. Freeman, fully aware of the spectacle inherent in an autopsy, always did his best to make the class entertaining. After other St. Elizabeths staff members gave summaries of the patients' case histories, Freeman would predict what the autopsy would reveal. "I was often in for a surprise, and so were the clinicians. Patients with mental disease are often poorly equipped to discuss their symptoms. I used to remind the students that we had to rely on objective or laboratory findings rather than upon the history," Freeman recounted. "It's like veterinary medicine, I would say."

Using a favorite technique of stage performers, Freeman often asked students to assist him in the demonstration. Still unaccustomed to surgical tools, some handled the knives and saws awkwardly. One student once cut his own nose with a knife, and "it's a wonder nobody lost a finger," Freeman recalled. Often, when Freeman had excised the organs, he would hold them up—dribbling blood—so that the students in the back row could see.

On one occasion, Freeman found his Georgetown students lacking in attention. An uninteresting part of an autopsy was evidently boring them, because "I heard the unmistakable click of dice," Freeman reported. "Without raising my head I said loudly: 'Put them away! Two hundred and eight bones are enough in this case!'"

At the same time he was on the Georgetown and U.S. Naval Medical School faculties, Freeman lined up yet another teaching job at George Washington University. Such multiple commitments at area universities were common among doctors working for the government. Just as at Georgetown, the medical school at GWU stood at the edge of ruin, and in fact the Council on Medical Education and Hospitals of the American Medical Association nearly stripped it of its accreditation in 1931. In filling its classes with students, it was much less selective than the best medical schools of the East, and many GWU students held part-time jobs and worked their way toward attaining their M.D.s. When he approached GWU, Freeman found the school "long on titles and short on cash," but he regarded the prospect of teaching there as an honor that could serve as "a crowbar to pry open a place in Philadelphia later." Freeman got the appointment and found himself a full professor of neuropathology. Initially, Freeman spent an hour weekly showing his GWU students photos and slides of neuropathological specimens, but the students struck him as unengaged. Then Freeman hit upon the idea of bringing in living patients to add immediacy to the discussion of specimens from patients with similar case histories. "The effect was electric. The students immediately became interested; the rustle of papers and the hiss of whispered conversation ceased," Freeman wrote.

The students felt so energized by Freeman's class that they petitioned GWU to have Freeman replace another faculty member who taught neurology by reading from a textbook while sitting on a desk. In the fall of 1926, Freeman was named professor of neurology and chair of the neurology department. "This proved an excellent challenge to me, and I went through the chronic wards of the hospital choosing patients who would demonstrate most clearly what I wanted to show to the students," he noted.

As Georgetown and GWU grew into direct competitors for students and funding, Freeman faced the necessity of choosing one over the other. Georgetown offered him a full professorship in neurology, but he cast his lot with GWU, which had made him a department chair and gave him more satisfying work. But Freeman had already taken advantage of his Georgetown contacts to start graduate work there in pathology. He chose to focus on biometrical studies in psychiatry, and Eugene Whitmore agreed to serve as his faculty sponsor. In earning his M.S. and Ph.D. degrees (he attained the latter in 1931), Freeman knew ahead of time that his "final examination" would be a raw oyster–eating contest with Whitmore at Wearly's restaurant in Washington. Freeman won.

Thus credentialed, Freeman put more energy into a project that had interested him ever since his arrival in Washington: the writing of a comprehensive textbook on neuropathology. While Freeman believed that the field of pathology was well served with adequate teaching books, "it has jestingly been said that pathologists know all about all of the diseases of all the organs—except the most important." The brain had received scant attention from authors of pathology texts, brain examinations were rarely conducted during autopsy, and only one neuropathology book already existed in English. Freeman hoped his book would fill the gap. It was a difficult undertaking. Much of the best research in the cellular function of the brain during the 1920s and 1930s was happening in Europe, and Freeman was committing himself to an intensive study of works in several languages. In addition, research on cellular function was advancing rapidly. Any author of the kind of textbook Freeman envisioned would see his words grow obsolete in a short time.

As usual, Freeman pursued his work on the book at a blistering pace. (A dozen years later, he wrote a poem that ended with the couplet, "Pace? It's hard on the nerves. / Let's go!") Between his responsibilities at St. Elizabeths, his private practice, his teaching, his involvement with his family, and his attendance at medical meetings, he researched and wrote several chapters. Often awakening at 4:00 A.M., he wrote in early-morning bursts, which he followed with his day at St. Elizabeths and then visits with patients well into the evening. When he finally returned home and dropped into bed, the sounds of the traffic along nearby Connecticut Avenue kept him awake. But he would again get out of bed early to begin another arduous day. He heightened the pace of his work on the book toward the end, continuing to work even while recovering from a bruised chest he suffered after a road accident in which his car was struck by another and flipped onto its side. At last, in 1932, the book was done.

After Freeman sent his manuscript to W. B. Saunders Company, a Philadelphia publishing house that had previously issued his grandfather Keen's six-volume textbook on surgery, he learned that the publisher had recently received two other neuropathology texts for consideration. Freeman's manuscript was the one that earned the offer of publication. The other two books were later published by other firms and would eventually diminish Freeman's sales. Not long after Freeman completed the index while sick in bed, his book was published in 1933.

*Neuropathology: The Anatomic Foundation of Nervous Diseases* was written in what turned out to be an uncharacteristically detached and nonanecdotal manner, a style Freeman would not use when he wrote his books on psychosurgery in the years to come. It continued Freeman's interest in discovering tangible differences between the nervous systems and the brains of healthy people and those with mental illness. He concluded that paranoia, psychopathic disorders, and some varieties of mental deficiency left no visible mark on the brains of their victims. In patients with manic depression, however, Freeman noticed differences—specifically alterations in the shape of ganglion cell nuclei—between those who died in manic and those who expired in depressed states. Most strikingly, Freeman declared that pathological studies of cadavers showed that schizophrenia was somehow linked to a deficiency of iron in the brain, and he hypothesized that patients lacking sufficient iron in their brain could not properly utilize oxygen.

It was Freeman's first published book, and he took pride in seeing the fruits of his labor so beautifully presented between covers. "Friends tell me that the chapter on neurosyphilis is still the best in the English language, and I agree with them," he wrote years later. That pride revisited him when he traveled to London in 1935. "One of the most delicious thrills I ever received was finding my book on neuropathology in the Library of the Royal College of Physicians, with the notation: 'Not to be taken from the Library.'"

Long hours and stress from writing the book, accompanied by lack of sleep, had left Freeman a mental wreck. Sixty-nine years earlier, his grandfather W. W. Keen had experienced a similar collapse at the close of his demanding service as an army doctor. Freeman had what he called a "nervous breakdown." It felt like "my brain was on fire," he wrote, and in his delirious mental state he designed a fantasy house, shuffled his way through his daily responsibilities, and composed a macabre poem that he titled "Psychological Plagues." Part of the thirteen-stanza poem read: "burn the candle of Man's life / at both ends. / Spare not his frontal lobes where / foresight / and ambition / and self-control / are centered, / But tire them out by constant work. / Turn the fruits of victory into ashes of / fatigue in his mouth."

The driven and energetic Walter Freeman that his family and friends knew had vanished. How to bring him back? Through talk therapy? Through an examination of his problems? What Freeman prescribed for himself was what he often recommended for his own patients with nervous stress: recreation and distraction. He took a month off from Blackburn Laboratory and booked passage on an ocean liner to France. After a week near Etretat on the Normandy coast, Freeman reboarded the same ocean liner for its return to the United States.

Aboard the liner, Freeman breathed in the salt air and watched the movement of the ocean as he returned home. He had gained some important information about his own limitations. Clearly, he had driven himself too hard while writing the book. He discovered that he must pay attention to insomnia, which foretold worse things to come if he didn't attend to its warning. "The sea voyage cleared the atmosphere," he wrote. "I deliberately set aside all plans for the future until the last evening at sea." And on that night he resolved to play more golf, never again spend a whole summer in Washington, and quit his job at St. Elizabeths.

Later, Freeman dashed off an article on his lessons from the breakdown, which he titled "Danger Signals: On the Advantages of a Nervous Breakdown and a Few Neurotic Symptoms in Certain Men Under 40." The article characterized such breakdowns as blessings in disguise that could help victims avoid worse disasters ahead, including diabetes and strokes, if only they heeded the warning to slow down. Freeman ignored his own advice, however; he pushed himself harder and harder and in fact developed diabetes and suffered a minor stroke in later years. But he did take an immediate step to prevent insomnia in the future. He began swallowing Nembutal, a barbiturate sedative, which he continued using for most of the rest of his life. "I am dependent on it but do not consider myself addicted since I have rarely needed more than three capsules a night," he wrote. Freeman evidently did not regard his use of Nembutal as a dependency because his dosage and the frequency with which he took it did not increase.

In a sense, Freeman had left St. Elizabeths just in time. The Great Depression brought financial hardships to most government-run psychiatric hospitals, and St. Elizabeths was no exception. Patients would suffer from lower staffing levels, deteriorating care, bad food, and, most of all, the nihilism of many doctors, who knew of nothing they could do to help patients aside from those suffering from the often treatable condition of neurosyphilis. In 1933, 43 percent of all patients who escaped the grip of hospitalization did so not by being discharged but by dying. The American Psychiatric Association did not organize a generally accepted nomenclature of psychiatric diseases until that same year. Patients living in psychiatric institutions had doubled in number during the previous three decades.

Psychoanalysis, quickly gaining adherents in the United States, offered little to deeply psychotic patients. Neuroscience was still too primitive to suggest any new treatments; Otto Loewi, a faculty member at the University of Graz, had first identified a neurotransmitter in the brain only a decade earlier. American psychiatry had as yet failed to develop its own methods and traditions for treating people whose mental problems left them deeply disabled.

Freeman shared none of the nihilism of his colleagues. Finances partly dictated his decision to leave St. Elizabeths. His private practice was growing, and he was simply too busy to continue at the hospital on the meager salary that had actually declined during his nine years of employment there. But Freeman did not let go of his clinical teaching at the hospital until after the outbreak of World War II, even as his contacts and friendships at St. Elizabeths waned.

There was another aspect to Freeman's decision to leave the hospital: while other Washington doctors respected some of the research and medical work that went on there, the setting of a large psychiatric institution simply struck some of them as a weird and unscientific place for a medical practice. By extension, they believed, strange and unprofessional doctors must work there. Many on the St. Elizabeths staff reciprocated by avoiding contact with their peers and resenting their opinions. "Giving up a job of $5,000 a year in the Depression cost some uneasiness but it was a question either of giving it up or surrendering the growing practice and the close professional contacts with my colleagues," explained Freeman, who at all costs wanted other doctors to see him as professional and scientifically inclined.

And Freeman highly valued his contacts. A few years after joining the St. Elizabeths staff, he attended the International Congress on Mental Hygiene, which was convening in Washington. Freeman had no interest in most of the sessions, which focused on various aspects of behavioral psychiatry, but instead haunted the social events, where he could meet and exchange business cards with the foreign visitors. He added their names to his "card catalogue" of influential people in neurology and psychiatry. Starting in the mid-1920s, he bucked the norm at St. Elizabeths by taking on responsibilities in a staggering number of professional societies, panels, and associations. Among the first was the American Society of Neuropathologists, which Freeman helped found as an informal group to share specimens and slides, and over which he briefly presided. He also took part in early activities of the American Academy of Neurology, the Society for Biological Psychiatry, the Washington Society of Pathologists, the Washington Society of Mental and Nervous Diseases, and the American Medical Association's Section on Nervous and Mental Diseases.

Freeman was elected secretary of the latter group in 1927 and became its chair four years later. At the 1931 AMA convention in his hometown of Philadelphia, he not only delivered an address titled "Psychochemistry" to his fellow section members but won a bronze medal for a multimedia exhibit he prepared with the neuropathologist Karl Langenstrass. The exhibit cleverly presented brain specimens sealed inside large watch crystals—a display technique that Freeman had picked up from exhibitors at a previous AMA convention—combined with medical histories offered on pull-down window shade rollers and with screenings of films showing the movements, reflexes, and gait of neurological patients. Freeman discovered that medical exhibits not only directed attention to his research interests but allowed him to advertise himself to his peers. "While papers are more enduring, since they get published eventually, they lack the personal touch," he wrote. "I made many more acquaintances and became much better known through exhibits than through papers." Many of the physicians drawn by the 1931 exhibit and those that followed entered their names and addresses in Freeman's guest book; from there they were carefully added to his catalogue of contacts who would receive reprints of Freeman's published papers. Later, when Freeman was trying to popularize and legitimize lobotomy as a treatment for psychiatric illness, he would remember his earlier successes with medical exhibits.

Another group that attracted Freeman's interest was the Medical Society of the District of Columbia, which he had first tried to join when he came on staff at St. Elizabeths. A clause in the society's bylaws, however, barred membership to full-time government employees. In 1926, now with his own private practice to supplement his government service, Freeman reapplied. His application sparked a heated debate on the merits of offering membership to doctors who juggled government jobs and private practices; eventually the organization changed its laws to allow admittance to any white physician in the District. When Freeman became president of the society a quarter-century on, he worked to overturn the remaining barriers to the admittance of doctors who were not white.

During the 1930s, Freeman began eight years of work as a member of the Mental Health Commission of the District of Columbia, a group established to examine and make commitment recommendations for new patients in the psychiatric service of Gallinger Hospital, which was later renamed D.C. General Hospital. Freeman and the other psychiatrists on the commission occasionally visited patients' homes in order to learn more about their ability to care for themselves, and one such visit resulted in an unusual acquisition. "I was a few minutes early at the house of an aged lady who could no longer be managed at home," he recounted. "It was an undistinguished row house, somewhat dingy, with Victorian period furniture; on the wall was a magnificent painting." The artwork, a rendering of the Madonna painted on a wood panel, was owned by the patient's companion, who had resolved to sell it. She said it was the work of Giacomo Carucci da Pontormo, a fifteenth-century master. Freeman was determined to buy it, and it remained in the Freeman home for decades.

Meanwhile, Freeman's forebears were shifting stages as well. William Williams Keen had long remained mentally sharp; well into his nineties, he was studying such diverse topics as allergies, anthropology, chemistry, physics, and astronomy. His intellectual activity ended when a stroke partially paralyzed him and muddled his thinking so that he was left "disoriented and querulous, asking constantly to be taken home," Freeman wrote. Freeman's son Franklin remembered visiting his great-grandfather with his siblings after the stroke. "He was propped up in this big Victorian bed, with a big headboard and footboard, like the bed in Lincoln's White House room. He was pink and bald with a fringe of white hair and a black moustache. . . . I shook his hand very gravely, and then we were ushered into the dining room to get chicken noodle soup with bread and butter balls." Pushed in a wheelchair, at the age of ninety-four he attended the 1931 convention of the American Medical Association, the organization over which he had presided decades earlier. Freeman, along with his mother and his brother Norman, witnessed the loud ovation that greeted Keen when he appeared on stage before the assembly.

Bedridden, he died a year later, in 1932. His oldest grandchild left little record of his feelings on the passing of this medical giant who had shown him the exciting potential of a medical career and had helped him obtain his first real job as a doctor. "A full, rich life" was all he wrote.

Corinne Keen Freeman soon followed her father in death. During Freeman's years at St. Elizabeths, his mother had finally let go of what he called "her devotion-to-duty theme for me," her expectation that her oldest son should dote upon her. After several years of living in Paris during the 1920s, she had returned to Philadelphia and involved herself in civic and medical organizations. Freeman admired her energy but felt no deep affection for her. When they traveled together in 1929 to attend the American Medical Association's annual meeting in Portland, Oregon, it "was the first time that my mother and I could get along without disagreement." At age sixty-four, just after her election as the president of the women's auxiliary of the AMA, she began feeling pain in her abdomen. This daughter, widow, and mother of physicians waited several days before seeking medical help. An operation to remove her intestinal obstruction failed. "My eyes were moist when I saw her fighting the oxygen tent," Freeman remembered, "but dry when she died." The deaths of his grandfather and mother provided Freeman with inheritances that helped his family weather the worst of the Depression. They also cleared the way for Freeman to form an enduring professional and personal relationship with a doctor, new to Washington, who would collaborate with him on the most important medical experiment of Freeman's career.

## CHAPTER 5

# A PERFECT PARTNER

GEORGE WASHINGTON UNIVERSITY would serve as Walter Freeman's professional headquarters until 1954. Founded in 1821 as Columbian College, the school began offering medical courses within its first few years and launched the first medical department in the District of Columbia. An early proponent of clinical studies in the medical curriculum, GWU opened a hospital during the 1890s, and in this aging facility, Freeman saw many of his first patients as a full-time professor. At GWU, he struck a lucrative balance between institutional work and private practice. He would see his family grow to six children and would lose one to a terrible tragedy. He would embark upon a wild, controversial journey—the refinement and promotion of lobotomy—and as a result would become one of the most famous physicians in America.

Freeman quickly set to work at building a functioning neurological laboratory at GWU. He had no chance to duplicate the wealth of resources and equipment at St. Elizabeths, but he soon pieced together enough equipment to furnish a fourth-floor, one-room lab that gave work to one technician. Freeman assumed the duties of lab photographer. When his work required equipment or skills unavailable in his own modest laboratory, his colleagues at St. Elizabeths, Johns Hopkins, and the U.S. Naval Hospital were usually willing to help.

Confining himself to the lab, however, was not Freeman's style. It was not long before his six-foot-tall figure grew familiar to GWU students and faculty. Believing that students learned best with plenty of visual stimulation, he adopted a striking way of dressing and presenting himself, an odd mix of tradition and fashion. His goatee and glasses, round-rimmed and clouded with scratches, evoked the dusty image of the old-fashioned denizen of the laboratory. But his cane and wide-brimmed hat called to mind something of the stylish rake. Dangling from the vest pocket of his three-piece suit was the "penis ring" he had appropriated a decade earlier. To him the ring was a mark of distinction, an accessory unique to Walter Freeman. It was as if in his appearance he demanded respect both for the place he held in medical tradition as a third-generation doctor, the grandson of W. W. Keen, and an innovator who wanted to use his place in that tradition to set medicine on its head.

Once at GWU, Freeman heightened his national reputation with his work on the American Board of Psychiatry and Neurology, an organization he helped found in order to establish and maintain professional standards in those two medical specialties. Psychiatrists and neurologists needed to certify their peers for competency through examinations, interviews, and scrutiny of professional credentials, just as members of other medical specialties had been doing for many years. In December 1934 Freeman was one of twelve neurologists and psychiatrists who met in New York City to organize the certification process. Freeman thought that the even division of neurologists and psychiatrists at the meeting—six of each—created a misleading impression that the two specialties were easily delineated, when in fact many neurologists treated psychiatric patients, and vice versa.

Initially, everyone on the board was older than Freeman by an average of nineteen years, and he thought them "a rather crabby bunch." The group included a hospital psychiatrist, several teachers of psychiatry and neurology, and a couple of pure clinicians. Also involved was Adolf Meyer, the dean of American psychiatry and the first proponent of an examining board, who sat at the long meeting table sipping cups of highly sweetened coffee and occasionally offering confusing summations of the opinions of others. "If one listened intently to [Meyer's] introduction, it was better to stop listening then and there, since this was followed by elaborations, arguments pro and con, side issues of pertinence, and revolvements of his thinking that became so involved that his final summing up sometimes bore little relation to his original thesis," Freeman noted. "Only those accustomed to hearing him argue with himself were able to follow through to the end."

Predictably, the group first got hung up on the name of their board—or, more specifically, whether neurology or psychiatry should come first in the name. Eventually, psychiatry triumphed because of the far greater number of physicians practicing in that specialty. When the time came to elect officers, Freeman won the position of secretary, which had been reserved for a representative from the AMA's Section on Neurology and Psychiatry. Freeman cast the deciding vote for himself. He would remain secretary for twelve years, and when he finally surrendered the position in 1946, his successor would declare, "Walter, *you* are the American Son of a Bitch of Psychiatry and Neurology, emeritus."

The board agreed to start examining doctors in 1935. At first, the board grandfathered in physicians who completed medical school by 1919 if their credentials appeared adequate. Some pre-1920 graduates still had to undergo examination if their record failed to impress the board, including one seventy-two-year-old candidate who passed his exam. "Sad to relate, he died within a short time, but he was enormously gratified at his success—and so was the board," Freeman noted. Freeman and board president Lloyd Ziegler missed the grandfather provision by a single year, but the board granted them certification without the exam anyway. Understandably, some candidates resented these exceptions to the rules. "The fact that we were nominated [to the board] by the Section on Nervous and Mental Diseases of the American Medical Association and had been professors in medical schools for several years made no difference" to the critics, Freeman observed. Of course, many candidates for certification who had to take the examination and some who failed it were also longtime members of medical school faculties.

Freeman took an active role in designing the assessment, which emphasized oral responses and observation of the candidate's clinical examination of patients. Psychiatrists and neurologists had to show a working knowledge of both specialties. The goal was "to certify candidates who could safely be entrusted with responsibility for patients under their care," Freeman wrote. "This was one of the main reasons for ensuring that the candidate in psychiatry could recognize an organic condition and that a candidate in neurology could recognize a functional condition." As secretary of the board, Freeman had the responsibility of visiting the sites around the country where examinations would be performed, ensuring that the facilities and testing materials were ready, and recruiting examiners. Freeman avoided recruiting examiners who were "brusque, sarcastic, rigid, and especially those who insisted in too great detail upon their own views, who examined on the basis of their own specialized researches." When Freeman himself conducted the neurological portion of an examination, he considered himself a strict evaluator, especially when a psychiatrist faced him. "I expected of the candidates some recollection of what they had learned in medical school, refreshed by recent reviews," he declared, and when Freeman had tested a candidate, "he knew he had been examined." Not all candidates appreciated Freeman's strictness; one, after receiving his failure notice, arrived at Freeman's office in a fury and wanted to punch him out, but Freeman was not there.

The examinations were difficult and intimidating; many doctors failed to pass. Among the earliest to be examined were about thirty candidates from the Phipps Psychiatric Clinic, which Adolf Meyer directed. All except two failed. The board's rejection of one prominent psychoanalyst caused a furor. Dexter Bullard, the superintendent of Chestnut Lodge, a psychiatric hospital in Rockville, Maryland, and a frequent referrer of neurological patients to Freeman, strongly believed that psychoanalysts should not touch their patients. So when his board inquisitor asked him to make a neurological examination, Bullard flatly refused. Unsympathetic to psychoanalysis and its practitioners, Freeman viewed Bullard's position not as an ethical choice but as an example of a physician reacting "strangely under the stress of examination." Given a failing mark, Bullard became an enemy of the board's activities and later a vocal critic of psychosurgery.

There was no uproar when another unsuccessful candidate killed himself years later. "My main recollection of him is the sweating, the tremor and the disorganized examination that he performed. My colleagues on the Board agreed that he was impossible," Freeman wrote. "A sad ending for all the time, money and energy expended in his attempt to become a certified psychiatrist." Some successfully certified candidates later killed themselves as well, a fact that ignited in Freeman a long-burning interest in the disproportionately high number of psychiatrists who committed suicide.

Meanwhile, Freeman continued work on two lines of research that had first attracted his interest at St. Elizabeths. With Herbert Schoenfeld, a surgeon interested in neurological work, he began experimenting with various ways of photographically capturing images of the brain's ventricular system, which collects the liquid metabolic products of the brain and spinal cord. With these images in hand, for example, a physician could view an outline of a patient's cortex, determine the nature of a tumor, and decide whether to perform surgery. An earlier method of ventriculography had replaced some ventricular fluid with air, which was visible in X-ray photographs. "When appropriate patients appeared, I persuaded

Schoenfeld to inject a small quantity of thorotrast, a colloidal thorium dioxide into the ventricles," noted Freeman, who found the resulting photographs revealing and spectacular. With Schoenfeld, he wrote papers on thorotrast ventriculography, which was slow to catch on because the injected chemical agent was radioactive and thus raised the risk of malignant tumors as well as produced inflammation and irritation. Unfortunately for these partners in research, another procedure to chart the brain was being developed in Portugal by an ambitious neurologist named Egas Moniz. His method, cerebral angiography, mapped the organ's blood vessels by using thorotrast and later a much safer organic iodide solution.

Freeman's second avenue of research was an investigation of the origins of multiple sclerosis (MS), a highly debilitating neurological disease that often attacks adults in the prime of their lives; it sometimes results in paralysis and otherwise produces problems with muscle coordination, vision, and speech. Since the condition is incurable, Freeman believed that the best hope for its victims was to strengthen the ability of their bodies to repair damaged nerve fibers. His experiences with two patients suggested a possible treatment. The first, Wilford Wright, became paralyzed and was diagnosed with MS soon after his graduation from the Wharton School of Business. After a brief recovery, the disease returned with a fury and limited Wright's ability to talk, see, and walk. There was not much that Freeman could offer Wright—fever and vitamin therapy had proven absolutely ineffective for other MS patients—so he simply recommended bed rest. As Wright's condition deteriorated, "I sent him to the Veterans Home in Bay Pines, Florida, to die," Freeman wrote. To Freeman's astonishment, Wright gradually began showing improvement. The patient began typing letters to Freeman in 1933, and, still unable to walk, he learned how to pedal around on an oversized tricycle. Eventually, Wright pedaled all the way from Florida to Washington, a six-week trip he made in part to visit Freeman, and he later tricycled all around North America. In time, he regained his ability to walk.

The second patient was Lillian Murphree, a government employee whose MS left her a nearly blind quadriplegic. Freeman sent her to Florida, where she underwent a recovery similar to Wright's, yet even more spectacular. A few years in Florida enabled her completely to throw off her paralysis. She went on to become a fashion model and charm school proprietor in Texas. "How she can pirouette on the stage is almost unthinkable to me when I recall how desperately ill she was at the beginning," Freeman wrote. He concluded that, combined with bed rest, relocation to the warm climate of the South was partly responsible for these patients' victory over MS. "Whether the effect is due to fewer head colds or digestive upsets, or chilling, or to the more leisurely tempo of life, nobody knows," he observed.

Freeman seems not to have seriously considered their recoveries as anomalies or as phenomena explainable by other reasons. He promoted his views on the benefits of the South to members of the Multiple Sclerosis Society's medical board, on which he served for ten years, but few others gave credence to the limited anecdotal evidence that he offered. He launched an MS treatment center at GWU Hospital and attended various conferences exploring the causes and treatment of the disease, but his interest in the disease had peaked once he reached his conclusions, and it eventually fizzled. By his own admission, Freeman had little to contribute to MS research except to advance his own hypothesis. He had latched onto an unorthodox and controversial treatment for a desperately serious disease, defended it by recounting the positive experiences of his patients, and ridden the hypothesis as far as it could carry him. It did not carry him far in this instance. In just a few years, however, Freeman would find another controversial wave to ride, and that one had enough momentum and complexity to carry him for the remainder of his life.

In order to catch that wave Freeman needed a partner. Fortunately, the perfect partner arrived at GWU in 1935, a neurosurgeon named James Winston Watts. Once again, Grandfather W. W. Keen had made an important, albeit indirect, contribution to Freeman's career. Back at the turn of the twentieth century, when Keen assisted the young surgeon Harvey Cushing, he had helped propel Cushing to launch the new medical specialty of neurosurgery. Cushing went on to teach at Harvard Medical School for many years, and he and his neurosurgical disciples grew into some of America's greatest medical heroes—doctors who could diagnose and safely treat tumors and other brain disorders that had formerly spelled a death sentence. As the specialty matured, it increasingly attracted brilliant and hardworking young surgeons. Watts was one of those drawn to the excitement and promise of neurosurgery.

For years, two general surgeons with an interest and limited experience in neurosurgery had taught this young specialty at the GWU Medical School. As chair of the neurology department, Freeman wanted a neurosurgical specialist on the faculty, someone whose foundation in neurology as well as surgery would make him a close collaborator in teaching and research. Watts had actually initiated contact with Freeman after hearing Freeman speak on neuropathology at the annual conference of the American Neurological Association during the summer of 1933. Freeman's magnetic bearing at the lecture podium and the vigor of his speech had entranced the young neurosurgeon. Watts first sighted Freeman "strolling down the boardwalk wearing a beard (in the days before beards were in fashion), a Texas sombrero, and carrying a cane," the younger man later wrote. "I heard him lecture, and in contrast to the usual scientific paper or discussion, read in a monotone, Freeman was articulate, forceful and inclined to be dramatic."

The two men met and spoke at the conference. Several months afterward, Watts, then teaching neurosurgery at the University of Pennsylvania Hospital in Philadelphia and working in the service of Freeman's old mentor, Charles Frazier, discovered that Freeman had not forgotten their meeting. One morning, "I had left for the hospital in time for an eight o'clock operation," Watts recalled. "A few minutes later, my wife, Julia, answered a ring at the door in her nightgown, thinking I had forgotten something. There was Dr. Freeman, who had driven up from Washington that morning, walked around the city for a half hour to pass the time and still arrived at the ungodly hour of 7:30 A.M."

Freeman was in the midst of his impatient search for a neurosurgeon colleague. Watts, in his mind, had quickly become the best candidate. Born in 1904 in Lynchburg, Virginia, Watts had medicine and money in his blood. His maternal grandfather had been a family physician, and his father was a banker who described his occupation as "capitalist." Watts graduated from the Virginia Military Institute and studied medicine at the University of Virginia. He served his neurosurgery residency in Chicago. After a year of study in Europe (where he took advantage of the chance to examine Lenin's brain in Germany), he settled in New Haven in 1932 to join the research fellows at Yale University's three-year-old Laboratory of Primate Neurophysiology, the nation's first lab that focused on the links between brain physiology and neurological disorders. It was one of the few labs in the United States exploring experimental neurology. The Yale lab hoped to shed light on the mysteries of the human nervous system by experimenting on the brains of monkeys and chimps, by studying the effects of cutting or removing parts of their brains.

At Yale, Watts had built a friendship with John Fulton, the brilliant physiologist who directed the laboratory. Fulton, a native of St. Paul,

Minnesota, had earned a Rhodes scholarship to Oxford before studying with Harvey Cushing and becoming the youngest full professor at Yale. His large head, clean-shaven face, and well-tailored suits became familiar sights at medical assemblies around the world. Married to a wealthy heiress and living in Mill Rock, an estate at the edge of New Haven, Fulton indulged his passion for gourmet food, costly wines, and rare scientific books while juggling countless projects and experiments at Yale. A prodigiously organized man (and a childless one), he published more than 130 articles and books during the 1930s alone.

Over the next several years, Watts and Fulton would spend weekends together with their wives, attend medical conventions in each other's company, and collaborate on research. Once, while Watts was attending a medical meeting with Fulton, one of the physiologist's newest chimps made a public display of her attraction to the young neurosurgeon. The chimp dashed away from Fulton "and vaulted over three or four rows of seats, jumped in my lap and put her arms around my neck," Watts recalled. "People in the audience were alarmed at this animal, but Professor Fulton said calmly, 'Don't worry, he's a friend of the chimpanzee's!'" This affectionate chimp was named Becky, and she would later contribute to the development of psychosurgery and lobotomy. Watts's patients often displayed a similar partiality to the genial Virginian, and he became known as a kindly surgeon who paid unusual attention to the comfort of those in his care.

Fulton had supervised Watts's research studying the effects of lesions in the frontal lobes on the workings of the intestinal tract, which Freeman had read and admired. Previously, many neurologists believed that the frontal lobes, even in monkeys, concerned themselves only with the highest mental functions such as decision making. Through these experiences, Watts had grown interested in far more than the technique of operating on the brain and other parts of the nervous system. He had developed a deep appreciation and knowledge of neurological diagnosis, neuropathology, and segments of practice that overlapped with biological psychiatry. This orientation led Freeman to regard the neurosurgeon as good partner material and to aggressively court him, despite their differences in personality and practicing style. Watts was a deliberate and nonconfrontative man who enjoyed taking an afternoon nap and admitted that he took "nearly everything seriously." As a medical student, he once wrapped a handkerchief around his neck and pretended he had a sore throat when he realized with mortification he had come to class without wearing a necktie. In later years, Watts lamented the decline of formality in dress and manners.

Watts also hesitated to undertake unnecessary risks in the operating room. When he grew convinced, for instance, that coughing and sneezing hospital personnel contributed to the spread of infection during surgery, he insisted on keeping them away from patients. "I've had fights with operating room supervisors about that," he declared. "I've thrown assistants out, nurses out, antagonized the operating room supervisors, but I simply wouldn't permit it." Freeman, on the other hand, breezily broke convention and cared little for what he considered the unnecessary formalities of surgery, such as antisepsis and draping.

Watts wanted to return to the South, or at least to a southern-flavored city, and Washington, despite its lack of medical renown for anything other than ophthalmology, appealed to him. "I did want a University appointment," Watts later said. "And I wanted a place where I was either the chief or had a good chance to be the chief. . . . I'd gotten tired of assisting other people and I thought I'd like a position of responsibility."

Freeman's first move was to invite Watts to come to Washington to lecture at the Medical Society of the District of Columbia on the frontal lobes and their functions. On a tour of the campus, GWU's hospital thoroughly underwhelmed the visitor. Then containing just one hundred beds, it was nondescript and poorly furnished. The lobby had seating for only four people on a single bench. The X-ray facilities particularly alarmed him; he noticed that from the ceiling dangled stray wires at head level whose electrical charges made one's hair stand up on end. But at a party that the Freemans hosted for Watts and his wife, Julia, as well as in all their other meetings, Watts felt impressed by Freeman's energy and brilliance. A couple of months later, Watts succumbed to Freeman's persuasion and agreed to join the GWU medical faculty as an associate professor of neurosurgery. He began teaching at the start of the 1935–1936 academic year.

Once at GWU, as the two men jointly taught many clinical and lecture courses, Watts received a full dose of his colleague's indefatigable personality. "I don't use the word[s] 'brilliant' or 'genius' very often, but he bordered on both of them," Watts recalled. He found Freeman stimulating and hardworking, with only the neurologist's endless energy a fault. "The worst thing about Walter Freeman was that he never got tired. . . . That was [what] made him the most difficult," Watts admitted. In the classroom, Watts considered Freeman "a great showman. Gosh, the students loved his lectures and demonstrations." Freeman, in fact, had evolved into one of the most colorful and popular instructors on the GWU campus. In the gloomy and tight days of the Great Depression, Freeman's lectures sometimes substituted for entertainment. Medical students "often brought their 'dates' (it was cheaper than going to the movies)," Watts noted. "After examining a patient with a stroke or Parkinson's disease, he would march up to the blackboard, pick up a piece of chalk in each hand, draw a coronal section of the brain, using both hands simultaneously and sketch in the internal capsule and basal ganglia, dotting in the pathologic lesion. The students were fascinated." Freeman admitted that he rehearsed for up to a half-hour before class time to keep his ambidextrous sketching sharp.

Eccentricity followed Freeman into the classroom, where he employed unorthodox methods to signal his impatience or keep the students attentive. Watts recalled watching his colleague stare intently as a medical student would haltingly relate a case history, when "there would be a sudden sweep of his hand; the students became alert, then he would open his hand slowly and allow the fly to escape." Freeman knew the power of unpredictable happenings in the classroom and noted with pride the time "I set fire to my coat pocket by failing to empty my pipe before I put it away." "Students notice those things." Another time Freeman nearly set the classroom ablaze during a Saturday afternoon clinic in which he emptied the pipe into a ventilator shaft. "The smoke began coming out at an alarming rate, but the fire was soon controlled," Freeman remembered.

Freeman worked hard to entertain in the classroom. Using live patients, he gave demonstrations of hypnosis, hysteria, and the behavioral effects of sodium amytal. He once wrote a paper about the art of showmanship. "I wrote that what the teacher had to say did not have to be important, indeed did not even have to be true, but it had to be interesting," he observed. "When a woman with Cushing's syndrome was being presented, I called attention to the hirsutism by pulling my trouser leg up to the knee and placing my leg beside hers. I showed the class how to perform cistern puncture, saying, 'Awfully simple if it goes right—but simply awful if it doesn't.'" He retained this drive to entertain for years, though he sometimes crossed the boundaries of professional decency. In 1941, for instance, while leading a neurological clinic at Western State Hospital in Washington State, he staged a tour de force for fellow physicians. Using a patient to display the infantile behaviors of senile dementia, "I pulled from my hip pocket a nursing bottle full of warm milk and fed it to the greedy old lady. That's a picture they'll not soon forget," he wrote in a letter to Marjorie. "She fumbled around with it and tried to get the whole bottle in her mouth, just as our babies used to do. And then I gave her the bowl of my pipe to suck on and she did the same thing. I'll say she was demented!" As a teacher, Freeman aimed to retain the attention of his audience at all costs, for a drifting classroom could signify a shrinking of his public stature. Spiller and Frazier had shown him that great men seize the limelight without regard for the human casualties of their actions.

At home, the Freeman family continued to grow. Two more sons, Keen and Randy, joined the brood in 1934 and 1936, respectively, to complete the family at six children. (Marjorie also miscarried a pregnancy when Randy was about four.) Freeman tried hard to spend more time with his children, especially after he learned that their governess "sewed up the pockets of the boys and made the children walk two by two when on an outing," he recalled. To Freeman, placing such restrictions on kids was not the way to encourage curiosity and their adventurous spirit. He and Marjorie took the family on day trips to nearly every Washington-area spot that gave them the chance to learn and explore, including Rock Creek Park, the National Zoo, Great Falls, and Glen Echo. They went on weeklong summer outings to camping grounds along Chesapeake Bay and the Potomac River. Freeman wanted his children to gain fluency in French, so at the age of nine his daughter, Lorne, became the first Freeman child to spend a block of time away from the family. After passing a summer with friends in Paris, she returned home not only able to speak French but also speaking English with a French accent.

Freeman's recollections of his weeks in the woods with his father still resonated, however, and he wanted to embark upon some extended trips with his own children. These trips would both satisfy Freeman's need to break free of Washington at least once each summer and build the character and abilities of the kids. Marjorie did not look forward to the discomforts that lay ahead, and she rarely agreed to go along.

The first trip—one of many that Freeman would take with his children over the next dozen years—was in 1932. Packing his Ford with the kids, Freeman drove to his aunt Dodie's house in Vermont. He took the family on the cog railway to the top of Mount Washington, where the notoriously variable weather pelted them with rain, hail, and snow. They made a rough descent with the youngest, four-year-old Paul, clinging to his father's shoulders. "I think this was expecting too much of the children," Freeman noted, "but it showed what they could do under stress and set the scene for many later trips that proved no more taxing. I believe that outings such as these, getting the children away from their home base and creature comforts, established a firmer basis of friendship and cooperation among us. Furthermore, they led to greater self-reliance, so that as they grew up to adulthood, they were able to step into their proper roles." For the remainder of his life, Freeman felt proud of his role in exposing his children to the challenges of the outdoors precisely when they were most impressionable.

Besides these treks with his family, little gave Freeman greater pleasure than his periodic trips to Europe to attend medical conferences, mingle with his colleagues from around the world, and add new names to his address book. Often, Marjorie accompanied him, and they usually extended their time abroad for side trips to visit friends and see the sights. The first of these overseas excursions had come in 1931, when they traveled to Bern, Switzerland, for the First International Neurological Congress. For Freeman, a highlight of this trip was his opportunity to meet the legendary Russian physiologist Ivan Pavlov, whose work on the conditioned reflex in dogs had brought him international fame. Like Freeman's grandfather, W. W. Keen, the eighty-two-year-old Pavlov was remarkably energetic in his old age, and Freeman described the white-bearded physiologist as "a rather short, erect old man whose gestures and movements were as lively as those of a man thirty years younger." Disappointingly, the translations of Pavlov's talks and papers "did little to clarify his thoughts, at least to me," Freeman noted.

Four years later, Freeman was eager to attend the Second International Neurological Congress in London. Marjorie agreed to go, too, and she gathered up several gowns that she had bought in Europe four years before, but she presumably was not thrilled when she learned that Freeman was planning on transporting strange cargo. He had consented to help the cancer investigator Charles Oberling, whose family was hosting their daughter, Lorne, in France and who was researching the viral transmission of cancerous malignancies, in the transportation of a shipment of infected chickens. The Freemans left their children with friends, accepted the chickens from the Rockefeller Institute, and booked transatlantic passage from Baltimore on the *Normandie*. This would be the most important overseas trip of Freeman's life, but it began with a dull ship voyage. Freeman had to content himself with studying the many "well upholstered women in bathing suits" on board. "Their cheeks, chins, breasts, bellies, hips and thighs wobbled like jelly. I was fascinated," he later remembered, and added that he met nobody interesting on board.

They disembarked in France, with Freeman carrying the crate of infected chickens under his arm. Hours later, after clearing customs and taking a train to Paris, they were able to exchange the birds with Oberling for their daughter. They spent a few days in Paris before crossing the channel for the congress. Freeman brought with him an exhibit on the use of thorotrast in ventriculography, with photos and transparencies, which he planned to display in London to supplement a paper that he would deliver at the congress.

The congress attracted a wide range of neurological luminaries from around the world, including Pavlov in a return appearance just a year before his death. Wearing a stylish hat and toting his cane, Freeman took the opportunity to investigate London on foot. There were also side trips to Oxford and other attractions, but for Freeman the greatest draws of all were the events of the congress at University College.

Freeman wasted no time in setting up his display in the congress exhibit hall and soon met the occupant of the neighboring exhibit stall, Egas Moniz of Portugal, who had brought along a similar presentation detailing his work using thorotrast in cerebral angiography. It was eight years since Freeman had first heard of Moniz's invention of the imaging technique and four years since he had heard the Portuguese neurologist describe angiography at the first congress in Bern. Moniz, sixty years old at the time of his visit to London, was not aging well. He had ulcerated and swollen hands from a thirty-six-year-long battle with gout, a puffy and excessively wrinkled face, a withdrawn demeanor, and the air of someone constantly in pain. Although the two neurologists advocated competing methods of visually revealing the brain, they developed a rapport. They conversed in French. "Moniz seemed to me a kindly old gentleman, who had made a great discovery and could now rest on his laurels," recalled Freeman, who was impressed by Moniz's conviction in his own ideas. He would soon learn that Moniz had no intention of going out to pasture, and that Moniz was germinating ideas that would greatly influence Freeman's own.

For many attendees, a highlight of the congress was a full-day symposium on the function of the frontal lobes of the brain. The Canadian neurosurgeon Wilder Penfield recounted his surgery on a patient whose giant brain tumor necessitated the removal of her frontal lobes. Amazingly the patient was Penfield's own sister. Richard Brickner of the New York Neurological Institute presented a paper on Joe A., a patient of the neurosurgeon Walter Dandy. After building a successful career as a stockbroker, Joe A. had suffered headaches of increasing intensity, had become forgetful, and had fallen into a coma. Dandy found a meningial tumor in the frontal lobes and had to remove them almost in their entirety. Joe A. became Brickner's patient about a year after surgery. According to Brickner's assessment, Joe A. lost self-restraint and his capacity for complex thinking along with his frontal lobes. He became a braggart (especially in exaggeratedly describing his sex life), sometimes threw childish tantrums, told school yard jokes, and grew uninterested in keeping himself clean and clothed. But in situations that demanded little social sophistication, Joe A. could get along, and Brickner found that people meeting him casually rarely noticed anything awry. The patient's intellectual abilities seemed intact.

Moniz, whose command of English was fair, must have listened to these two reports with heightened interest, in addition to the French neurologist Henri Claude's opening comments giving an overview of the personality changes that accompany frontal lobe injury. He may have concluded from Penfield's and Brickner's presentations that damage to the frontal lobes caused changes that were more irksome than debilitating and that patients who lacked frontal lobes could still function and exercise their intelligence. In this sense, Moniz may have believed, they were better off than many mental patients he had encountered who could not function and contribute to society.

A separate session on frontal lobe physiology was moderated by John Fulton, Watts's mentor at the Laboratory of Primate Physiology at Yale. Here, an experimental psychologist stepped to the front of the assembly to deliver another report on frontal lobe function. He was Carlyle Jacobsen, who, starting in 1933, had worked with Fulton on experiments involving Becky, the chimpanzee that had jumped into Watts's lap, and another chimp named Lucy. (Previously these animals had been used in investigations of the common cold at Johns Hopkins University.) Over a period of time, Becky and Lucy had been trained to perform a series of tasks, continually growing in complexity, in order to receive rewards of food. In one set of experiments designed by Jacobsen, the chimps had to recall the location of food hidden beneath cups. In another set, Becky and Lucy had to locate and, using sticks of varying lengths, retrieve bowls of fruit that were placed on platforms out of their reach.

In the course of these experiments, it became clear that Becky and Lucy were not temperamentally identical. Becky frequently showed frustration and emotional disturbances as she learned and sometimes failed at her tasks; she displayed a behavioral trait that Jacobsen described as an "experimental neurosis," caused by her struggle to complete her tasks. Lucy had a calmer personality and took her setbacks in stride. After training the chimps, the researchers removed the frontal lobes of the animals' brains in order to determine the effect of this loss on their problem-solving abilities. Jacobsen found that the chimps seemed to have only a slightly harder time solving their experimental challenges without their frontal lobes. As an aside, however, he mentioned that the animals experienced changes in their handling of the experimental anxiety. Becky, formerly easily frustrated, grew less upset from her missteps after her operation; her "neurosis" had seemingly vanished. Lucy, however, displayed the opposite reaction: she showed much more anger and frustration postoperatively than she had with her frontal lobes intact.

What happened immediately after Jacobsen's presentation remains one of the most debated sequences of events in the history of psychiatric medicine, a scene passed down in conflicting recollections and interpretations. At the center of the dispute is the question of when Moniz formed the idea of conducting brain operations to rid patients of psychotic symptoms. Did hearing about Becky's indifference to her anxiety after the loss of her frontal lobes suddenly give Moniz the idea to treat mentally ill humans with brain-damaging surgery? If so, why did he ignore the experiences of Lucy, whose feelings of anxiety were stronger after her operation? Or had the notion of psychosurgery been percolating in Moniz's mind for some time before his arrival in London? The one indisputable fact is that promptly upon returning home, Moniz began planning his own human experiments in frontal lobe leucotomy, surgeries designed to treat psychiatric illness.

In later years, Fulton repeatedly declared his conviction that the inspiration for leucotomy came to Moniz at the moment of Jacobsen's talk at the frontal lobe symposium in London. Of course, it was in Fulton's own interest to position this work as the launching pad for an important new medical procedure, and one medical historian has raised the possibility that Fulton invented the inspiration Moniz drew from Jacobsen's talk as a myth-making "parable." As Fulton told the story, "Doctor Moniz arose and asked if frontal lobe removal prevents the development of experimental neuroses in animals and eliminates frustrational behavior, why would it not be feasible to relieve anxiety states in man by surgical means?" Believing that Moniz was proposing the complete removal of the frontal lobes in humans, a risky and difficult procedure, Fulton said the suggestion left him feeling "a little startled." (Fulton later added, a bit regretfully, that "since the Moniz procedure had been suggested by the results which we reported in 1935, I had felt in a measure responsible for a procedure that had had widespread adoption before the basic physiological mechanisms had been adequately elucidated.")

Freeman, who was also present at the frontal lobe session but never gave an eyewitness account of Moniz's questions and seems not to have heard them if they were ever spoken, believed Fulton's interpretation of events. Freeman wrote that only Moniz "envisaged the possibilities" of the issues his questions had raised. "Certainly I did not," Freeman noted. But in 1950 when he questioned Moniz about the importance of Jacobsen's presentation in the development of psychosurgery, Moniz peevishly responded that he could not remember what, if anything, he had asked of Fulton at the London conference, and that "I couldn't find any reference to this subject in any of my papers. I think this subject is of no importance therefore." Moniz then recounted how he had been interested in methods of disrupting neural connections in mental patients as early as 1932. "Naturally I studied the problem in various aspects and all questions direct and indirect on the subject interested me deeply. I was always thinking of this daring operation and all elements were important for me." Moniz seemed to be saying that Fulton's and Jacobsen's research might have played a minor contributing role in the first leucotomies, but then again it might not have played any role at all. In his first paper reporting his experimental surgeries, published in 1936, Moniz did not mention the Fulton–Jacobsen research.

For Freeman, the London conference had been a professional bust in one sense. "My paper and exhibit were soon forgotten," he wrote. "They represented about three years of very occasional opportunities and those who had worked with thorotrast were quick to point out the dangers of its use in the brain." Moniz's cerebral angiography was the method that neurosurgeons would adopt in order to create photographic images of the brain at work. But in another sense—one that Freeman would not understand for many months—the congress was to determine the course for the rest of his career. For he would remain in contact with the Portuguese neurologist for the next two decades. When Moniz began undertaking brain operations to treat mental illness in November 1935, Freeman was one of the first to hear about them and to champion them. And Freeman would make sure that he and Watts would be the first in North America to replicate Moniz's surgeries on their own.

Egas Moniz, as it turned out, was not the name given to him at birth. Born in Avanca, Portugal, in 1874 into a wealthy family with nine-hundred-year-old roots in the country, he was christened António Caetano de Abreu Freire. Active in the political left in his years as a university student, he began writing liberal pamphlets under the pen name of Egas Moniz, a Portuguese patriot who helped halt the Muslim invasion of the Iberian Peninsula in the twelfth century. The young man's adoption of the name of Moniz would be like an American political activist today using the name of Tom Paine.

Moniz completed medical school at the University of Coimbra in 1902 and continued his studies in Paris with advanced training in neurology. There his research on neurological aspects of human sexuality drew the attention of the monarchy at home, and Moniz's monograph was for a time banned from publication in Portugal. (It would later go through nineteen published editions.) In 1911 he began teaching at the medical school in Lisbon. That same year, the end of the monarchy and the establishment of a republican government in Portugal reignited Moniz's political passions, and he won election to the national parliament. He made a steep ascent into the upper ranks of the Portuguese government, where he served as minister to Spain, minister of foreign affairs, and his nation's signatory of the Treaty of Versailles, which formally ended World War I.

Throughout his work in politics and government, during which he became one of the world's most politically powerful physicians, Moniz did not abandon medicine. Like W. W. Keen, turning war into a research opportunity, he made significant contributions to the understanding of neurological injuries sustained in battle. The ascension to power in Portugal of a rightist regime in 1920, followed by a military coup six years later, forced Moniz out of politics and made it possible for him to pour his energies into his pursuit of a method to produce photographic images of the vessels of the brain. After injecting many different substances into the carotid arteries of dogs and cadavers, he attained success in 1927 with a solution of sodium iodide. (He later advocated thorotrast and other substances as better contrast media.) The effect of this new technique of brain imaging was dramatic. "Here was a new method for exploration of the brain without opening the skull," Freeman wrote. "Soon Moniz described not only the normal pattern of the arteries and veins, but also their abnormalities in cases of tumors, cysts, abscesses, thromboses and aneurysms." This invention of cerebral angiography, which Moniz zealously claimed through aggressive publication of his findings, earned Moniz his first nomination for a Nobel Prize in medicine in 1928. Another nomination for the prize followed five years later.

Despite his discomfort and disability from gout, in his late fifties Moniz became a prolific neurological investigator, publishing or copublishing thirty scientific papers in 1933 alone. He was afflicted with an intense passion for the unknown. His professional interests ranged from hypnotism to psychoanalysis. He later wrote that during the early 1930s he grew interested in psychiatric illnesses and new approaches to treating them. Moniz was not a psychiatrist, but like many other neurologists of the time, he frequently saw patients with psychiatric problems. He strongly believed that biologically based, somatic treatments, not behaviorally based therapies, offered the best hope for these patients. Over the next two years, he later claimed, he developed a theory that characterized some mental illnesses as fixed delusions caused by the presence of similarly fixed neurological circuitry in the brain. In people with mental disease, these circuits were in constant activity that produced obsessive and harmful behavior. If there were a way to disrupt those neural circuits, he believed, psychotic behaviors might be eliminated. Moniz had begun to focus his attention on the frontal lobes as the home of those incorrigibly fixed neural pathways for two reasons: the frontal lobes had no other clearly understood role, and they were thought to be somehow involved in intellectual functions. Before attending the London conference in 1935, however, Moniz published nothing that even hinted at his thinking along these lines.

Only after returning home from London did Moniz begin to act upon his ideas. "For awhile, he stayed at his rococo hideaway, the Casa do Marinheiro, whose red-damasked study and stately grounds afforded him the privacy he needed to plan his next step," wrote the psychosurgery historian

David Shutts. Then, after he had decided to go ahead with his history-making operations, Moniz had to find someone whose hands would follow his instructions. By this time, gout had rendered Moniz's hands incapable of manipulating surgical instruments with any accuracy. In Lisbon, he found the Oxford-trained neurosurgeon Almeida Lima to be a willing partner. Handsome, well dressed, and dark-haired, Lima was, like James Watts, a professional descendant of Harvey Cushing. Together Moniz and Lima determined to damage as little of the gray matter of the frontal lobes as possible by focusing instead on severing white neural fibers that allowed communication between the prefrontal lobes and other regions of the brain. (Leucotomy derives from the Greek term *leukos*, meaning "white"—thus "cutting white matter.")

Moniz and Lima made one practice attempt on the brain of a cadaver. Then, on November 12, 1935, operating at the Santa Marta Hospital in Lisbon, they journeyed into the new world of psychosurgery by injecting 0.2 cubic centimeter of alcohol into the frontal lobes of a patient. Wherever nerve fibers contacted the alcohol, they dehydrated and died. Since the alcohol did not produce permanent results and Moniz feared that traces of alcohol were damaging untargeted parts of the brain, he adopted another surgical technique. He began using a device called a leucotome, adapted from a widely used surgical instrument called a trocar, which had at one end a wire loop that extruded when Moniz pressed a plunger. After Lima cut burr holes in the top of the patient's skull directly over the frontal lobes, Moniz inserted the leucotome so that, as best as he could determine, the end of the tool abutted the superior, middle, and inferior gyri of the prefrontal lobes of the brain directly behind the forehead. He extruded the loop and, rotating the tool, cut 1-centimeter-wide cores of brain tissue without harming the surrounding tissue. Eventually, Moniz directed Lima to cut as many as four entry holes above each frontal lobe.

Moniz maintained that although the operation did not eliminate his patients' delusions and hallucinations, it diminished their emotional responses to those psychotic symptoms. The cutting of the leucotome made them less agitated and volatile and, he hoped, more able to take on the duties of a normal life. Furthermore, he asserted that leucotomy in no way hurt the patient's intellect. He failed to offer evidence to support this last declaration.

Most of Moniz's patients were institutionalized at the Bombarda Asylum in Portugal. The first patient to receive the alcohol injection, a middle-aged woman suffering from agitated depression, was one of the most miserable patients in the hospital. After leucotomy, she reported some relief and was cured, according to Moniz, although she never left the hospital. Seven other patients diagnosed with agitated depression or paranoid schizophrenia, among the hospital's most hopeless cases, received the alcohol injections next, and sometimes more than once. From their experiences Moniz concluded that the best results from leucotomy could be expected from patients with disordered emotions or feelings, rather than those with such psychotic symptoms as hallucinations. Moniz's ninth leucotomy patient was the first to undergo surgery with the leucotome. The greater precision achieved with the new instrument produced less unintended damage to the brain, Moniz believed, but Sobral Cid, the director of the Bombarda Asylum and a proponent of psychotherapy who held a dim view of somatic psychiatric treatments, refused to make any more patients available to Moniz. The last six patients in Moniz's initial leucotomy series of twenty came from other hospitals.

Moniz wasted little time in bringing his surgeries and their results to the attention of his medical colleagues. His reputation as the innovator of cerebral angiography and his status as the head of the neurology department and the dean of medicine at the University of Lisbon gave him easy access to the medical societies and their journals. Less than four months after the first operation, on March 3, 1936, Moniz reported on his first twenty leucotomy patients before the Academy of Medicine in Paris. He judged 35 percent of the patients cured, another 35 percent helped, and 30 percent unimproved by leucotomy. Given the mental deterioration of his leucotomy patients before operation, these statistics, if accurate, were impressive. Within weeks he published an article in a French medical journal and followed that with the completion of a lengthy monograph detailing the procedure and its results. Republication of these works in many languages came soon afterward. The monograph, in which Moniz coined the term *psychosurgery*, came out simultaneously in several languages.

Very quickly the author faced a firestorm of criticism, "which, if not organized, at least was widespread," noted an English contemporary. Psychiatrists were outraged over the incursion of the neurologist Moniz and the neurosurgeon Lima into the evaluation and treatment of psychiatric illnesses. Many questioned the crudity of Moniz's underlying theory of mental illness and wondered how the destruction of healthy brain tissue could be considered an ethically sound treatment for psychiatric disease or anything else. Others criticized Moniz's presentation of only anecdotal evidence to buttress his assertion that leucotomy was effective and safe, since in his rush to publish, Moniz had not allowed time to follow up on his patients beyond the first few weeks after surgery. Moniz had not examined the seven patients he had pronounced cured, for instance, beyond eleven days after the operation. And clearly some of the patients in the unimproved category were worse off than they had been before surgery.

Some of the most damning criticism, however, focused on Moniz's failure, or reluctance, to investigate whether his operations had provided support for his hypothesis on the link between fixed psychiatric behavior and fixed neural connections in the brain. Sobral Cid dismissively called Moniz's speculations "pure cerebral mythology."

Moniz, of course, was not the first person to suggest that mental disorders might respond to brain surgery. Trephining, the practice of opening large holes in the skull among the Incas and other ancient cultures, dates back to prehistoric times and may have served as a way to relieve illness or affect behavior. "Since no written document corroborating these claims is available (on account of the fact that the Incas had no writing), little of use can be said in this respect," one medical historian has noted. One trephined skull discovered in France and carbon-dated to 5100 B.C. gave evidence that the patient not only survived the procedure but also lived past age fifty, which is very elderly in Neolithic terms.

Contrary to common belief, Phineas Gage, a railroad worker whose frontal lobes were pierced in 1848 by a tamping iron during an explosion near Duttonsville, Massachusetts, is not an important figure in the development of psychosurgery. Gage survived the accident and lived for another eleven years. Although Gage underwent a change in personality as a result of his brain damage—from a responsible and energetic worker to an impatient, stubborn, and rude man who could not hold down a job—Moniz and the other psychosurgeons who followed him rarely cited this case as a significant factor in the development of their theories of brain function or their procedures. In his book, *An Odd Kind of Fame: Stories of Phineas Gage*, Malcolm Macmillan convincingly shows that Gage's connection with the rise of psychosurgery is "very minor, if not entirely mythical."

Meanwhile, nineteenth-century anatomists like Franz Josef Gall, Pierre Flourens, and Pierre Paul Broca investigated the role of the brain as the seat of thinking and emotion and examined the localization of specific controls and functions in discrete areas of the brain. Into this arena stepped Swiss psychiatrist Gottlieb Burckhardt, whose brain operations in 1888 and 1889 are discussed in the second chapter of this book. But even before Burckhardt had planned his surgeries to relieve symptoms of schizophrenia and other forms of psychosis, several British surgeons were attempting operations designed to treat mental problems.

In 1887 two English physicians reported on a head injury patient who periodically became violent and shouted out to imaginary people. Recalling that they had found damage to the angular gyrus region of the brain during an autopsy of an earlier patient with a similar history, they probed the same area of the living patient's brain. They found nothing unusual, and the patient never suffered another attack. In 1889 William Macewen, another English surgeon, explored the brain of another patient afflicted with violent behavior after a head injury. He found a piece of bone penetrating the brain. After removing it, Macewen reported that the patient's homicidal behavior ended. That same year and in 1890, papers appeared in British medical journals describing patients with neurosyphilis who recovered after surgeons drained their brains of yellowish fluid. Another physician reported a nearly identical case in New York State in 1890.

Other operations intended to relieve cranial pressure in neurosyphilitic people continued in Britain and the United States through the early 1890s, with one of the participating surgeons finally concluding that the procedure "is at the moment of no material benefit whatsoever." In 1895 Emory Lamphear wrote of his operation on the brain of a patient with psychiatric symptoms resulting from cerebral hemorrhage. The patient had suffered from "a band-like sensation around the head" and homicidal impulses directed at his wife, among other troubles. The patient's behavioral problems ended after Lamphear opened his skull, drained a pint of brain fragments and fluid from the third left frontal convolution, and filled the resulting cavity with saline solution. Whether they represented beneficial treatment or spontaneous remission, these and the earlier cases raised "the question of the propriety of excising cortical areas in insanity as well as of epilepsy, when certain subjective phenomena such as hallucinations of sight and hearing can be given a local habitation in the brain," noted a contemporary of these surgeons, C. K. Mills.

The psychiatric historian Edward Shorter explained that these surgical experiments produced "a climate of meddlesomeness" and a willingness to tamper with the brain that was widespread among physicians between the nineteenth and twentieth centuries; it was "ticking away just beneath the surface of medicine." Possibly as early as 1900, Ludwig Puusepp made his own contribution to that climate. The Estonian surgeon reported in 1937 that years earlier he had operated on the brains of three patients with manic depression by severing the neural links between the frontal and parietal lobes. The surgery had done no good, but Puusepp said he later performed leucotomy-like operations on fourteen additional patients with generally good results and few debilitating side effects. François Ody, a Swiss disciple of Harvey Cushing, claimed that he had performed leucotomies before Moniz, but he did not publish his results until 1938.

Not long after Puusepp's first experiments, several neurosurgeons began removing large sections of the brain to treat gangrene, abscesses, seizures, and tumors. Initially, when the patients survived, their surgeons rarely paid more than fleeting attention to any changes in behavior, thought processes, or emotional expression. One pioneer in this work, the famed American neurosurgeon Walter Dandy, noted of a patient in 1922 only that the removal of a frontal lobe had occurred "without any observable mental or other after-effect." Later, engaged in the removal of sizable sections of the brain, Dandy and other neurosurgeons paid a great deal of attention to the neurological effects of these operations, but the psychological impact of the surgeries remained unclear because of the lack of reliable tests to measure psychological functioning. Not until the late 1920s and early 1930s did the development of psychometric testing enable investigators to subject patients with parts of the brain removed to thorough mental examination. Joe A., whose case Brickner detailed at the 1935 London conference, was one of the first such surgical patients whose psychological and emotional aftereffects were closely examined and widely reported.

By 1935, however, many physicians found absurdity, if not danger, in the whole idea of leucotomy. William Sargant, a British neurosurgeon who later promoted psychosurgery in his country, laughed aloud when he first heard Moniz's claims. "This was too much even for me to swallow," he recalled. "Nobody then knew what were the functions of the frontal lobes; how on earth could such a complicated psychopathological illness as schizophrenia, with all its multiple and complicated psychopathological causes be helped simply by destroying certain tracts of the frontal lobes? It sounded utterly ridiculous."

But Moniz had many admirers. Two investigators began psychosurgical operations in Romania in late 1936. Near Turin, Italy, Emilio Rizzatti had performed many leucotomies by 1937, as had G. Sai in

Trieste. There were other admirers in the Western Hemisphere. Soon after the publication of the first papers on leucotomy, a Brazilian psychiatrist named Mattias Pimenta attempted the procedure on four patients in Sao Paulo. He thus performed the first modern psychosurgeries in the Americas. Soon afterward, Ramirez Corria leucotomized several patients in Cuba.

Moniz's most vocal supporter across the Atlantic, however, was Walter Freeman. Stirred by their meeting at the London conference, Freeman initiated a correspondence with Moniz in May 1936. His first communication to the Portuguese physician was a seventy-four-word fan letter sent soon after news spread of Moniz's leucotomies, praising the elder doctor's "magnificent researches." It included this sentence: "I enjoyed particularly your recent work on the reduction of psychotic symptoms following operation on the frontal lobe and I am going to recommend a trial of this procedure in certain cases that come under my care." Freeman added that he had been considering a similar procedure "for some time past," a fabrication probably designed to impress Moniz. Moniz replied in unidiomatic English that "I was very satisfied with your disposition to make the prefrontal operation in certain psychoses" and included the address of the French instrument maker who produced his leucotome. George Washington University soon ordered several. Moniz then sent Freeman an autographed copy of his book with a note that "I shall be very satisfied to get your informations after having made some prefrontal leucotomies." Moniz had inscribed the book, "To Dr. Walter Freeman of Washington, with kindest regards from Egas Moniz." Freeman's bond with Moniz was deepening both emotionally and professionally.

Moniz's ideas, and especially the decisiveness with which the Portuguese neurologist had acted upon them, resonated deeply within Freeman. Despite his recent appointment and assumption of significant responsibility at GWU, Freeman was dissatisfied with his professional accomplishments, and he yearned to make an indelible mark on the treatment of the mentally ill. "I recognized that I had done nothing important in either explaining mental disorder nor treating it," he wrote. For years he had followed all of the recent somatic treatments for psychiatric illness with great interest, including therapies involving cyanide, carbon dioxide, and sodium amytal. He had missed the boat not only in failing to pioneer these therapies but also in watching others win credit for the later development of insulin and metrazol shock therapy.

Put simply, Walter Freeman wanted a piece of the somatic therapy action. Nonbiological psychiatry did not attract him at all. "I could never get interested in the speculations of my psychiatric colleagues as to mental mechanisms in the production of psychoses," he observed; an attitude that may explain why the unsubstantiated theory behind Moniz's operations did not concern Freeman. On the contrary, when Freeman first read of the leucotomies in Portugal, he heard trumpets sound. "Here was something tangible, something that an organicist like myself could understand and appreciate. A vision of the future unfolded," he wrote. It showed a future in which Freeman could advance to the forefront of his profession, a new phase of life for which he had long prepared himself. This was to be Walter Freeman's period of great achievement, and psychosurgery would carry him through it.

To their university colleagues, however, Freeman and Watts detailed a more cautious reaction to Moniz's published work. Many other treatments such as sleep therapy had previously arrived and ultimately disappointed. "Yet careful study of Moniz' case reports and a review of the literature upon the functions of the frontal lobe reveal that prefrontal lobotomy rests upon a firm physiological basis," they declared.

If Moniz's early work in leucotomy excited Freeman, the man behind the work fascinated Freeman just as much. He found himself struck by "the scientific genius of the man." He regarded Moniz's creative leap in inventing this form of psychosurgery at age sixty-one as "a remarkable feat of intellectual ability, the mark of genius." Moniz, in fact, represented what Freeman wanted to become: a scientific renaissance man, impervious to criticism, whose strikingly original arcs of thought would bring him international acclaim despite his location outside the world's most prestigious centers of neurological research and learning.

A month after his first letter to Moniz, Freeman wrote again with an offer to arrange English translations of Moniz's book on leucotomy and the Portuguese version of his medical journal account of his first twenty patients. Freeman clearly wished to be Moniz's right-hand man in America. His efforts brought results. With Freeman's recommendation, publisher Charles Thomas agreed to produce the book, and Moniz approved of the arrangement.

In August, Freeman again contacted Moniz, reporting that he had reviewed Moniz's book for the *Archives of Neurology and Psychiatry*. "It is truly a fascinating study and I feel certain will prove epoch making in the scientific approach to mental disorders," he said in the letter. Moniz responded to the review with warmth; he added that his most recent series of leucotomies brought improvement in patients who had shown symptoms of schizophrenia within the previous two years.

At Yale, John Fulton had also heard the reports of Moniz's work. One of Fulton's associates observed that few times had an experimental lab finding so quickly made an impact in the operating room. Surely relieved that Moniz was not performing lobectomies—the removal of large sections of frontal lobe tissue—Fulton exchanged his views on the operations with Freeman, who also discussed the leucotomies with James Watts. Also an organicist, Watts shared Freeman's biological views on the causes and treatment of psychiatric illness, and he and Freeman could discuss the neurological significance of psychosurgery as equals. Like the carrier of an intellectual virus, Freeman infected Watts with his enthusiasm and determination to revolutionize the treatment of psychiatric illness. Perhaps influenced by Fulton's approval of the surgeries, Watts "also was ready" to make a plan for performing psychosurgeries. "I was interested in the frontal lobes, and naturally I saw [Becky and Lucy] at Fulton and Jacobsen's primate lab, and I'd seen the animals change as a result of bilateral, frontal lobe operations," Watts explained. "And it seemed like, as we read Moniz' [writings], that this was a way of treating patients who would otherwise be disabled indefinitely."

In Freeman's view, nothing could stop the spread of psychosurgery across the globe. He wanted to be the one who introduced it to America. Admissions to psychiatric hospitals were growing by 80 percent a year, with 432,000 patients filling the wards of state mental hospitals by 1936. Freeman "was concerned about the lack of therapy for mental patients," Watts recalled years later. "He was disturbed because [patients] were allowed to sit around and deteriorate. He did not blame anyone but wanted to do something to correct the situation."

Once Freeman and Watts had decided jointly to carry out the first psychosurgeries in the United States, they set their plan in motion. They examined the brains of cadavers to find safe pathways from the skull to the anatomical coordinates that Moniz had identified in the prefrontal lobes. They practiced sectioning the brains with a tool resembling a bread knife. The leucotomes arrived from France in July 1936, and Freeman and Watts became familiar with the use of these new instruments on brain tissue. Inspired by Harvey Cushing's thorough reporting on the outcome of his treatments for brain tumors, they devised a plan for follow-up on their psychosurgery patients. Freeman and Watts also gave thought to the legal and ethical consequences of operating on mentally incompetent patients. Permission to operate had to come from the patient, if possible; a spouse or other nearest relative; and another relative, such as a sibling, a parent, or a child. "Now sometimes it'd be five members of the family involved," Watts remembered. "And if anybody objected, I didn't do it. I had enough trouble without looking for it. If somebody raised too many questions, I just said, 'Sorry, rather not do it.'"

They felt nearly ready to look for a patient that month, but before the advent of widespread air conditioning, summer's peak was no time to operate on brains in Washington. Even if the patient escaped infection from sweat dripping from the surgeons, the sweltering heat could produce dangerous swelling. The first psychosurgery patient would have to wait for the cooler days of September.

With their entry into psychosurgery on temporary hold, Freeman began planning a big family vacation. He bought a York Senior Cruiser camper trailer, which had bunk beds for the children, a sofa bed for himself, closets, an icebox, a stove, a sink, linoleum floors, and activity tables. He pulled it with his 1935 V-8 Ford. From that moment on, nearly all of the family's voyages into the great outdoors would happen aboard this trailer or others like it. Freeman planned out an inaugural trip to Yellowstone National Park. Soon he and the kids, who now ranged in age from eight to eleven (baby Keen was left at home with Marjorie, who was already pregnant with their sixth child, Randy), fell into a workable routine. They ate breakfasts and dinners in the trailer and took their lunches in restaurants on the road. Freeman usually parked the trailer overnight at a campground near a lake or pool. "I used to say that the children cared little for the sights, but they knew every swimming place on the way to and from Yellowstone," he wrote.

After surviving the treacherously winding mountain roads of West Virginia, the campers found their next trip highlights in Colorado. They motored to Pike's Peak, Estes Park, Leadville, and the desert outside Grand Junction. "We were beyond civilization," Freeman remembered happily. When they at last reached Yellowstone, Freeman had to cajole the children away from their comic books to show them the sights. But despite the unpredictable ways in which the youngsters responded to the natural beauty around them—"they seemed just as much impressed by a squirrel as a deer," Freeman remarked—their maturation that came from experience on the road greatly gratified their father. They all learned how to muck out the trailer and wash their clothes. Lorne caught her first trout and fed the family when Freeman briefly fell ill.

The summer of 1936 was one of the hottest in the recorded history of the Plains and Upper Midwest, and on the way home the family drove through a sticky layer of air baking the state of South Dakota. Young Walter nearly got heatstroke; they battled flat tires and other mechanical problems, and the children fought off the flies and mosquitoes. "We arrived home lean, tanned and refreshed, agreeing that the trip was the best ever and intending to take another next year," Freeman wrote. There would be another trip the next summer and for many summers that followed. Getting away from it all, changing scenery, indulging in physical activity, and temporarily depriving themselves of the comforts of home all contributed to family harmony and the mental health of the children, Freeman believed. The worst thing they could do during a summer was to stew in Washington and think about their minor dissatisfactions with life. Just before starting something new, he believed, it was good to take a break.

# Chapter 6

# Refining Lobotomy

Three weeks after Alice Hammatt's psychosurgery in September 1936, which Freeman regarded as successful, he and Watts found another candidate for Moniz's leucotomy technique. She was a fifty-nine-year-old bookkeeper named Emma Ager who suffered from agitated depression, like Hammatt, and had spent nearly six months in bed. In addition, she experienced hallucinations and held a "fear of being poisoned [along with] constant mourning and weeping." Freeman was startled when the leucotome broke midway through the procedure, but otherwise the surgery proceeded without complications.

For the first time, Freeman photographed a psychosurgery patient before and after the operation—a practice he would continue to follow for decades—documenting Ager's transformation from a downcast and anxious-appearing woman to one looking more attentive and calm. Two weeks after the operation, Freeman noted that "the patient has had no complaints at all and she mentions her delusions in a rather disinterested manner, has not yet recovered from her confusion, and is unable to remember the names of her doctor or of her nurse. She is cheerful, however, although entirely lacking in spontaneity." She returned to her job after a two-month recovery. Ager held her bookkeeping position for eight years before retiring, but her employer then asked her to return to work for an additional two years. When Freeman examined the seventy-two-year-old patient in 1949, he found her "quite outspoken but not rude. She is keeping house, reads the Bible, attends church regularly, and, according to her daughter, has definite 'spiritual values.'"

Another psychosurgery patient followed on October 14, 1936, a thirty-three-year-old concrete worker named Linwood Roberts who had received outpatient treatment the previous year at the George Washington

University Hospital for a variety of complaints including "nervousness, worry, palpitation, throbbing in the head, fear of falling, and ideas of suicide," Freeman and Watts wrote. Reared in North Carolina, Roberts had moved to Washington, could not find steady work in construction, made few friends, and grew depressed. His mother began caring for him when he started losing weight, had trouble sleeping, and continually felt dizzy and weak. Soon confined to bed, he stopped speaking. "He had spent eighteen months in bed," noted his case history, "most of the time with his hand on his chest, waiting for his heart to stop beating."

Freeman examined Roberts and found the patient "had a solemn worried expression on his face and rarely smiled. He did not volunteer any information." Unexpectedly, psychosurgery turned him into a compulsive talker. Just after his operation, Roberts surprised his doctors with continuous chatter. Other patients complained about his unstoppable discourses, and the hospital staff had to shush him repeatedly. "You know, I have been talking a blue streak all day," he said. "I feel as if all that nervousness and worry were gone. You know I am anxious to get back to work again." He returned home, undertook a reconditioning regime of rowing on the Potomac River four hours a day, and eventually found work as a high school janitor. A dozen years after his leucotomy, he was still employed. "He makes his home with his mother and seems to have no great ambition to get married and raise a family, but he is well thought of by his employers and has few complaints," Freeman wrote.

Freeman and Watts's next psychosurgery patient had struggled against her obsessive-compulsive behavior and fear of contamination for more than thirty years. By the time she submitted to surgery, her illness had reduced her to "washing her hands as long as an hour at a time, scrubbing the toilet seat an hour, and burning two overcoats because they were touched by friends," the doctors noted. After her leucotomy on October 20, 1936, she experienced a short-lived improvement. She charmed her physicians with her wit and admitted, "I realize my fears were foolish. Even before [the] operation I knew that if everyone was like I was there would be no work done in the world. Yet I could not help it." She felt no postsurgical mental confusion and was soon discharged from the hospital to an anxiety-free life of going to the movies and visiting friends. That life lasted six weeks. One day her son found her "dressed in a bathrobe with her hands up the opposite sleeves, her hair disarranged, her face drawn in anxiety and her ideas of contamination as insistent as before." What exactly had caused her temporary improvement or her relapse, Freeman and Watts could not say.

Their fifth leucotomy produced their first immediate failure. The patient, a forty-seven-year-old woman who before surgery had attempted suicide by inhaling gas and had spent half a day in a state of unconsciousness, was probably brain damaged as well as suffering from depression at the time of surgery. To complicate her woes, Freeman and Watts cut several blood vessels in her brain during the operation and as a result she suffered seizures and bladder incontinence. Leucotomy did not diminish her anxiety. A month after surgery, Freeman found that she "is still quite confused, inattentive, and forgets easily."

The doctors pressed on. Their sixth leucotomy patient, operated on November 5, 1936, was Mildred Spurlin, a thirty-two-year-old secretary suffering from schizophrenia. Freeman and Watts knew that Moniz's surgeries on schizophrenic patients had produced few successes. In this case, Freeman decided to evaluate the location and effectiveness of the cuts by using his favorite brain-imaging medium: injected thorotrast, known to be radioactive and carcinogenic, which showed up in the X-rays of the patient's brain. Not only did this patient suffer no apparent harm from the thorotrast, she also recovered from her psychological withdrawal. Although she continued experiencing hallucinations, Freeman believed that she paid less attention to them. She suffered a temporary relapse when a callous acquaintance ran a hand along her hair, which had been cropped for surgery, and called her a "fuzzy-wuzzy." But she struggled to recover. Seven months after surgery, she was "in splendid condition," well enough to return to work, but when she discovered she had lost her job, she had a recurrence of her schizophrenic symptoms and received insulin shock therapy at St. Elizabeths Hospital. She remained there for more than thirteen years, "grossly obese, untidy and mannerless. Her eyes were dull and her voice was toneless. She never complained. She ate greedily and hoarded trash, stayed by herself and slept soundly."

Freeman and Watts had already operated on half a dozen mental patients, but by early November, Egas Moniz had heard nothing of the work of his protégés. His interest in Freeman's work was keen because lack of cooperation from referring physicians was making it difficult for Moniz to locate new patients for his series of leucotomies. When Freeman's letter describing them finally arrived—the continuation of a frequent correspondence in French and English between the two psychosurgeons—Moniz offered encouragement and advice in his unpracticed English. "New cases, the choice of the patients, the continuation in the way of this new treatment, ever inoffensive, will be the basis for arriving to sure conclusions and to get all indications of the method still far from being determined," he wrote. The letters that followed reveal the extent to which Freeman wished to ingratiate himself to the elder physician. For Freeman considered Moniz a genius, "a true pioneer," he wrote, "in that he has not rested upon his arms with the accomplishment of one stroke of genius in neurologic diagnosis [cerebral angiography], but that he has also synthesized, and has given to the world a very important means of allaying mental distress."

When Moniz wrote that he had made some changes to his leucotomy technique, Freeman replied that he and Watts would promptly apply those alterations to their future operations. Freeman offered to send one of Moniz's articles to the *Journal of the American Medical Association* "with a note to the editor whom I know personally, requesting an early printing." When the journal rejected the paper on the grounds that it was too specialized and technical, Freeman successfully placed it in the *American Journal of Psychiatry*—where the editor overcame his reservations that Moniz's "case histories are not very impressive and his deductions perhaps not very critical"—and sent Moniz one hundred copies. Freeman expressed the hope—which was not realized—to visit Moniz in Lisbon during the summer of 1937. "I trust that by that time the procedure of prefrontal leucotomy will be even more firmly established than it is now." Freeman cemented his adulation by contributing an essay in Moniz's honor to an issue of a Portuguese medical journal celebrating the tenth anniversary of the invention of cerebral angiography. "All hail to its originator!" Freeman wrote.

A few weeks after their first leucotomy, they had presented the results to their colleagues in the District of Columbia. (They later published a paper on psychosurgery in the journal of the local medical society.) Now they felt ready to tell the world of their work. During the third week of November, they traveled to Baltimore to describe their cases to a potentially skeptical audience: members of the Southern Medical Association, assembling for their annual meeting. These doctors represented a variety of medical specialties as well as a spectrum of philosophies and approaches among those practicing psychiatry. Sessions in Baltimore focusing on psychiatric illness and brain disorders drew a mix of neurologists, neurosurgeons, neurologically minded psychiatrists, and psychiatrists engaging patients in

Freudian psychotherapy, which had rapidly grown in popularity during the previous decade. Brain function was a topic of high interest just as it was at the London Conference of 1935. One of the other conference presenters in Baltimore, Spafford Ackerly, was going to read a paper about a patient with a brain tumor who had undergone the removal of both frontal lobes, just like the patient who Brickner reported on in London.

Unlike most other presenters at the conference, Freeman appreciated the power of the popular press to heighten the interest and receptivity of his audience. Decades earlier, Grandfather W. W. Keen had energetically penned articles and letters to the editor for popular magazines and daily newspapers in his campaign to build public support for animal vivisection and other issues he championed. Freeman's experiments with oxygen as a treatment for mental disorders had landed him in the pages of *Time* magazine several years earlier. He had long ago learned that "the public press has one great advantage that is not possessed by the scientific periodicals, owning the ability to announce their discovery within a day or two after the time that it is presented at the scientific gathering," he wrote.

As the Southern Medical Association conference neared, Freeman devised a plan to generate publicity and support for leucotomy. After the sixth operation, "I called Tom Henry, science writer for the [*Washington Evening Star*], had him come to George Washington University Hospital, explained what we were doing and what theories had been advanced, persuaded him to see patients before and after, and to witness an operation," Freeman wrote. Henry made quick use of his opportunity by publishing a story on November 20 that maintained that lobotomy "probably constitutes one of the greatest surgical innovations of this generation. . . . It seems unbelievable that controllable sorrow could be changed into normal resignation with an auger and a knife." The result was just what Freeman wanted. "As was to be expected," he recalled, "there was considerable journalistic interest when Watts and I arrived in Baltimore."

Pleased with the excitement of the newspaper reporters, Freeman nevertheless felt an obligation to make sure that Henry received the meatiest information on his leucotomies. When four other reporters ambushed him shortly before he was scheduled to present his paper on November 21, Freeman was unprepared but quickly rose to the occasion. He delivered an extemporaneous oration on Phineas Gage's famous "crowbar" accident of the previous century, the anatomy of the frontal lobes, frontal lobe diseases, brain injuries resulting from war wounds, Fulton and Jacobsen's research on monkeys, and Moniz's leucotomies of 1935—all topics that allowed him to dance around his own psychosurgeries without actually describing them. "So with diversion and delaying tactics I escaped," Freeman wrote, "and, at the same time, gave Mr. Henry his chance. He had come in during the time the interview was progressing and first looked a little shocked but caught on as I went into detailed consideration of inconsequentialities."

And before he knew it, Freeman was standing at the conference podium, looking tense and reading his portion of the paper. His audience listened carefully as he recounted Moniz's contributions to psychosurgery, the technique of the operation, and the first six cases he and Watts had treated. He showed his "before and after" photographic portraits of patients and speculated on the emotional states manifest in their expressions. "We are able to say, with Moniz, that no patient has died and none has been made worse," Freeman told the audience. "All of our patients have returned home, and some of them are no longer in need of nursing care. All of them are more comfortable, having been relieved of certain symptoms that previously had been very troublesome. It is as if the 'sting' of the psychosis had been drawn." But he carefully avoided asserting that he had cured anyone. A relief of symptoms—anxiety, worry, apprehension, nervous tension, and insomnia—was the most he claimed.

And for the first time, Freeman uttered the word *lobotomy* in public. The new term, he and Watts believed, more accurately situated their surgeries in the frontal lobes of the brain, where the functioning of cells was changed by the severing of neural fibers. Never again would they refer to their operations as leucotomies—a semantic distinction that separated their work from Moniz's and more easily allowed them to stand alone as innovators when they later modified and abandoned Moniz's technique.

When Freeman concluded his reading, the audience could not keep quiet. Spafford Ackerly, the Kentucky neurosurgeon who reported on frontal lobe excision at the conference, declared, "This is a startling paper. I believe it will go down in medical history as a noted example of therapeutic courage." Ackerly's statement was one of the few supportive ones. Others, especially psychiatrists, were horrified by Freeman's description of intentionally damaging healthy brain tissue. One, a New York City psychiatrist named Joseph Wortis, repeated the contentions made a month earlier of Freeman's Washington colleagues that any benefit to the six patients could easily be attributed to the shock of surgery. "I have seen mental patients in Bellevue Hospital who have become more normal after such a shock as the fracture of a leg," he said. (Less than a year later, Wortis would become the first American physician to administer insulin shock therapy to mental patients.)

Freeman had a ready response. Surgical shock that benefited mental patients was rare, he said. In addition, his patients had not undergone shock during or after lobotomy, as shown by their steady pulse and respiration rates, blood pressure, and body temperature. Then the truly derisive comments began. Audience members questioned the ethical soundness of causing injury to patients, of tampering with their personality, when doctors were sworn to heal them. Freeman and Watts were inviting legal trouble, they added. Freeman acknowledged that their patients *had* lost something: the emotional drive of their neurotic and psychotic ideas. Ackerly's patient and others who had undergone lobectomies to remove the threat of brain tumors had lost far more brain matter without the destruction of their lives. Freeman's defense swayed few in the audience, however, and he and Watts teetered on the brink of losing credibility and respect.

At this point, one of the most influential figures in the history of American psychiatry took the floor. He was Adolf Meyer, seventy years old, a longtime member of the faculty of Johns Hopkins University Medical School, balding, and wearing a bushy goatee that gave his face a narrow look. After his birth in Switzerland and studies of neurology and psychiatry in Germany, England, France, and Scotland, Meyer's American career had begun in 1893 as the staff pathologist at the Illinois Eastern Hospital for the Insane. He had lamented the unscientific approach in the United States of most asylums and schools that taught psychiatry, and he devoted his professional life to forging a new kind of psychiatry—one that unified all psychiatric practitioners—from asylum doctors to neurologists—under a single specialty that valued treatments to both psyche and brain. In addition, he worked to upgrade psychiatric training in medical schools. "I am not antagonistic to this work, but find it very interesting," Meyer said from the audience in Baltimore. "I have some of those hesitations about it that are mentioned by other discussants, but I am inclined to think that there are more possibilities in this operation than appear on the surface."

Meyer went on to discount the probability that shock was responsible for the diminishment of psychiatric symptoms in the lobotomy patients. Instead, he attributed the benefits to "things we are learning concerning the frontal lobes and their role in the functioning of the personality." Meyer cautioned Freeman and Watts, and all present, not to lead the public to hope that an operation can remove worries and anxiety. "The work should be in the hands of those who are willing and ready to heed the necessary indications for such a responsible step, and to follow up scrupulously the experience with each case," he concluded. "The available facts are sufficient to justify the procedure in the hands of responsible persons, but it is important that the public should not be drawn into any unwarrantable expectations. At the hands of Dr. Freeman and Dr. Watts I know these conditions will be lived up to."

Meyer's words, resonating with authority, had an electrifying effect. It was a powerful vote of confidence for the two GWU doctors and their new treatment. Here was the man who had introduced factual investigation to American psychiatry and had championed the importance of collecting a complete patient history before administering treatment. He had outlined a new holistic approach to psychiatry, "one that might join the separate fields of the asylum physician, the elite neurologist, and the general practitioner into a single medical specialty," noted the medical historian Jack Pressman. He had formerly presided over both the American Psychiatric Association and the American Neurological Association. Freeman, who knew Meyer only through some fleeting meetings and their joint work on the American Board of Psychiatry and Neurology, was impressed by the attainments of the man who had stood up to support him. "Meyer studied the *person*, drawing his ideas from the patient in his surroundings rather than establishing a theory . . . and then attempting to 'explain' the phenomena on the basis of the theory," Freeman wrote years later. Freeman never anticipated Meyer's support.

After Meyer spoke, the audience in Baltimore settled down and offered no more criticism. Freeman, in particular, took to heart Meyer's directive to "follow up scrupulously" on the lives of his lobotomy patients, and made of it "a mission that has lasted for over thirty years," as he wrote at the close of his career. "Had it not been for his sympathetic and helpful discussion, the advance of lobotomy would probably have been much slower than it was," Freeman observed.

Buoyed by Meyer's endorsement, Freeman considered his Baltimore report well received. He dispatched a typescript of it to Moniz, who sent his congratulations. "The results that you and Dr. Watts have got in your patients were very remarkable. . . . We can make cuts in the brain of man that produce sometimes cures or improvements of the patients. . . . In other words," Moniz noted, "the new idea opens, I think, the human brain to the investigations which were not possible till now." He warned Freeman, however, of the fierce battle that lay ahead. Psychologically oriented psychiatrists, those who believed that patients' traumas and life experiences led to mental disease, would not accept brain surgery as a legitimate treatment for psychiatric illness. "I think that we are on the right road; our surgical orientation will be brought further in other parts of the brain; but we shall have great difficulties to overcome," he concluded.

Freeman's letters to Moniz voiced his expectation that some foundation would come forward to financially support further psychosurgery research. But Freeman did not receive any such support in the coming months or, for that matter, ever. His repeated appeals for funding from the U.S. Public Health Service and other agencies produced nothing. "As a fund-raiser I was a flop," he later concluded. He blamed part of his lack of success "on the prejudice, general among psychiatrists, against mutilation of the brain; part was due to the fact that Watts and I were in private practice and not in full time research; part may have been due to the locus of the research in Washington," Freeman wrote.

Starting immediately after the Baltimore conference, though, intrigued visitors came calling. Richard Brickner, who had reported on Dandy's patient, Joe A., in London, arrived in Washington to watch Freeman and Watts perform a lobotomy and to examine four of the previous patients. "He has already obtained a leucotome and will start work in New York in the near future," Freeman told Moniz. "He was very much impressed by our results." Another luminary from the London conference, John Fulton, dropped by with his colleague, Henry Viets. Freeman informed Moniz that Fulton and Viets "are greatly pleased with the results and are with us that it is the most important advance in psychiatry that has come about in a long time." In a few years, Fulton's enthusiasm over Freeman's methods would cool.

Thus encouraged, Freeman and Watts set an ambitious goal to complete an initial series of twenty lobotomies, just as Moniz had, by the end of 1936. They operated on a new patient only two days after returning from Baltimore and wielded their leucotomes at a steady pace—sometimes every other day—during the next several weeks. The results were very mixed: several patients experienced no long-term improvement, including their first lobotomy patient to carry a diagnosis of manic depression, but a few attained a dramatic recovery by the doctors' standards. One, a hypochondriac who had recently grown agitated and bad tempered, ended her physical complaints and astounded everyone with her cheery behavior. A twenty-five-year-old woman with catatonic schizophrenia appeared to awaken from her stillness right on the operating table. A disastrous outcome was the case of a suicidal woman, sixty years old, who died after the operation from a large hemorrhage of the brain.

For Freeman, the most memorable of the patients of December 1936 was Paul K. Hennessy, the fifteenth case, with whom Freeman kept in contact for more than twenty years. A Rhodes scholar and an attorney working for the government, Hennessy had appeared before Freeman a few days before Christmas. Hennessy was anxious and plagued by feelings of inferiority, with a diagnosis of "severe psychoneurosis complicated by alcoholism," and the doctors decided to perform a lobotomy, which produced a startling mishap. "During the operation, while I was drilling through his skull, the bone gave way and the drill went all the way in," Freeman wrote. "There was a gush of blood and my heart sank." Remarkably, Hennessy seemed to suffer no complications from this accident. He felt so well after surgery that on Christmas Eve "he dressed and put on his hat over the bandages, walked out of the hospital, and proceeded to celebrate." When they heard of his escape, Freeman and Watts went out in search of him, mostly in the bars of downtown Washington. The timing was poor regardless of the holiday—Marjorie was in Columbia Hospital giving birth to Freeman's sixth child. Not until they located Hennessy, intoxicated in a saloon on early Christmas morning, did Freeman at last have the chance to hold his newborn son, Randy.

After his recovery from the lobotomy, Hennessy went into the army and was court-martialed three times; he received acquittals in the first two cases and a discharge after the third. In 1957 Hennessy reported to Freeman that he traveled extensively as a manager of fund-raising campaigns for hospitals and had recently raised half a million dollars for a Canadian hospital. He proudly declared that his job required good people skills, public speaking talents, and, most of all, a resistance to needless anxiety. Hennessy attributed his success to his ability to "keep my sense of humor and not worry about what I can't help." Alcoholics Anonymous had helped as well. Evidently, whatever excess tension he had in 1936 had been removed for good.

The last patient in the 1936 series of psychosurgeries, a sixty-year-old woman diagnosed with agitated depression, provided Freeman and Watts with their most discouraging experience so far. The physicians decided to try cutting eighteen cores in the patient's brain instead of the usual twelve. After taking nine cores from the right side of the frontal lobe, they injected thorotrast and took X-rays in an effort to detect hemorrhage. Finding none, they completed the remaining nine cores. The patient appeared to have withstood the operation, but within a few hours she became paralyzed on her left side and lost her ability to speak. She fell into a coma and died six days after surgery. "This, of course, has distressed us greatly since the outlook in the case of this particular patient was so good," Freeman wrote to Moniz. Although the patient's family refused to give permission for an autopsy, Freeman and Watts determined through thorotrast injection that massive hemorrhaging had occurred near the site of some of the cores.

Freeman and Watts met their end-of-the-year goal of twenty lobotomies but at the cost of a life. Despite the fatality, Freeman judged at the start of 1937 that lobotomy was proving to be a good treatment. Of the first twenty patients, "three are fully recovered, two of them being back at their previous occupations; many of the other cases are too recent . . . for us to decide what their future will be," he wrote.

Their own future in the operating room was open to question as well. Throughout these initial operations, Freeman and Watts had settled upon a division of labor during surgery that was highly unusual, but it satisfied the interests and matched the talents of the two men. Normal protocol dictated that Watts, the neurosurgeon, was in charge. But Freeman's mastery of brain anatomy, his knowledge of Moniz's technique, and his drive to advance lobotomy into the mainstream of psychiatric treatment made him the physician in control in many respects. He admired those who controlled the operating room, and he took pride in the surgical innovations and prowess of his grandfather, W. W. Keen. (Freeman was delighted after he allowed a full beard to grow during the summer of 1938 and a colleague "produced a photograph of my grandfather and pointed out the striking resemblance. Even I could see it," he wrote.) From the first lobotomy on Alice Hammatt, Freeman—who never studied surgery in any detail, never worked as a surgical resident, and had no credentials as a neurosurgeon—had wielded the leucotome in partnership with Watts. Technically, by the rules of George Washington University Hospital and most other medical centers in the United States, Freeman had no business picking up a leucotome or any other surgeon's tool in the operating room. "I had a feeling that he wanted to be the neurosurgeon himself, and he really enjoyed being referred to by patients and in the press as a neurosurgeon," Watts later recalled.

On the day of one early lobotomy, however, Watts had a cold and felt too infectious to work in the operating room. By the rules of the hospital, the surgery should have been postponed—an unacceptable choice given the doctors' aggressive operating schedule. Freeman took sole charge of the operation, an action that prompted a complaint from Daniel Barton, a GWU hospital surgeon who happened to look in. Barton brought the matter to the dean of the medical school and the medical director of the hospital. The hospital's decision: Freeman was "not a surgeon and if he wants to operate he'll have to apply for surgical privileges," Watts later remembered. Lacking surgical training, Freeman faced certain rejection in any such application, and at this point in his career he would not consider placing his ambitions on hold for the several years it would take to become a surgeon. Freeman took no more solo flights in the operating rooms of George Washington University Hospital, but this was just the first volley in a long battle over his surgical credentials. It would run for the remainder of his career.

The doctors had spent the last quarter of 1936 performing lobotomies in a race against the clock. (Ten years later, a review of these first twenty patients showed that there were fourteen survivors, four of whom were employed, four who were running households, four who were living at home in the care of their families, and two who were confined to hospitals.) In 1937, however, they slowed their pace. Freeman had not forgotten Adolf Meyer's advice to carefully follow up on the progress of their lobotomy patients, and he and Watts operated on only twelve new patients that year. They devoted time to further improving their technique, by replacing the wire stylet in the Moniz leucotome with a more effective cutting blade. They worked to promote psychosurgery in the medical community and among the public. Freeman often lamented that critics could not comprehend the great potential for good in the procedure. "We may recall the response of Faraday when he had demonstrated the principle of electromotive force to an admiring audience," he once wrote. "A lady asked what it might be good for. 'Madam,' he replied, 'of what use is a newborn baby?'"

Freeman decided to answer the question of the value of psychosurgery at the 1937 annual meeting of the American Medical Association, which brought more than seven thousand physicians to Atlantic City from throughout the United States. Freeman reserved a booth in the exhibit hall.

He and Watts pulled together an eye-catching exhibit designed to stand out among the hundreds of other displays and to draw members of the press. They displayed their thorotrast-dyed X-rays, comparative photo studies of lobotomy patients, and—most strikingly—two live monkeys, one of which had been lobotomized. With their booth set up a day before the AMA meeting was scheduled to start, they awaited attention. As the convention's registrar of exhibits led the press around the hall with no intention of highlighting the lobotomy display, the Freeman–Watts materials worked their magic. It did not take long. On June 6, 1937, the *New York Times* published an article by William Laurence describing their operations with the headline "Surgery Used on the Soul-Sick." With only a scattering of accuracy, Laurence announced the arrival of "surgery of the soul," a "new surgical technique, known as 'psycho-surgery,' which, it is claimed, cuts away sick parts of the human personality, and transforms wild animals into gentle creatures in the course of a few hours." Lobotomy had "changed the apprehensive, anxious and hostile creature of the jungle into creatures as gentle as the organ grinder's monkey."

Laurence had no way of knowing that the lobotomized monkey had died from a cerebral hemorrhage even before the article appeared in the newspaper. No matter, declared Freeman, "Watts and I had made the headlines even though we did not get an award. He and I worked hard on that one and talked ourselves hoarse to the visitors." Other laudatory reports on the lobotomy exhibit appeared in news publications of cities as distant as Paris and Shanghai. "Our work was roundly criticized [at the convention] by a number of individuals but met with a warm reception by persons who, in our opinion, knew something definite about the problem," Freeman wrote. Freeman noted with satisfaction that his display had also succeeded in inspiring a newspaper advertisement that read, "Doctors say that worry can be cut out of the brain with a knife. There is a better way. GO TO CHURCH." Freeman exhibited at every AMA convention from 1937 through 1946, and he continually sharpened his techniques of gaining attention, sometimes by using a clacker to produce a loud sound to attract crowds and shouting like a carnival barker.

At the same time, Freeman was trying to find a large mental hospital that would allow him and Watts to operate on its patients. As early as October 1936, Freeman had hoped to begin the first large-scale series of lobotomies on the nearly limitless group of candidates at St. Elizabeths Hospital. William A. White, the hospital superintendent and Freeman's former boss, was sick and unable to discuss the matter for several months after Freeman conceived the idea. When he at last broached the subject with White, Freeman found the superintendent skeptical, "although I am rapidly winning over some of his younger colleagues." He remained optimistic that permission to lobotomize patients at St. Elizabeths would soon come.

It never came while White was director. White questioned the ability of a mentally ill patient to competently authorize a hazardous procedure like a lobotomy, and he knew that patients' families might have ulterior motives for allowing the surgery. He explained that "relatives not infrequently desire the death of patients in hospitals. I do not mean that they do this consciously, although I have no doubt that they do in many cases, but that they do so in the back of their heads there is no question, because these sick people cause them a tremendous amount of trouble." Given the incompetence of patients and the conflicts experienced by families, life-threatening surgery should be avoided, White maintained. He called lobotomy a "spurious and irresponsible" treatment. Freeman and Watts argued that it was responsible and in fact ethical to pursue a treatment that held hope for patients whose condition was regarded as otherwise untreatable. But White's final word was not negotiable: "Freeman, it will be a hell of a long time before I let you operate on any of my patients."

Had they based their expectations on the American reaction to the leucotomies that Moniz began performing in 1935, Freeman and Watts might have anticipated little opposition to lobotomy. Few medical journals in the United States commented on Moniz's work at all. What commentary did appear was mild. "Such a radical procedure is not to be widely recommended at the present time," observed the writer of one submission to the *New England Journal of Medicine*, speaking of Moniz's surgeries. "The operation, however, is one that might rightfully be considered in a patient with long standing agitated depression, if done by a skillful neurosurgeon, in a properly equipped hospital. It may mean a better future for certain patients with chronic mental disease. . . . The operation is based, however, on sound psychological observation and is a much more rational procedure than many that have been suggested in the past for the surgical relief of mental disease."

Freeman, a student of medical history as well as brain anatomy, knew the past too well to suppose that lobotomy would make its appearance without controversy. For hundreds of years of Western medicine, a progression of practitioners who concerned themselves with illnesses of the mind—spiritualists, alienists, neurologists, psychiatrists, psychoanalysts, and psychologists—had debated whether mental disease originates in malfunctions of the brain or in an individual's imperfect adjustments to life events and traumas. The two camps have battled each other without pause with one and then the other scrambling to the top.

During the earliest years of modern psychiatry dating back to the eighteenth century, the organicists—those who believed that psychiatric illnesses resulted from biological disorders of the brain—sat on top. In 1758 William Battie theorized that compression and blockage of the blood vessels and nerves feeding the brain caused the delusions and other symptoms of psychiatric disease in his mad patients. He lacked any scientific evidence to back up his suppositions but hoped future research would fill the gaps of knowledge. Johann Riel further refined these notions, speculating that "irritation" of the brain caused psychiatric illness. He treated his patients with such painful procedures as the application of hot irons, which was designed to draw interior irritants to the surface of the skin. Other early psychiatrists observed that certain mental afflictions appeared to run in families, and some hypothesized that each generation would bring more frequent and severe cases of madness. The study of eugenics and the practice of involuntary sterilization—both popular in the United States during the 1920s and 1930s—were the twentieth-century outgrowths of this line of thinking.

The nineteenth century saw the maturation of biologically based research on mental disorders. Neurologists, more often than psychiatrists, focused their investigations on drugs, the anatomy of animal and human brains, and the relationship between mental states and the workings of our highest organ. The microscope, affording the opportunity to detail the structures of nerves and brain matter, dominated the European study of psychiatric disease by the 1880s. Carl Wernicke, who attempted to identify specific regions of the brain that were responsible for the onset of certain psychiatric symptoms, and Jean-Marie Charcot, a respected French pathologist who spent years trying to find the biological origins of a set of disorders he grouped together under the name of "hysteria," brought this first era of the primacy of organic psychiatry to an end. Both failed in their quests—in Charcot's case because his target, hysteria, proved imaginary. By the end of the nineteenth century, the biological psychiatrists had built a vast literature explicating the anatomy of the brain and the functioning of the nervous system, but they had succeeded in identifying the causes of and developing treatments for only a few illnesses with psychiatric symptoms, including neurosyphilis.

Along with this evolution of biologically based psychiatry came an approach to diagnosing, treating, and categorizing mental illnesses from a social and psychological orientation. In nineteenth-century Germany, a movement of "romantic psychiatrists" began applying moral instruction to the treatment of their patients. In the process, they made a point of discussing with patients their feelings and experiences. Jean-Etienne Esquirol, who practiced medicine in the early nineteenth century, studied the relationship between psychiatric illness and gender, age, and occupation. For decades, though, there was little advancement in the study of patients' experiences as a key to their problems.

At the close of the nineteenth century, Emil Kraepelin was among the first to examine large numbers of case histories of psychiatric patients and to draw from that information a division of mental illnesses based on the progress and general traits of the diseases. Kraepelin made the experiences of patients paramount in his studies, rather than developing ideas on abnormal functioning of the brain that could never advance beyond hypothesis. Most importantly, he divided mental illnesses without an obvious biological origin into two categories: disorders affecting the mood of the patient (the so-called affective diseases, producing depression, anxiety, or other changes in disposition) and those having no such influence on mood, such as schizophrenia. With this categorical simplification inspired by his observation of patients, Kraepelin swept away countless groups, classes, and designations of diseases that the biological psychiatrists had devised, or merely speculated upon, on the basis of their interpretation of mental symptoms. Slightly modified, Kraepelin's division of psychiatric diseases remains in use today. It allows practitioners to approach mental illness as a combination of biological malfunction and psychological response to the events of life.

Starting in the 1890s, however, the balance between the biological and psychological wings of psychiatry shifted. Biology lost its dominance, thanks to Sigmund Freud and his adherents. Like Freeman, Freud was originally a neurologist with a passion for the microscope and the anatomy of the brain. In time, however, he grew convinced that the ideas of patients, not their brain function, caused most mental problems. Freud believed that neuroses—relatively bearable psychiatric problems that arise from daily life and do not require hospitalization—sprang from the repressed sexual memories and fantasies of childhood. Only through psychoanalysis, a lengthy process of investigation in which the patient recognizes and relives the attitudes of the past through a relationship with a therapist who serves as a surrogate, caring parent, could the neuroses be resolved. The observation of symptoms and the diagnosis of disorders dimmed in importance; the biological workings of the brain were not considered. Freud, in fact, favored removing the practice of psychoanalysis from the medical profession.

As early as 1894, American physicians began reading Freud's published work. When Freud made his first visit to the United States in 1909, his ideas dazzled many North American physicians. A psychoanalytic society was launched in New York two years later. Analysis quickly captured the imagination of American psychiatrists, who had no tradition of professional innovation or of employment in any setting other than the lowly insane asylum. Propelled by the financial opportunities offered by analysis, the percentage of American psychiatrists in private practice quadrupled between 1917 and 1933. With the rise of Nazism in Europe, scores of prominent psychoanalysts—many of them Jewish—fled to the United States. This infusion of talent and energy added to the momentum of analysis in American psychiatry. No other nation of the world so warmly accepted and practiced psychoanalysis. "If in the end analysis won, it was not necessarily because of the power of Freud's ideas but because analysis opened the road to private practice," noted the psychiatry historian Edward Shorter.

But analysis had not yet won when Freeman and Watts performed their first lobotomy in 1936. For one thing, Freud had little to offer to people suffering from psychoses, the debilitating mental illnesses such as schizophrenia, psychotic depression, and manic depression that most often landed patients in mental hospitals. Even if psychoses could be dealt with through analysis, the lengthy course of treatment made its application in the crowded wards of America's psychiatric hospitals, which sometimes housed thousands of patients, impractical. Only in a few institutions clustered around Washington, D.C., and Baltimore, were such physicians as Adolf Meyer, Dexter Bullard, and Henry Stack Sullivan attempting to treat psychotic patients with analysis.

Freeman thought these efforts to apply Freudian techniques to deeply afflicted mental patients were doomed, and he doubted the efficacy of analysis in general, even in cases of neurosis. "Insight is a terrible weapon, and few know how to use it constructively," he wrote. "When we realize, really get to know what stinkers we are, it takes only a little depression to tip the scales in favor of suicide." Regardless of any improvements that analysts produced in their patients, Freeman believed, "I am forced to the conclusion that the application of the [psychoanalytical] theories has less bearing upon the management of the patient-in-distress than the close personal relationship that they achieve. Whatever their theory, its effectiveness is highly correlated with the personality of the therapist." In other words, Freeman thought that doctors who convey their genuine concern for their patients, regardless of their therapeutic orientation, would get the best results.

If psychoanalysts could offer little help for seriously psychotic people, biologically oriented psychiatrists were determined to develop new treatments. The physiologist John Fulton, writing several years after the introduction of psychosurgery, summed up their position. He called the neural network in the brain "the matrix of the mind" and asserted that "interrupting the vicious circles which form the basis of abnormal states" could successfully treat psychiatric disorders. "Mental disease in fact can be looked upon as little more than [the brain's] faulty interpretation of sensory data," he wrote, "and if it is possible to help the mentally ill to interpret their sensations more correctly, it is clearly our duty as physicians to do so, whatever may be the means employed to achieve this end."

Psychosurgery was just one of the new therapies that treated mental diseases as brain disorders, and the reported successes of each therapy generated enthusiasm for the others. A modest revival of the organicist spirit was underway—one unlike earlier efforts to subject the patient's body to various soothing treatments for mental illness, such as immersion in spa water. The new radical techniques of the 1930s followed more in the tradition of centuries-old attempts to "shock" patients into recovery, such as bowssening, the practice of giving the patient alternating blows on the head and submersions in water, and hydrotherapy (portrayed in the novel and movie *The Snake Pit*), in which the patient is immobilized in a freezing bath.

The therapies devised during the 1930s were designed to shock the brain. Today the mention of these treatments—Metrazol convulsion, insulin coma, electroshock—chills many people, especially psychiatric patients, but in their day these therapies, which were brutal to watch and often horrendous to endure, presented an alternative to indefinite confinement in a ghastly mental hospital. The new treatments did not cure many people—with the exception of electroconvulsive therapy, which is still widely used to treat serious depression—but they held out the chance of a cure and they gave hospital staffs *something* to do for psychotic patients. No longer did mental hospital employees have to think of themselves only as custodians. As Edward Shorter noted, shock treatments "gave psychiatrists powerful new therapies in a field dominated for half a century by nihilistic hopelessness, and it is against that sense of despair that they must be set."

Through the "shock" of an outside agent, these therapies all produced unconsciousness or convulsions. The end of the coma or seizures, believed Freeman, brought "restoration of more effective, flexible, adaptive activities of the brain. Theories as to how this is brought about are numerous, and attempts to obtain the beneficial results without too great a danger to the patient are equally so." The first of these treatments to appear was insulin coma therapy, which an Austrian psychiatrist named Manfred Sakel applied to schizophrenic patients in 1934. A dangerous procedure demanding close monitoring for days at a time, insulin treatment produced prolonged unconsciousness by raising the patient's insulin concentration to near-fatal levels and thus reducing the amount of glucose in the bloodstream. Sometimes patients also went into convulsions, but Sakel considered seizures undesirable because of the memory problems that often followed. Sakel—who claimed to be a direct descendant of the twelfth-century Jewish philosopher Moses Maimonides—was an unfriendly, misanthropic man who felt no need for closeness with others. "He had a withdrawn, rather defensive attitude; he never married; he lived, apparently only for his work; and he resented the intrusions of others," wrote Freeman, who met Sakel twice. At one point Sakel claimed that 88 percent of his schizophrenic patients improved with insulin, an assertion widely disbelieved. Even so, few other treatments produced any improvement in schizophrenics, and a hundred American hospitals eventually built insulin therapy units. These wards were quiet, eerie places, with rows of comatose patients arranged in beds, and attended to by staff trying to keep them from dying from hypoglycemia.

Next came a convulsive therapy using Metrazol, a drug similar to camphor. In 1934 Ladislas von Meduna, a Hungarian-born psychiatrist who loved to play bridge, became the first to use Metrazol intravenously on schizophrenic patients after coming to the conclusion that convulsions produced by epilepsy seemed to help such patients recover their mental health. Why not, he reasoned, induce convulsions and see what happens? Meduna immigrated to America soon afterward, and many psychiatric hospitals adopted his therapy through the early 1940s. Freeman met Meduna several times—his son, Paul, served part of his psychiatry internship under Meduna—and was struck by his eyes, "which were level and penetrating." A gigantic drawback to Metrazol was the feeling of intense fear and anxiety it produced in patients just before the onset of convulsions. Two American psychiatrists who experimented with Metrazol noted "the fleeting but quite definite and almost animal-like expression of fear that appears just before the first tonic convulsive tightening of the body. . . . It is almost haunting in character and stays with the observer throughout the attack and afterward. . . . It seems as though one is carried back in time and sees in [Metrazol subjects] fear appearing at a lower biological level," wrote Samuel Clark and Frank Norbury. "One does not see this expression often at the onset of convulsions due to other causes."

Europe was also the birthplace of the third major shock treatment, electroshock therapy. In the early 1930s, the Italian psychiatrists Ugo Cerletti and Lucio Bini began using electric current to induce seizures in dogs. But intimidated by the specter of the electric chair and distracted by the successes reported with insulin and Metrazol, Cerletti delayed six years before submitting human subjects to electrically produced convulsions. His first patient, a thirty-nine-year-old schizophrenic man found wandering around Rome and suffering from hallucinations, gave the psychiatrist a vivid demonstration of the power of electric current when applied to the brain: "He started to sing abruptly at the top of his voice, then he quieted down," Cerletti wrote. "Naturally, we, who were conducting the experiment, were under great emotional strain and felt that we had already taken quite a risk. . . . All at once, the patient, who had evidently been following our conversation, said clearly and solemnly, without his usual gibberish: 'Not another one! It's murder!' [In another account, the patient's outburst is translated from the Italian as, 'Look out! The first is pestiferous, the second mortiferous.'] I confess that such explicit admonition under such circumstances, and so emphatic and commanding, coming from a person whose enigmatic jargon had until then been very difficult to understand, shook my determination to carry on with the experiment." Nevertheless Cerletti administered another 1.5-second shock, then asked the patient for his impressions. "I don't know. Maybe I was asleep," the patient replied.

While electroshock did not cure schizophrenia, it often diminished the most immobilizing symptoms of that disease as well as other mental disorders. Patients did not dread the convulsions as much as with Metrazol—most could not recall their electroconvulsive treatments at all—and electricity did not bring them to death's door, as with insulin. Electroshock, which arrived in the United States at the end of the 1930s, replaced insulin and Metrazol therapy in some hospitals due to its ease of use.

Although patients undergoing electroshock rarely died, they frequently suffered broken bones resulting from the severity of their convulsions, despite the squads of hospital staff that piled atop patients during treatments to hold them down. Freeman helped find a solution to this problem. In 1934 he began treating an explorer named Richard Gill who was severely disabled by multiple sclerosis. One day Freeman astonished Gill by speculating that curare, a deadly compound used by South American tribesmen to kill game, might provide relief for spastic disorders like multiple sclerosis.

As a graduate student in Paris, Freeman had seen curare used to treat certain spastic conditions. Gill, who had hunted with curare-tipped blowgun darts during his healthier days, well knew that curare kills by relaxing and paralyzing the victim's skeletal muscles, thus causing an agonizing asphyxia. He overcame any disinclination to expose himself to such a deadly poison and was able to procure a supply of curare from Ecuador. After several pharmaceutical companies turned down Gill's request to investigate the curative properties of the compound, Freeman made some inquiries and came up with the name of Abram Bennett, a psychiatrist in Omaha. For a time, Bennett had been experimenting with curare on spastic children. Freeman put him in touch with Gill. Although curare never played a significant role in the treatment of multiple sclerosis, Bennett went on to show that in small doses curare induced a paralysis of the head and neck muscles that could reduce the spasms of patients undergoing electroshock. Until 1952, when a safer drug replaced it, curare helped end the incidence of bone fractures as a side effect of electroconvulsive therapy.

Freeman viewed the sudden interest in the shock therapies as a good thing, because "it prevented a runaway enthusiasm for psychosurgery that might have done a great deal of harm." While doctors set up shock units in hospitals around the country, lobotomy was allowed to evolve gradually, a few patients at a time. Freeman initially mocked the theory behind insulin coma therapy, but he later grew more interested in it, writing in

1937 that insulin treatment appears "sufficiently meritorious to stand out as treatment endeavors in the very difficult field of psychiatry." Later that summer, Freeman tried both insulin and Metrazol on patients, becoming the first physician in the Washington area to experiment with the shock therapies. This readiness to test the newest treatments was in keeping with the character of a man whose drive to break new ground in medicine coexisted with a conviction that manipulating cells and neural pathways could quiet the mentally diseased brain.

In his search for candidates for Metrazol convulsive therapy, he thought of his seventy-year-old aunt, Florence Keen. Although it is surprising that he chose an elderly relative for a procedure so disagreeable to patients, he was concerned by Florence's mental state, which had deteriorated in the five years since the death of her father, W. W. Keen. Florence had lived with her father and run his household for decades, and in his absence she became chronically depressed. With his surgeon brother, Norman, acting as assistant, Freeman began Florence's first treatment session by injecting Metrazol into a vein. Within ten seconds, "she began twitching, then opened her mouth widely, arched her back, stiffened out in a tonic convulsion that lasted about 20 seconds, followed by clonic movements [muscular contractions] for another 25 seconds," Freeman later recalled. "Then she relaxed with no respiratory movements for many seconds." After turning blue, Florence gasped and resumed breathing. "Gradually, the color returned to her face—also to my brother's. 'Jesus!' he said, and wiped his brow." Five more Metrazol treatments over the next two weeks left Florence less depressed but with a damaged memory. When Florence died ten years later, "she remembered me in her will, but I doubt if she ever forgave me," Freeman noted.

Not long afterward, Freeman discarded insulin and Metrazol from his repertoire of treatments. Electroshock, which he began using in 1938—again scoring a first in the nation's capital—remained in. He acknowledged that electroconvulsions proved a poor therapy for schizophrenic patients; compared with lobotomy, they were not sufficiently damaging to the brain, even when he tried electric jolts "up to 100 times the duration of current that is suggested without doing much to influence the course of the disease." (Likewise, his main complaint against insulin and Metrazol treatments was that they did not produce enough permanent damage to the brain. "Maybe it will be shown that a mentally ill patient can think more clearly and more constructively with less brain in actual operation," he wrote.) For patients suffering from anxiety, depression, and other affective disorders, however, he believed that electroshock was a good choice. "Electroshock treatment seems to be the answer for many people who get tangled up in their thoughts, get to thinking in circles, unable to make up their minds, harassed by doubts and fears and driven to desperation," Freeman wrote in 1945. In his experience, it improved these patients quickly, often in just a few days, and it was well suited to government workers stressed by overtime hours, "the civilian war casualties, as it were."

One patient in particular, however, did not appreciate Freeman's technique as an electroshock practitioner. During the early 1940s and before the widespread use of curare and other muscle relaxants, he decided to treat a patient with electroconvulsion in his private office. Normally his secretary or a relative of the patient would assist by holding the patient down during the convulsions, but in this case the secretary declined and the patient's husband, disabled by a heart condition, could not help. Freeman decided to administer the treatment unaided. During the first convulsion, the patient broke both legs. "I gave her an injection of morphine and left her too soon to see some patients at the hospital," he wrote. "When I returned the fat was in the fire. The woman was writhing in pain and her husband was outraged." The patient sued, asking for $50,000 in damages. Freeman settled with her out of court. The incident left Freeman with "a mixture of regret, self-justification, anger at my secretary for her refusal to help, self-condemnation, humiliation at the disapproving attitude of my colleagues, and confidence that the matter would be settled and the damages covered by insurance. When the matter was settled, I offered to shake hands with the patient and her husband, but was glared at." The patient's subsequent relapse into depression convinced Freeman that he should have performed a lobotomy on her instead of administering electroshock.

# Chapter 7

# The Lines of Battle

As early as December 1936, Moniz warned Freeman that lobotomy would place him in the tempest of the long-running historical conflict between the biological and behavioral camps of psychiatry. "The psychiatric doctors professing the classic and psychological school that divorced the mind from the brain or at least keep some restrictions between the organ and the function, will have a certain resistance to [your] organic orientation, for a long time," he wrote. Thirteen years later, Moniz elaborated: "In the depth of the problem is the old question between those who wish to see the problem through the brain or without this organ of the mind. . . . For us the progress of psychiatry will be the anatomical study of the brain and the operative trials. It will be the only way to [make] a medical specialty of psychiatry. The other will be the philosophical and psychological way, good for nothing in scientific progress."

These battle lines did not move much during Freeman's lifetime. One of the most aggravating of the Freeman–Watts patients, in fact, was a twenty-eight-year-old woman debilitated by obsessions including a terror of shoes, whose two lobotomies in September 1937 allowed her to resume her life as a graduate student in mathematics at Columbia University. Freeman caught up with her five years later, when she had completed her requirements for a Ph.D. and was teaching at a university in Puerto Rico. The patient, however, attributed her recovery to postoperative psychoanalysis, not to psychosurgery. "We are not concerned here with trying to evaluate the degree of improvement due to lobotomy and that due to analysis," Freeman and Watts noted. "We merely wish to point out that after an extensive bilateral prefrontal lobotomy, this woman resumed her activities after having been disabled for eight years, and demonstrated a high degree of intelligence in the pursuit of her graduate studies."

Many of the physicians involved viewed the conflict between organic and psychological treatments as an all-out war. The English neurosurgeon William Sargant called the early lobotomies "a preliminary skirmish in a surgical attack on the supposed 'soul' of man" and observed at the 1939 annual meeting of the American Psychiatric Association that members "felt so insulted by [Freeman's] attempt to treat otherwise incurable mental disorders with the knife that some would almost have used their own on him at the least excuse." Decades later, Sargant found the basis of much of the opposition to psychosurgery little changed. He was struck by the "moral repugnance this operation still arouses among some psychiatrists who are too prone to deny mental disorders any real physiological basis—believing them to be mainly psychological, environmental or spiritual in origin! And when they are asked why they object so much . . . the old bogy of 'personality deterioration,' long ago minimized by new techniques, is still produced to frighten patients and relatives into choosing other less certain treatment alternatives, or doing nothing at all even if the patient is suffering terribly as well as having been totally incapacitated in a mental hospital for years on end."

As might be expected, medical ethics frequently entered into the controversy over psychosurgery, a procedure that altered personality in patients whose judgment and ability to consent to surgery was sometimes dubious to begin with. Not even Freeman and Watts claimed that the brain tissue they damaged was unhealthy; they believed that their lobotomy patients would lead better lives when some parts of their brain were unable to communicate with others. To many critics, however, it was unethical to harm healthy tissue, even in the hope that beneficial results would follow. Even worse, some detractors declared that playing with a patient's personality amounted to tampering with his or her human essence. Lobotomy threatened to alter a patient's emotions, sense of altruism, and sense of humor—all traits that separate humans from animals.

One such critic was Dexter Bullard, a psychoanalytically oriented psychiatrist whose spirited attacks on Freeman over the course of many years started with an attempt to shout down Freeman during his presentation before the Medical Society of the District of Columbia in 1936. In the summer of 1941, Bullard and his wife gave Marjorie Freeman a rare opportunity to defend her husband's work. When the Bullards met Marjorie at a social event, they indignantly laid into one of Freeman's recent articles. "Then Anne [Bullard] said, 'But Walter should have waited and not have such an article come out until he had been successful,'" Marjorie reported to her husband. "I asked her if she had read the article, and when she said yes, I replied, 'Well, read it again with more discerning eyes,' with my sweetest and cattiest smile. They were both so bitter that it would seem to indicate an amount of jealousy which pleased me enormously." (Bullard and Freeman maintained an odd friendship despite their philosophical differences and sometimes traveled together to medical conferences. Freeman once accepted Bullard's invitation to drive with him to an AMA event in Cincinnati but later canceled his plans after feeling strong premonitions of disaster. Bullard set off in his Packard, overturned the car on a slippery highway in the Appalachians, and had his hand impaled by a shard of the windshield.)

At one medical meeting, Harold Singer, the editor-in-chief of the *American Journal of Neurology*, sarcastically asked Freeman why he didn't simply take the easier path of cutting the cerebellum located in the brain stem. Whereas Singer was suggesting that damaging any part of a patient's brain would produce results similar to cuts in the frontal lobes, Freeman and Watts firmly believed that their work, with that of Moniz, demonstrated that the frontal lobes played a special role in regulating tension and anxiety. "Which is better," Watts once replied, "to damage the brain a bit and get the patient out of the hospital or [to] do nothing?"

Some religious critics may have answered that it is better to do nothing. Catholic doctrine of the time, for instance, maintained that people are not entitled to employ or alter their bodies for any purpose other than "natural uses," and that doctors may not mutilate their patients' bodily parts "except when no other provision can be made for the good of the whole body." Eventually, the Catholic Church decided that lobotomy was "morally justifiable when medically indicated as the proper treatment of serious mental illness or intractable pain. . . . These operations are not justifiable when less extreme remedies are reasonably available or in cases in which the probability of harm to the patient outweighs the hope of benefit to him." Freeman and Watts noted that lobotomy patients often curtailed their attendance at houses of worship because of inertia and lack of motivation, but they refuted arguments that psychosurgery erased spirituality from the soul. Watts recalled that in an informal survey of twenty lobotomy patients, eighteen said they believed in God, one was undecided, and one denied God's existence. "Now, I don't think that we took God away from these people," he concluded.

At medical conferences around the country, Freeman spoke and gave papers on the early results of psychosurgery. In general, he wrote, he found his audiences "much interested but still lacking conviction." Sometimes his listeners became hostile. At a meeting of the Chicago Neurological Society, Freeman presented a paper on the new procedure. Opponents, including the psychiatrist Loyal Davis, the father of the future first lady Nancy Davis Reagan, called lobotomy lazy, mutilating, and criminal. "One of my friends observed that it was well that the stem of my pipe was strong, otherwise, I would have bitten it off," Freeman recalled.

As time passed, Freeman perceived that the general receptivity of his audiences changed. Giving a public lecture in 1944 in Hartford, Connecticut, Freeman found an audience open to hearing about the benefits of lobotomy performed on a large scale. When asked what lobotomy could do for the estimated five hundred thousand to six hundred thousand chronic schizophrenics confined to American psychiatric hospitals, Freeman declared that psychosurgery could return up to sixty thousand of them to a productive life in society. Apparently nobody present questioned this startling prediction.

The previous year, during a trip to Toronto made at the invitation of the government of Ontario, Freeman operated—apparently without Watts's assistance—on a man whose psychiatric condition had made him mute for three years. The patient "came out of his stupor on the operating table and . . . began talking in an animated fashion before he left the operating room," Freeman wrote. This demonstration, along with a lecture he gave later in the day, made a strong impression, Freeman believed. His lecture audience, packed all the way into an outside hallway, hardly moved. "I must have hypnotized them or something." Speaking invitations came from everywhere: nursing groups, neurological conferences, church assemblies, and community organizations. "I try to bring in a little vulgarity always, because I find that there is nothing that makes people wake up like a belly laugh," he declared. "It sort of relaxes the hypnotic tension, and does away with the snores. I think I am a pretty good speaker now, after long years of practice." Partly on the basis of his reception from such audiences, he judged that lobotomy "is slowly gaining against psychiatric prejudice, and it pays the income tax, so that is about all that can be expected of it."

To the further aggravation of his opponents, Freeman refused to acknowledge that their assaults affected him personally. "My grandfather,

W. W. Keen who introduced antisepsis into the United States was called a liar by his contemporaries, and even I was accused by one of my friends of criminal activities in following out leucotomies," he wrote in 1946. "But at least there was no official obstruction, and all of us, working with our private patients had only the natural laws of torts and damages to contend with." He calmly explained to one antagonist, the psychoanalyst Smith Ely Jelliffe, that "I know that your opinion will be shared by many who see no justification in a mutilating operation upon a part of the brain, the highest part, indeed, but I maintain that in some of the cases that have come to operation, the patients have thereafter been able to think more constructively with less brain." Writing sixteen years later, in 1956, he told a supporter, "If I have been attacked on occasion I am sure there is nothing personal in it, and I think all but one or two of the attackers are personal friends of mine although we differ markedly in our attitude toward this particular subject. If they had shown any more significant results in their attempts to improve similar patients I would go along with them much more readily."

Support from their peers, among them some of the biggest names in neurosurgery and neurology, gave Freeman and Watts their biggest weapons against the assaults of critics. In 1939 William Sargant visited Washington with a colleague and headed straight to GWU to evaluate three lobotomy patients. The first, an alcoholic (not identified in Sargant's account but possibly Paul Hennessy), struck Sargant as a failure. The patient disclosed that "he could now drink half his ordinary amount of whisky and get twice as tight! . . . Obviously, [the lobotomy] had done his alcoholism no good and perhaps a great deal of harm: he was much less tense but of course did badly later on as he continued drinking heavily." The second patient, a schizophrenic, was no longer troubled by the voices she continued to hear. The final patient Sargant met, a "chronic melancholic," reported that she felt much better since her operation, but she noticed a pronounced change in her personality—a tendency to be grouchy, impulsively angry, and self-absorbed. Nevertheless, Sargant concluded, "Freeman and Watts had confirmed the value of an operation which, for the first time in my experience, seemed able to relieve chronic tension and anxiety in whatever psychiatric setting it presented itself. With each of these three very different psychiatric conditions, the chronic anxiety and intense fears of impending disaster had diminished or disappeared."

Not all neurosurgeons viewed lobotomy in such a positive light; Wilder Penfield, the Canadian physician who had spoken at the 1935 London conference about his excisions of the frontal lobes of patients with brain tumors, feared that psychosurgery could produce convulsive seizures. "Walter," he told Freeman during the first few years of lobotomy, "don't you realize that you're doing a very dangerous thing?"

At last, five years after the first lobotomies that Freeman and Watts performed at GWU, the American Medical Association weighed in on the propriety of psychosurgery as a treatment for psychiatric illness. The group sided with critics alarmed at the dangers of damaging healthy brain tissue and the risk of producing convulsive seizures. "In our present state of ignorance concerning the frontal lobes, there is ample evidence of the serious defects produced by their removal in non-psychotic persons," the AMA statement read. "It is inconceivable that any procedure which effectively destroys the function of this portion of the brain could possibly restore the person concerned to a wholly normal state." (This was the only instance during the 1930s or 1940s in which a medical organization took a meaningful public stand against lobotomy.) Freeman, however, never proposed lobotomy as a means to restore sick patients to healthy wholeness, or even as a way to eliminate the symptoms of mental illness. He and Watts repeatedly promoted lobotomy only as a treatment of last resort—one that would dull the "sting" of mental disease—and they carefully reported on the aftereffects of the surgery. For many patients, they maintained, the price paid in changes to the personality was worth it. "Psychosurgery was a form of human salvage, not rescue," noted the psychiatric historian Jack Pressman. Freeman and Watts gained some reassurance by the first British psychosurgeries in 1941, an important milestone for a nation with traditionally conservative psychiatric practices.

Through their use of thorotrast dye, X-rays, and conversations with patients, the doctors had made a curious discovery. In general, patients who displayed the greatest changes and difficulties after lobotomy—who showed such side effects as apathy, indifference, and uncontrollable repetition of words and gestures—lost more of their symptoms of anxiety and depression than those who retained alertness and normal behavior. "We noticed that considerable dullness and unresponsiveness is a prelude to more satisfactory recovery from depressed and agitated states," Freeman noted. That eight of the first twenty patients had relapsed led Freeman and Watts to conclude that a comfortable recovery was a sign that the patient's key neural connections had not been severed. Two of the early lobotomy patients, Mildred Spurlin (the sixth case) and Ella Hanshew (the twelfth case), would be found at St. Elizabeths Hospital when Freeman checked on them twenty-four years later. (Freeman did not consider Hanshew's case a total failure, however, because she stayed out of psychiatric hospitals from 1936 to 1951. "She has been spared some fifteen years of institutional life, and she enters the hospital at the age of thirty-nine instead of twenty-four," he wrote.)

By August 1937, nearly a year after venturing forth into psychosurgery, Freeman and Watts had reoperated on four patients, those "who sustained no improvement, or only a short remission," Freeman noted in a letter to Moniz. (Eventually eight of the first twenty patients required reoperation, and two underwent a third lobotomy.) "When a patient has improved following lobotomy, gone home, taken up housekeeping, or found employment and then has a relapse, it is reasonable to think that by cutting additional pathways in the frontal lobe the symptoms may be relieved again," the doctors later observed. One such patient was the woman known as Case 4, the obsessive-compulsive woman, operated on in October 1936, who feared contamination. In her reoperation on March 1, 1937, Freeman and Watts cut cores in the neural connections in the lower half of both frontal lobes. The patient's recovery was far more difficult than previously, and her behavior indicated that her thinking was confused and inattentive. The doctors saw her "shake her hands as though to rid them of germs, or go through washing movements of her hands," although she "did not seem distressed by her ideas of contamination." She went home in this state; at first a difficult burden for her family to handle, she gradually emerged from her mental fog sufficiently to help her daughter clean house. Freeman kept track of her for many years and noted that "a few of the cherished [obsessive-compulsive] mannerisms have persisted, but they have lost their importance to her."

There were many other reoperations in the months and years that followed. Some failed to produce even mixed results. Among those receiving repeat lobotomies in 1937 was a woman whose suicidal impulses did not subside after the first lobotomy—and she again attempted suicide by setting herself on fire after her second operation. The family of the woman known as Case 6, who had recovered from catatonic schizophrenia after her first lobotomy only to require hospitalization again the following year, initially refused to allow Freeman and Watts to perform a second operation but did give permission for her to receive shock by Metrazol and insulin therapy instead. When those treatments failed to produce a recovery after several years, the family at last authorized a second lobotomy. Although the patient emerged from this surgery less withdrawn, she was never again able to live without daily assistance from her family.

Perhaps the repeat operation that made Freeman and Watts the most anxious was the one performed on Case 10, a woman originally given a lobotomy in December 1936. Cutting their cores more deeply than in the first surgery, the doctors produced a fountain of blood that sent the patient's vital signs plummeting. Watts quickly cauterized a severed artery, but for weeks the woman suffered paralysis and could not walk. According to the psychosurgery historian David Shutts, this patient sued her physicians, and Freeman and Watts agreed to an out-of-court settlement of $2,500. "But such things," Freeman observed, "are more or less to be expected in a free country like ours." Freeman, in his follow-ups on this patient, noted that her paralysis eventually faded and she returned to work with, he observed, a "sterile" mental life. Surprisingly, this patient returned in 1941 for a third operation. The additional cuts to her brain diminished her emotional affect and left her, in Freeman's words, "indolent, petulant, puerile, and irresponsible." By 1956 Freeman had tallied three hundred of his patients who had undergone more than one lobotomy.

Freeman frequently puzzled over the delayed relapses that often required reoperation. Over time, he encountered several patients who had apparently been helped by lobotomy but who inexplicably experienced a return of their symptoms several years later. By the same token, he followed some patients who went through "delayed recoveries" and found relief from their symptoms long after their lobotomies—sometimes ten years later. "A recent letter from a patient who was hospitalized for eight years after her operation in 1940, and is now making her living in institutional work, reads in part, 'I do talk out of turn and I *must* correct it. *Will.* Am conscious of my personality fault. It is up to me to correct it and remain silent.'" He ignored the possibility that the recoveries were spontaneous or unrelated to the surgery, and he found comfort that "the relapse rate is almost balanced by the delayed recovery rate."

At a slower pace, Moniz continued his work in Portugal. By the spring of 1938, he had performed seventy-three leucotomies, with the last of these cases producing the first Portuguese fatality, a death from cerebral hemorrhage. "We were already too confident in ourselves," Moniz lamented as he noted that the holes in the skull and location of the cuts of neural tissue were inaccurately placed. Freeman updated Moniz on the first forty

Freeman–Watts lobotomies. Three patients, all of them dating from the first operations performed in 1936, had suffered convulsive seizures postsurgery. Four patients had died, although two of the deaths appeared unrelated to their psychosurgery. Five required care in psychiatric institutions. One was partly paralyzed on the left side because of the doctors' improper placement of the leucotome.

Of the first six cases treated in Washington under Moniz's technique, only three were still thought to be significantly improved: "One is substantially well and another, a schizophrenic, is living at home but still looks and talks like a schizophrenic although the family says she is better."

By this time, the signals of impending war had nearly smothered all important medical research in Europe, especially new investigations into psychosurgery. Although Freeman had not seen Moniz since the summer of 1935, his relationship with the Portuguese physician was growing closer. In 1939 Freeman was shocked to learn that a patient, a young man suffering from hormonal problems, had attacked and seriously injured Moniz. Contrary to some later accounts of the crime, the assailant had not received a leucotomy. "I knew at the fifth shot that I was being attacked," Moniz wrote in his memoirs. "I arose as well as I could and he shot me twice more. Realizing that my right hand was severely injured, I tried to hit him in the head with the ink stand." After his attacker fled, Moniz lay bleeding. "It was a poor lunatic," he told people who came to his aid. "Call my wife. I want to see her before I die. . . . Let me die here quietly. I've been mortally wounded. That crazy man plugged me with bullets. I couldn't resist." His blood sprayed the patient record he had been completing at his desk.

Moniz did not die, but he took four bullets, one of which could not be removed because of its closeness to the spine. He became partly paralyzed and never recovered use of one hand, which was already crippled by gout. The incident made a distressing impression on Freeman. Writing years later, he noted that those in his profession "run a certain risk of assassination by patients. It's an occupational hazard." Moniz recovered slowly. Pressure in Portugal from psychoanalytically oriented psychiatrists upset about a neurologist trespassing into the treatment of mental disease eventually forced Moniz to abandon his investigations into psychosurgery; he had operated on about one hundred patients. Moniz felt further marginalized by his government's disapproval of his declarations that Portugal should enter World War II against the Nazis instead of remaining neutral. He also believed that his Portuguese colleagues were envious of his successes. The acclaimed physician diverted his energies to new areas and even contributed to a book on boston, a card game similar to bridge. Moniz retired from his academic position in 1944 at age seventy-five. His final addition to the literature of psychosurgery was a preface he wrote to a book authored by his former medical partner, Almeida Lima.

As Freeman watched Moniz's contributions to psychosurgery trickle to an end, he and Watts did not let up on refining their technique. Put simply, they wanted better results. "We studied brains of patients who died after operation, as well as the architecture of normal brains," Freeman wrote. "We read volumes on anatomy, physiology, neurosurgery, experimental surgery, war injuries and their consequences. We enlisted the aid of friends who had experience similar to ours. Every few months I reported to Egas Moniz, to [John] Fulton and to others who were interested in psychosurgery." In particular, Freeman and Watts wanted to reduce the postoperative incidence of seizures, which plagued more than 10 percent of their lobotomy patients. Case 5, for example, who was operated on in 1936, died eight years later in the throes of convulsions after battling frequent seizures since her operation. Freeman and Watts wrote that seizures "are the most embarrassing sequel to prefrontal lobotomy."

When patients' symptoms failed to improve after surgery, Freeman believed that not enough neural fibers had been severed. In 1938 Freeman and Watts modified their approach to the frontal lobes, initially by moving the entry holes closer to the top of the skull near the center line. Patients given lobotomies by this approach, though, fared poorly and frequently suffered convulsions. Freeman and Watts then moved the entry holes to the sides of the skull, where they could make better use of the anatomical landmarks of the brain and the skull. Using a cannula, a narrow rod marked with calibrations, they first measured the diameter of the patient's brain. "During the operation, I stood at a distance of some six feet behind the head of the patient, with my eyes in the plane of the coronal suture and the sphenoidal ridge, so that I could guide Watts in making the incision as accurately as possible," Freeman wrote. Watts's great skills as a neurosurgeon enabled him to insert the cannula in the hole at one temple and carefully thread it through the brain so that the instrument appeared in the hole at the opposite temple, avoiding blood vessels and critical tissue. "That's pretty damn dramatic, you know," Watts said. "To be able to do that with that degree of accuracy is very, very good. And of course it always impressed [an audience]."

Once they knew the diameter of the patient's brain, Freeman and Watts took out a new instrument in their arsenal. They had abandoned Moniz's core-cutting leucotome, which often unintentionally damaged tissue in its path through the brain and failed to remove targeted tissue cleanly, in favor of the Killian periosteal elevator, originally designed to lift brain tissue. This instrument resembled a calibrated letter opener with a curved tip and a movable sidearm, and it severed fibers in clean cuts. They inserted this new leucotome into the brain to a distance of about half of the diameter. Then they swept the leucotome both upward and downward in the plane of the coronal suture, a joint traveling from the top of the skull to the temples. Using the entry hole at the opposite temple, the doctors repeated the procedure.

Henry Ator, a Washington machinist, manufactured the tools for Freeman and Watts, who had them imprinted with their names on the handles. Watts favored the new instrument—which he and Freeman often called a leucotome—because it allowed him to probe brain tissue while leaving untargeted regions unharmed. Watts became adept at using the new leucotome to creep along the walls of the cranial vault as he felt for the ridges and protrusions that gave him anatomical orientation to his position. He also honed his skills in using his tools to distinguish between the various types of tissue within the brain. "Whenever I felt resistance," Watts said, "I would stop swinging because I'd be afraid that I might be touching or hooking a blood vessel or something like that." Generally, the new leucotome passed through the brain "just like soft butter," Watts observed.

The change in instrument and location of the insertion holes, which resulted in a refined lobotomy that Freeman and Watts called the "precision method," appeared "to allow better control of the placing of the lesions than insertion of the leucotome from above," Freeman informed Moniz. Precise cutting would improve the outcomes, Freeman and Watts agreed, along with the consistent use of X-ray imaging to study the lesions and the selection of patients whose illness had the best chance of recovery with lobotomy. But the partners continually experimented with modified approaches to the brain, sometimes placing the entry holes a bit behind the plane of the coronal suture. In these cases, they "found that inertia, incontinence, and other indications of severe damage to the frontal lobe were enduring residuals," they wrote.

Over time, Freeman and Watts developed several different lobotomy techniques, each intended to treat patients of varying types. While they believed that the standard prefrontal lobotomy most often produced the best results, some patients required less, or more. In the *minimal lobotomy* procedure, they omitted any cuts to the neural pathways in the upper portion of the frontal lobes. They chose this procedure if the patient was elderly, had been ill for only a short period, or was in danger of functioning poorly with the loss of responsibility that the standard lobotomy produced.

Schizophrenic patients and those for whom the standard operation had not worked often received the *radical lobotomy*. In this more extensive surgery, Freeman and Watts targeted fibers behind the sphenoidal ridge, a bony landmark that previously marked the boundaries of judicious cutting. Any patient risked "a considerable sacrifice of personality organization" after radical lobotomy, Freeman and Watts believed. Each millimeter of cutting behind the sphenoidal ridge made it more difficult for patients socially to adjust after operation, and in these cases Watts relied the most heavily on Freeman's guidance and knowledge of brain anatomy. (Years earlier Moniz had observed that leucotomy lesions behind the sphenoidal ridge made patients lose their "conscience"—his term for considerate behavior toward others.) At its worst, the radical procedure produced patients who lay inert, unable to move because of a profound lack of initiative. "The bizarre 'china doll phenomenon' sometimes occurred in these patients," noted the psychosurgery historian David Shutts. "Freeman and Watts would urge the supine 'doll' to open his or her eyes, but the inert patient would remain tight-lidded. When the patient was moved into a sitting position, the eyes opened; when the patient was laid down again, they would close." These patients often went home without yet recovering spontaneous movement, and they required intensive care.

Freeman and Watts clearly needed a means to gauge the amount of damage their cuts were producing—preferably while the lobotomy was in progress. Around 1939 they began performing some surgeries under local anesthesia. Unlike patients under general anesthesia, those given only Novocain to numb the scalp at the sites of the burr holes could respond to questions, take tests designed to measure their mental acuity and determine the presence of delusions or tension, and describe to the doctors their sensations and feelings. Because Freeman and Watts believed that lobotomy was most effective when it resulted in temporary disorientation and mental dullness immediately after surgery, they hoped to use conversations with patients during the operation to gauge exactly when their cuts had caused enough but not too much damage.

When conscious during surgery, however, many patients felt increased fear and anxiety. (Freeman, who usually took charge of these dialogues with patients, was fond of asking them to recite the Lord's Prayer, a strategy that in some cases could only have heightened their fear.) "An operation under local anesthesia is always a somewhat trying experience to the patient," Freeman and Watts wrote. "This must be doubly so when the patient knows that his brain is being operated upon, and probably no less so in a patient who is preoccupied with abnormal fears, anxieties, worries, depressions, and the like. We must commend the hardihood that has enabled quite a number of our patients to undergo prefrontal lobotomy under local anesthesia, even though they could be assured in advance that the operation itself was relatively painless." Fully cognizant of their surroundings, these patients watched themselves being strapped to the operating table, shaved, and covered with sterile draping. During the surgery, they heard "the rattling of instruments, the noise of the suction apparatus, and the menacing spark of the electrocautery." A few times the patient's panicky response convinced Freeman and Watts to change course and administer general anesthesia. Occasionally, though, the powerful obsessions and anxieties of patients blocked out the business of the operating room. One patient continued chattering about her apprehension over not having attended church the day before, even as a hole in her skull was being bored—a process the physicians said produced "a grinding sound that is as distressing, or more so, than the drilling of a tooth."

In general, the physicians found the first half of the operation, in which Freeman and Watts severed neural connections in two of four target areas of the frontal lobes, did not greatly change the mental state of locally anesthetized patients. Anxiety and fear persisted, memory and mental acuity remained intact, and patients could state where they were and why they were there. The third set of cuts usually produced the first noticeable changes: tension diminished and "sometimes the replies indicate lack of self-consciousness even to the point of being witty." When Freeman once asked after the third stage of the lobotomy, "What's going through your mind?" the reply after a pregnant silence from the man on the operating table was "A knife."

One woman identified as Mrs. A., the eighty-third patient in the Freeman–Watts series of lobotomies, made a written record of her memories from the surgery. Freeman described her as an acerbic and stubborn woman, a college-educated mother of two children. In 1940 she had grown obsessed with her church minister; she believed that he possessed evil powers and wanted to kill her. Despite Metrazol shock treatment and daily exercise, she suffered from insomnia and grew more paranoiac. In May 1941 Freeman presented her with the choice of commitment to a mental hospital or undergoing lobotomy. She declared of the clergyman, "Damn him, damn his soul, damn him! It's all his fault! Damn, damn! DAMN!" Then she chose lobotomy, and Freeman and Watts operated on her the next day.

A week after the surgery, Mrs. A. began setting down her recollections of the operation. She wrote that after she received the Novocain in the operating room, Freeman arrived at her side and asked her to count backward from 100 by sevens. "Not the easiest thing in the world to do," she noted, "but it was interesting. Yeah, something interesting about the 7's." Then Freeman posed her questions about the significance of the numbers 1066, 1028, –437, 606, 666, 57, and 99 44/100. Her replies: "William the Conqueror entered England. Office number. Absolute zero. The number of the serum cure for syphilis. Seen in roadside signs: a fever cure for colds. Heinz–57 varieties. Percent pure: Ivory Soap." Her mental powers were obviously intact, but the drilling had just begun. Several times Freeman left her side to help direct Watts's probing. "Then more directions, more directions, and three snips. Scissors? Sounded like it. Well, it must be over. That tenseness practically left my back." Freeman then asked her to recite the Lord's Prayer, which she did in German. "I didn't want to sing at all, but I had to do a verse of 'America.'"

Mrs. A. concluded her written account with observations on her mental condition. "Whereas, I was tense, sometimes panicky, now I feel calm—'Let come what may!' Whereas I dreaded what the day may bring, or what I might experience the next day, now I have no fears or dreads. . . . What seemed like 'complications' are now merely incidents. 'Experiences' are made part of the plan of life. 'Tenseness' seems to be something that belonged to somebody different. No worries! My! What a change!" When Freeman visited Mrs. A. at home a month after the operation, he found her adjusting well to a resumption of her work as a homemaker. She would not discuss her former feelings of persecution. Freeman contacted her again a year after surgery; she was working as a sales representative for the Fuller Brush Company. Two years after the operation, Mrs. A. agreed to discuss her old delusions with him. "What a fool I made of myself," she told him. "I guess it was my particular way of going off balance. I got on one line of thought and then couldn't get off it."

In the case of Mrs. A., despite the good results they reported on her, Freeman and Watts seem not to have achieved their ideal level of disorientation during lobotomy. Other patients they carried to the brink. Transcripts of the conversations Freeman directed during surgery often took surreal turns:

Freeman: Who am I?
Patient: Dr. Walter Kaufman.
Freeman: What about that relative who was troubling you?
Patient: He was after me.
Freeman: Are you happy?
Patient: Yes.
Freeman: What is a widow?
Patient: She is related to a man, and he went with another woman, and he picked up the other woman, and that's all.
Freeman: What is the difference between a dwarf and a child?
Patient: It wouldn't hurt a dwarf and it would a child.

Or, in another case:

Freeman [after cuts made in the left side of the frontal lobe]: Do you feel any difference?
Patient: I feel there's a book on my face. They worked on the left side. There was a drawing when they put the knife in.
Freeman [later during the operation]: Does your conscience hurt?
Patient: I don't know where it is. It was down by my heart, but I can't feel it at all.

In all, sixty-six patients operated on under only local anesthesia, Freeman and Watts used these conversations in an attempt to apply a "disorientation yard stick" that could measure the degree to which the patient's brain connections had been damaged. Most patients, Freeman maintained, recovered from this disorientation and emerged from lobotomy without any significant impairment of their intellectual abilities. He cited the case of one psychosurgery patient whose intelligence quotient increased from 82 to 95 after operation. Instead of damaging intelligence, he wrote, psychosurgery hindered "the utilization of intelligence, and this led back again to the question of motivation and thus to the emotional facet." Others disagreed with Freeman's position on patients' intelligence, with one physician noting that it "would appear that the tests at present used are not sufficiently refined." He was correct, and one of the lasting effects of the psychosurgery era was to improve testing techniques used to evaluate the value of lobotomy and many other psychiatric treatments.

Even with their intellect intact, though, lobotomy patients frequently suffered postlobotomy complications. Like children, patients had to pass through stages of development in which their level of maturity grew. "Prefrontal lobotomy has the effect of a surgically induced childhood. . . . The best one can say is that it is an immature, sometimes infantile personality," Freeman and Watts wrote in an article they coauthored with a nurse. Freeman urged families to give patients dolls or teddy bears to play with immediately after surgery and recommended they then graduate to drawing with crayons, reading picture books, and watching movies. Some male patients exhibited a kind of childish sexuality often characterized by uninhibited masturbation that Freeman treated with synthesized estrogen. Nurses had to learn new techniques of care: kissing, cuddling, and even tickling—the latter an effective way to disarm overly amorous patients. In instances that demanded discipline, spanking "is as effective a method as we know, since it seems to recall childhood experiences of a similar nature," the psychosurgeons observed.

When patients returned home, families had to deal with hyperactivity or inertia, incontinence, and overeating and weight gain. In one case history, the doctors documented the postoperative symptoms of a patient who suffered from severe inertia, disorientation, and a prodigious hunger. "On one occasion the nurse supplied her with food to the extent of 4,000 calories at one sitting, topped off by 400 grams [about one pound] of chocolate candy," they wrote. "The patient vomited and was ready to eat again." (These side effects prompted Freeman's son Franklin to quip that "talking about a successful lobotomy was like talking about a successful automobile accident.")

Freeman and Watts also concerned themselves with the effects of lobotomy on functional disorders that so far had proved to be less treatable, especially schizophrenia. Over time, Freeman had formulated the conviction that the hallucinations, detachment from reality, and withdrawal into fantasy typical in schizophrenics resulted from their emotional obsession with the future and their desire to avoid it. Lobotomy, he believed, should help. (We now know that schizophrenia is a genetically influenced disorder of brain development usually acquired early in childhood.)

When Freeman and Watts at last received permission to perform psychosurgery at St. Elizabeths Hospital seven years after the death of Superintendent William A. White, they set their sights on what Freeman called "some of the most difficult, disturbed and destructive cases" at the institution. Freeman confessed to his family, "I have hungered for the material there." At St. Elizabeths, as elsewhere, schizophrenic patients were quickly rising in number, and the new superintendent, Winfred Overholser, succumbed to the pressure of these increased numbers. Over the next seven years, Freeman and Watts operated on about eighty St. Elizabeths patients. One of the most unusual of these patients suffered from the schizophrenic delusion that he was fated to hold up such massive objects as the Statue of Liberty. Just before his operation, he undertook the burden of holding up the world. "When I last saw him he was lying in bed holding the world with his right hand," his case history noted. "His arm was extended, his fingers were tightly clinched. The fingers are distorted by long, continued pressure. He was sweating profusely and his tongue is out between his teeth like one who is exerting himself to his utmost." When a doctor pressed on his arm, the patient exclaimed, "Oh God! Don't you think I am holding up enough already?" The result of this patient's lobotomy is unknown.

Freeman kept in touch with one of these St. Elizabeths patients, Oretha Henley, for many years and considered her treatment unquestionably successful. Henley, who stood over six feet tall and weighed about 240 pounds at the time of her lobotomy, had suffered from psychiatric problems for eighteen years; she was admitted to St. Elizabeths in 1938. "She is the largest woman I have ever seen and is the kind of person [of] which one remarks, 'There is not an ounce of fat on her body,'" a physician, probably Freeman, noted in her case file. "When she has to be transferred from one room to another it requires five or six attendants to transfer her. Because of her size, she cannot be allowed privileges which other patients have, because when she goes on a rampage it is impossible to handle her. A lobotomy was proposed in the hope of reducing her aggressive tendencies. The patient has done some very fine needle work and if these tendencies can be controlled it might be possible to discharge her from the hospital." Freeman and Watts operated on her on February 2, 1944.

After surgery, Freeman reported, he could teasingly grasp her around the throat, "twist her arm, tickle her in the ribs and slap her behind without eliciting anything more than a wide grin or a hoarse chuckle." Three months later, Henley had moved to the hospital's open ward. She responded to questions and behaved cooperatively. The hospital wanted to discharge her, but her husband remembered her old conduct well and feared her. She was still a St. Elizabeths patient in 1946, by which time her weight had increased to 294 pounds and her record described her as "a goodnatured and smiling person." Soon the hospital paroled her, and in 1949 she wrote to Watts from her home in Denver, "I shall never forget the nice thing that you did and how happy I feel this afternoon for you operating on my head."

Freeman grew increasingly interested in treating deeply disturbed patients. "Recently we have become even more radical in our attack on these chronic schizophrenics," he wrote to Moniz, "the incisions far back in the frontal lobes, with the result that there has been a very prolonged period of inertia and incontinence, followed sometimes by an excessive degree of childish and boisterous behavior that has been almost as difficult to control as the original schizophrenia, so we must say that the problem of schizophrenia is not yet solved, although progress is being made." He painted a brighter picture when discussing in public the effectiveness of lobotomy on schizophrenic patients, as when he told an audience in Hartford, Connecticut two months before his report to Moniz that about two-thirds of schizophrenic patients treated by lobotomy had been helped by the surgery.

Meanwhile, Europe braced for war in the summer of 1939. Freeman and Marjorie nevertheless set off for Copenhagen to attend the Third International Neurological Congress. They traveled on the *Europa*, an ocean liner filled with Germans returning home and a few brave, or perhaps foolhardy, Americans. After docking in Germany, the Freemans made a whirlwind tour of that country, for they knew that the window for tourism was rapidly closing. Yet they appeared to be little concerned over the effects of Nazism and the tensions that were carrying Europe to war. Freeman had always been patriotic, but he remained surprisingly apolitical. Watts, in fact, confessed that despite their long friendship and professional partnership, he never learned whether Freeman was a Democrat or a Republican, a liberal or a conservative. Franklin Freeman believed that his father "leaned Republican."

(During a 1942 investigation of Freeman's patriotism and political beliefs, agents of the Federal Bureau of Investigation interviewed many of his neighbors and medical associates. One interviewee, a staff member at St. Elizabeths Hospital, recalled that Freeman "once had shown some color films taken in Germany and had spoken of his admiration of some German civil accomplishments, that he 'very definitely' admired such accomplishments as super-highways and slum clearance and had remarked that the Germans had done a good job along that line. [The interviewee] said he had questioned the Subject upon several occasions about those remarks. He said the Subject replied that he did admire some of the improvements to the country itself but that he certainly did not approve of either the Nazi regime or the Nazi philosophy. [The interviewee] said he believed the Subject to be preoccupied completely by his professional work, nonchalantly indifferent to any ideologies. As an exceptionally ambitious professional man, Subject could not become zealous over any ideology or political regime, according to [the interviewee]." Although the FBI found no one who questioned Freeman's patriotism, some interviewees took advantage of the opportunity to criticize other aspects of Freeman's character. A former neighbor called him "indifferent, aloof, conceited, peculiar and eccentric." Another neighbor noted that "her late husband, who had been a patient of Subject, often remarked that he believed Subject was of German ancestry because of his militant and disciplinarian attitude." One informant disclosed that Freeman "always seemed preoccupied, that he sometimes spoke to neighbors and at other times ignored their efforts at conversation or greeting.")

Visiting the Berlin clinic of the neurologist Hugo Spatz, Freeman did discuss the impending war but found more diverting discussion in "a point of international controversy that was still going on," namely, the location of the brain lesion that produced Parkinson's disease. Generally, Freeman found German doctors to be strikingly uninterested in psychosurgery, perhaps because thousands of German mental patients had been exterminated over the previous several years, a measure that precluded any need for treatment. Elsewhere in the German capital, the Freemans drove along the boulevards and shopped. Freeman picked up a Leica camera and a pair of binoculars. "Marjorie did most of her Christmas shopping and used a newly bought large suitcase that had to be strapped on the back of the car to hold all the presents."

Over the next several days they visited Heidelberg, where Marjorie had attended school as a child; Nuremberg, where they toured the famous stadium built for Nazi rallies; and many other German cities. They observed soldiers lifting their arms in "Heil Hitler" salutes, government cars crisscrossing the country, and fighter planes arranged on airfields in neat formations. None of these sights inhibited them from enjoying lavish meals and ample drinks in their luxury hotels. At Hamburg on August 25 they dropped off their rental car, arranged for the shipment of their trunk loaded with presents, "and took the train for Copenhagen, expecting to return [to Germany] to sail back to New York."

As might be expected of a conference that mixed neurologists from Germany, Britain, Poland, Italy, the Baltic nations, and other countries embroiled in the international conflict, talk of war limited the depth of scientific discourse. Freeman was disappointed by the response to his seminar on lobotomy, "which aroused little interest. The split between neurology and psychiatry was greater in Europe than in the U.S., and neurosurgeons had little contact with mental patients." One member of Freeman's audience, however, later recalled that he had listened to the psychosurgeon's presentation with "astonishment and horror." With his new Leica in hand, Freeman added photographic portraits to the visiting cards he customarily collected at such gatherings. His relentless picture-taking embarrassed Marjorie.

Sweden was the final stop before their itinerary called for their departure for the United States, and Freeman had arranged social calls with several prominent physicians. It was supposed to be a weekend visit. "I saw more of them than I bargained for, and also many another doctor, for we woke up on Sunday morning with the news that Hitler had marched into Poland," he wrote. Immediately they tried to buy passage to the United States from Sweden. "The *Athena* was all booked—fortunately for us, because it was sunk," he remembered. When departure proved impossible, they settled down in Stockholm and dined with four other American refugees at a lavish banquet given by the neurologist Nils Antoni. Over the next three weeks they ran out of money but got by on funds wired from home and a loan from John Fulton, who was also stranded in Sweden. At home, Freeman's colleague Bob Groh cared for the children. At last, through a friend who had contacts at the Swedish Line, they secured space—a bridal suite that was decorated with frescos full of cupids and that had been given up by the novelist Thomas Mann—on an outbound ocean liner. The passenger list also included the cowboy actor Tom Mix. After a tense voyage, during which the ship was fully emblazoned with a protective showing of the neutral colors of Sweden, the Freemans returned home. The war would prevent Freeman from receiving the published proceedings of the Copenhagen conference, shipped overland through Siberia, for two years.

After their return, Freeman remembered the war years as "a time of work with little let-up." Military enlistments had drained GWU's department of neurology and neurosurgery of nearly all the staff except for Freeman and Watts, who were too old for military service, and Bob Groh, a neurologist who had survived polio and a bout with cancer that necessitated the amputation of his tongue. Resident physicians were scarce. "Therefore we have to fall back upon cripples, women and foreigners," observed Freeman, who began importing a skilled stream of residents from Latin America. (In a letter to his brother Norman, Freeman recorded a conversation between his two youngest sons and a new enlistee. "Are you going off to war and be killed?" Randy asked. "No, Randy," the enlistee replied, "I think I'll come back." Then Keen piped in, "Oh yeah? That's what you think.")

Freeman, as a consequence, frequently worked six-and-a-half-day weeks. At the same time, the demands of his private practice increased. "I found myself giving less and less care to more and more people, which is not good, but under the circumstances unavoidable. . . . I have at times seen as many as ten new patients in a day," he wrote during the war. Nevertheless, he and Watts managed to attend to a steady stream of private lobotomy patients in their offices in the LaSalle Apartment Building at Connecticut Avenue and L Street, a palatial penthouse in which patients waited in a fifty-foot-long living room, Freeman occupied the two-story-high dining room (he used the kitchen for examinations), and Watts had an office at the top of a spiral staircase. There they netted $12,300 (about $145,000 in current dollars) from forty-nine psychosurgeries in 1942.

Freeman looked forward to the end of the war, which would bring a return of the assistants, fellows, and residents who usually did much of the work he had to take on. By the summer of 1942, the strain of overwork, a familiar sensation from his "nervous breakdown" of a decade before, began to worry him. He was experiencing bad digestion and insomnia. He underwent a full physical without informing Marjorie of his suspicions of strain. "I don't believe in letting on to her since she is inclined to remind me of things, and then the old stomach gets to going again," he told his brother Norman. Before undergoing tests on his gallbladder and gastrointestinal system, which required dietary adjustments he was determined to keep secret from Marjorie, he kept out of her sight at mealtimes.

Unwittingly, Freeman was following in his father's footsteps—keeping his ailment and discomfort to himself in a misguided attempt to shield his family from emotional upheaval. As a result, "Marjorie got to thinking that I didn't like home and home meals any more, but after it was over I came back and upped my calories, and Marjorie is none the wiser," he reported.

Freeman reluctantly heeded the signs of overwork but refused to slow down much. He would not drop any of his many irons in the fire. Walking was his pressure valve: he occasionally strolled a dozen miles in an afternoon, an exercise regimen made easier by the gasoline rationing that eliminated nonessential driving. These impromptu hikes sometimes took him farther than he had planned. While attending a medical conference in California during the spring of 1943, "I got a bit of the itching foot," he declared, "and, having brought along a pair of hiking boots, proceeded to go and look for the ocean." After several hours on roads of declining quality, he found out that he was still fifteen miles from his goal. He finally ended his hike when darkness fell. In setting out on foot in times of stress, Freeman was following his own advice to his patients. He frequently prescribed physical activity—horseback riding, walking, even cold bathing—to combat mental fatigue. A regimen of exercise, he believed, produced better sleep and an end to the fatigue. "When their muscles are working," he said, "their brains are resting." And when brains rested, they could no longer obsess about their own problems and concerns.

During the war, Freeman lectured around the country on his methods of stress reduction. He feared that the end of the war would bring an epidemic of Americans feeling nervous energy but lacking a healthy outlet. "We must develop a sure cure for the American malady of 'spectatorism,'" he told an audience in Kansas City in 1944. "We pay to see our games played by experts, when we should be on the corner lots playing them ourselves. We should go out and bat a baseball or swing a golf club to release the energy we now release in war jobs and in bandage-making, canteen service, civilian defense duties and the like." He elaborated on this theme in some verse he wrote for his son, Franklin, who was about to enter the military in 1945:

It isn't the ease and the pleasures of life
That make you glow inside;
It's the struggle and battle and dangerous strife
That you take in fighting stride.

To sit in the bleachers and yell for the team
Is a sop for the man who is soft,
But get in the game and soon it will seem
That your spirits will soar aloft.
The bigger the game the greater the thrill
And the greatest of games is war,
When man is the quarry you're trying to kill
And death adds up the score.

Ineligible for military service, Freeman threw himself with even greater intensity into the highest-stakes game available to him: the salvaging of lives through lobotomy. Though he worked far from the battlefields of war, death would still blot his happiness in the months to come.

## Chapter 8

# Advance and Retreat

Life in the Freeman household at 4501 Linnean Avenue, Washington, D.C., settled into a wartime routine. Freeman frequently rose early to work and kept at it until late at night. Lorne, who had grown into a dignified young woman with a retinue of suitors, enrolled at Stanford University in Palo Alto. As Freeman emptied a big wardrobe trunk in the cellar for Lorne's use, he came across an unexpected memento of his childhood: an old sign lettered "Dr. Freeman" that used to hang outside the window of his father's medical office. Freeman told his family of this discovery without nostalgia or even affection, but the modest, long-forgotten sign pointed to the difference between his father's simple attainments as a physician and Freeman's own aspirations. The son needed no signs to direct the public to his door. His achievements made them come.

Lorne's parents took an interest in the romantic life of her collegiate years, with Freeman noting that one boyfriend, a serviceman facing a dishonorable discharge after "serious trouble," was a "psychopath" and Marjorie hoping that Lorne would limit her dates to officers. Her brother Paul worried that she would never find a suitable mate because of her devotion to her father. "She really looked up to my father and all his work," he remembered.

During the summers of the war years, with Lorne away and Walter, Franklin, and Paul frequently gone, the Freeman house felt empty. What the younger boys, Keen and Randy, lacked in numbers, however, they made up in noise. "Almost every hour there is a yell from one or the other," Freeman wrote, "and there follows for a variable period a series of yowls until I have to yell out, 'Kill him,' or some similar appropriate phrase. The fighting keeps on for a little while and then they are back peaceably doing their previous tasks again." Freeman enjoyed spending time with the boys:

cutting and cording wood in Rock Creek Park, engaging them in conversation, and having them join him for week-long camping trips in the Virginia and Maryland countryside. During one such trip with Walter, Franklin, and Paul in 1944, the family traveled a hundred miles to Hancock, Maryland, where they launched a boat in the Potomac River and leisurely let the water push them along. Soon they stove the boat into a rock and had to stop for repairs. They slept in their trailer, which strained to accommodate the long legs and outstretched arms of the group. Franklin spent one night on the floor of the trailer, with his long frame resting on a series of pillows and life preservers. "I recalled an old song we used to have on the Edison Phonograph . . . 'All de odder white trash sleepin' on de flo', Mammy only loves her boy' and sang it to Franklin as he stretched out seemingly endlessly," Freeman later related.

This expedition was a short version of the annual summer trailer trips Freeman had inaugurated during the mid-1930s. In 1937 the family had driven their car and trailer through New England to Prince Edward Island and New Brunswick. This trip marked a family milestone: the only night Marjorie ever spent in the trailer. The trip hit its nadir when a cow ate her purple and yellow polka-dotted bathing suit. The following year, the family went west and stopped in cities hosting meetings of medical organizations. Freeman maintained his reputation as an erratic driver by disobeying a highway flagman's signal and being dragged into court in southwestern Virginia. After visiting San Francisco, where he grew frustrated when his children demanded hamburgers and peanut butter sandwiches in a Chinese restaurant, Freeman vowed to teach them to eat more adventurously on their return home. Once a week he planned dinners that served as cultural tours of all corners of the inhabited world, and he rejoiced when he convinced the kids to eat frog legs—without disclosing to them the source of the meat. (This culinary strategy evidently worked, because his children willingly wolfed down melon soup and chow mein at a Chinese restaurant in Vancouver several years later, "even though the chow mein resembled a mass of boiled string.") There was no trip in 1939 because of Freeman's overseas travel, but in 1940 the trailer family traversed several U.S. national parks en route to the Canadian Rockies. On the way home, the car broke down and Freeman bought a new one on the spot in Eau Claire, Wisconsin.

The final trailer trip before the United States entered the war, in 1941, was one of professional importance to Freeman. The Freemans launched their trailer on July 31 and headed west for Portland, Oregon. Later they camped near Mount St. Helens in Washington state and Freeman hooked a dog in the ear while fishing. For one week of the trip he left the children at a lodge near Lake Chelan and traveled alone to Ft. Steilacoom, the site of Western State Hospital. The hospital was hosting a neurological conference, and Freeman, engaged as a presenter, displayed his brilliance as a teacher and performer. He dazzled his audience with his trick of drawing diagrams on a chalkboard with both his hands, he showed his customary assurance in using hospital patients to demonstrate neurological disorders, and he tried out a new technique to show the movements of patients' muscles: he dressed the inmates in bathing suits. "This turned out particularly well for the women, whose bright costumes led the audience to compare me with Balanchine and corps de ballet," Freeman wrote. In addition, "I also gave one lecture on lobotomy that was to have a far-reaching effect after World War II, but aroused little comment at the time." The lecture caught the attention of the administration of Western State, which would later become one of the first psychiatric hospitals in America to make significant use of Freeman's transorbital lobotomy procedure.

While still at Western State, Freeman received word that Marjorie had fallen and seriously injured her hip. He stocked up on gasoline, which was often difficult to obtain in the months before the U.S. entry into the war, picked up the children at Lake Chelan, and set off on "one of the fastest cross-country runs I've made." Each day traveling east, he and the kids awoke at 4:00 A.M., drove for two hours before eating breakfast, and made quick stops only twice more, for lunch and swimming, before finding a campground for the night. They covered five hundred miles a day and made an exhausted return to Washington. Marjorie's condition had improved, and her physician told Freeman that the family's arrival "was the most powerful tonic he ever knew." Freeman's detailed ledger for the 1941 trip showed that he spent a total of $1,291 while on the road, which averaged out to $3.31 per person per day.

Although they usually left Marjorie at home in the company of the two smallest boys, Freeman valued these trips as a form of therapy for him and his family—something akin to the walks and baths he prescribed for his patients. "There is something essentially healing about a long trip by car," he wrote. "I think it has something to do with the bombardment of the senses, sights, sounds, smells and the jouncing of the muscles, the feel of the wind.... Also in driving, things repeat themselves in a sort of refrain, nonsense rhymes, peculiar names and words, snatches of song. These take away the preoccupations of the days past and to come, and my spirits rise accordingly."

The trailer traveled less during the war years. Freeman and his sons visited Shenandoah National Park in Virginia for a month during the summer of 1942, with Freeman returning home to teach a once-a-week class at GWU. There in the mountains a wasp stung him while he was photographing its nest, and Freeman observed with curiosity his body's neurological response as the ample hairs on his arm grew erect like the quills of a porcupine. Marjorie joined them for the final week, which they spent in the greater comforts of a lodge, but Freeman complained that it was "difficult to get used to a bed again after a month in the trailer" with the boys. The following summer the trailer sat idle, despite Freeman's intense dislike of Washington summers. "Everything is damp with the dampness of the tropics," he wrote during the summer of 1944, "almost mildewing, the paper bulges here and there, the paneling in the basement is warped, the 'cello has fallen to pieces, and [only] an electric light inside the piano which has burned most of the summer has kept things dried out enough so that it still sounds well."

With more time at home, the family settled into a routine partly directed by the rationing and restrictions of the war. "This is a very masculine household," Freeman related in a round-robin letter to his extended family in May 1944. "We sit down to breakfast, six men with a woman to serve us, and the men are out of the house before the maid comes. Then at night we sit down again, six men and one woman. The furniture is even expressive of this, with couches to sprawl on and home work scattered liberally over various tables, chairs, desks and the floor." These furnishings groaned under the burden of the Freemen men—all six-footers who wore size 13 shoes except for the little boys. In 1944 Freeman won the Davidson Prize for his research on multiple sclerosis, and he tried to use the earnings to commission the building of new chairs. Months later he checked up on the work of the furniture maker, who "had drunk up the money, entered it falsely on his books and hadn't even started on the job. He hadn't even repaired the chair that was given him as a model," Freeman complained.

Walter, who began high school in 1942, was bright, adept at the sciences, and a bit awkward. His twin brother, Franklin, had grown into a joking, happy-go-lucky young man who did not take his studies very seriously. He was friendly and outgoing, so a couple of years later, while Frank was in training for the infantry, Freeman suggested that he get in touch with one of his lobotomy patients living near the training camp. "Maybe she can give you a good home-cooked meal," Freeman wrote his son. Paul, the shy and sensitive one who frequently wore a worried look on his face, was a dreamer who liked to spend time by himself gazing at the *National Geographic* maps that covered the wall in his room. Keen and Randy, despite their two-year age difference, were evenly matched for their frequent but quickly forgotten squabbles. Keen was an avid reader, and Randy seemed to coast on a single burst of energy that lasted from early morning to late in the evening. Together the family celebrated Christmas by decorating a tree with ornaments that included "a green Donald Duck that one of my patients had made out of string. Most life-like," Freeman noted. Wartime blackout regulations prevented them from decorating a tree outside the house. Once a week this large and energetic family invaded a Chinese restaurant on Connecticut Avenue to give the cook, Marjorie, a break.

Marjorie had reached twenty years of service with the U.S. Tariff Commission. Her morale deflated when her supervisors hinted that she might find a new job elsewhere. Tariffs, after all, were largely irrelevant in a time of world war. It was about this time that she began to indulge in excessive drinking, a possible cause of the fall that had injured her hip. By mid-1943 she was weary of the insecurity of her job and of the stresses of wartime life. "Marjorie is tired and worn out with the heat, the family cares, the maid troubles, the rationing troubles and she has been compelled to neglect quite a part of her office work during the summer," Freeman observed. But convinced that staying at home would bore her, he hoped she would not stop working. At the end of the year, she decided to take a leave of absence from government service instead. She began shopping, cleaning house, and making meals more often—all activities that she formerly disliked. She took dancing lessons and developed an infatuation for her young Swedish instructor.

Overall, she found her absence from government employment a relief. In a year, she decided never to go back to work again. Her departure from government service failed to end her hard work. In 1945 she developed a hernia from all the time she spent strenuously cleaning windows and scrubbing floors. "It seems that for some weeks she had been reaching and lifting too much, yanking out stuck screens and wrestling with trunks and furniture," Freeman observed. Domestic help was extremely hard to find during the war—the available assistants "were leftovers, too incompetent to hold even Government jobs," Freeman noted—but at the GWU neurology clinic Freeman had met a woman with epilepsy who was eager to work. "If I can get Mother accustomed to her fits she might make a good person to have around," he hoped.

Slowly but steadily the happiness drained from their marriage. Freeman had a limited appreciation of the demands of motherhood and sometimes quoted his mother's assertion that since a single child took a mother's full time, then six more children could do no more. The Freemans still functioned well as a social pair, frequently hosting parties at which Freeman enjoyed mixing a signature cocktail from orange peel, sugar, and medicinal alcohol. But Marjorie, though she admired her husband's professional accomplishments, resented his commitment to psychosurgery and the long working hours it demanded. "I think she was not exactly on his cheering bench when he got into lobotomy," recalled son Franklin, describing her as being often critical of her husband's excessive devotion to the procedure.

The war years—the time in which their oldest children neared adulthood—did not see the eradication of their love, however. "Always when you go away I not only miss you but I definitely feel the need of you in more ways than one," Marjorie wrote Freeman during one of his trips in 1945. Then a subtle note of resignation crept into her note. "But at any rate you have given me six gorgeous children, who are all very fine people and a great comfort and source of happiness to me."

Freeman worked long and hard, his labors both a cause of and a response to his marital problems. The family's finances were now secure, so money was not his motivation. He bought a Dictaphone and used it at all hours. "Occasionally when I wake up in the middle of the night and think of all those things I have to report on I sneak down for a couple of hours and then get back to bed for my beauty sleep," he confessed to his brother Norman in 1943, a year in which he often worked sixteen-hour days. Sometimes his travel schedule was brutal. During a five-day stretch in 1944, Freeman went to Philadelphia to help direct examinations for the American Board of Psychiatry and Neurology, gave a paper at the New York Academy of Sciences, spoke at a conference of the American Psychiatric Association, and delivered another paper to a meeting of neurosurgeons.

Freeman and Watts knew by 1940 that they needed some means of exposing larger audiences to the rationale and benefits of psychosurgery. That spring they began work on a motion picture intended to document actual lobotomies. Technical problems—including shooting the film at the wrong speed and recording the sound on unwieldy, large disks—plagued the project. The doctors gave up and began afresh on another movie a few months later. This one they completed and deemed good enough to exhibit at medical conferences and before physicians interested in learning the technique.

At about the same time, Freeman focused his attention on a project he and Watts regarded to be absolutely necessary to the popularization of psychosurgery: the production of a monograph that would explain the technique of lobotomy, outline its effects, and recount the course of the first eighty patient histories. So far, the medical literature in English included only a single work on lobotomy other than their own articles. With a book proposal in hand, they approached the Macmillan Company, which declined to become involved. A Macmillan representative wrote that "the sale for a work of this kind was so limited, i.e., perhaps 200 or 300 copies, that the unit cost of a small edition would seem so far beyond the bounds of reason." Watts later learned the real reason for the rejection: the publisher "considered the subject matter of the book too controversial, and for this reason would be unable to publish it."

Other publishers came to mind, including W. B. Saunders, which had issued Freeman's first book. In 1940, however, Freeman and Watts proposed the book to Charles Thomas of Springfield, Illinois, a prominent publisher of neurological textbooks and books on other medical topics. In November of that year, Thomas offered to publish the book. Opponents of Freeman and Watts quickly let Thomas know that they took a dim view of the authors and their treatment of the mentally ill. "There is considerable opposition to our publishing this book," Thomas wrote several months later. "But there is just as much enthusiasm for our producing it, too, by authorities equally as good as those who oppose the matter. We will give the book our best effort and our best work and endeavor to put it over strongly."

Freeman and Watts divided the content of the book. Watts took on most of the text dealing with the surgical technique, and Freeman handled psychosurgical theory and case histories. They began the slow process of writing. "Library [research] time had to be stolen from practice, and the early mornings from a warm bed," Freeman wrote. "Most of it was hammered out on the typewriter between 4 A.M. and 7 A.M., after which there was a day's work to be done." He wrote in a style that would become standard for him in the following decades—chatty, anecdotal, punctuated with striking and sometimes inappropriate turns of phrase, and novelistic in the sense that Freeman often tried to convey information about the emotional quality of his patients' lives as well as verifiable data on their mental and physical states before and after surgery. In the book, for example, he described one patient after lobotomy as having "one of the most disagreeable, dissatisfied expressions on her face that I have ever seen. Her lips are pouting; there is a furrow between her eyebrows, deep pouches under her eyes, and she looks at everyone with cold, fishy, blue eyes. In addition to this, she is deaf and altogether an unlovely specimen." Freeman and Watts spent much of the latter part of 1941 proofreading, checking facts, selecting and placing the illustrations, and completing the index.

The completed book, which was titled *Psychosurgery: Intelligence, Emotion, and Social Behavior Following Prefrontal Lobotomy for Mental Disorders*, began with a surreal excerpt from Harvey Cushing's *The Life of Sir William Osler*, a biography of a noted English surgeon. In it, Osler relates a dream he had of Cushing operating on a patient: Cushing cuts a large hole over the patient's cerebellum, inserts a whipping cream blender, and turns the crank. The patient recovers well and buys three oranges from a street vendor selling fruit in the operating room. "The great principle . . . in cerebral surgery is to create a commotion by which the association paths are restored," Osler writes.

In later pages, the authors made official their departure from Moniz's ideas on the causes of psychoses. Fixed ideas embedded in unchanging synaptic patterns in the brain's neural communications did not produce psychiatric disease, they maintained. In place of Moniz's hypothesis, they wrote of the relationship between the frontal lobes of the brain, representing the cognitive region of the personality, and the thalamus, representing the emotional center. Healthy people achieved a balance in the influence of each. In mentally disordered patients, the emotional signals from the thalamus arrived with such strength and urgency that they obsessed the patient's thinking and affected his or her actions to a damaging degree. Freeman and Watts speculated that foresight, a normal function of the frontal lobes, metamorphosed into anxiety and tension when too greatly influenced by emotional signals from the thalamus. Lobotomy then did not force the brain to establish new paths of communications, as Moniz proposed, but muted the effect of the impulses from the thalamus. They found some support for their thalamus-based theories when they conducted postmortem examinations of lobotomized patients who had died some time after their surgeries. In some brains, they found that the thalamus had degenerated in areas that formerly had received neural signals from the severed tracts in the frontal lobes, and they concluded that this deterioration was in part responsible for the patient's blunted emotional response.

Freeman and Watts derived many of their ideas about the thalamus from the research of James Papez. In 1937 Papez hypothesized that the thalamus serves as the brain's emotional router, a transmitter of emotional stimuli to the frontal lobes for cognitive appraisal and response. Today neuroscientists still regard the thalamus as an essential component of the limbic system, the neural network that receives and handles emotional response.

Freeman and Watts's book also included data from some of the first efforts anywhere to measure the effect of psychosurgery on patients' self-image. Freeman engaged Thelma Hunt, a psychologist, to devise the tests. She found a way to measure what she called the "self-regarding span," the amount of time patients would talk about themselves. Her stopwatch trials showed that psychiatric patients who had recovered without lobotomy could discuss their own lives for about nine minutes. In contrast, patients treated by standard lobotomy could talk for only four minutes, and those who had received minimal and radical lobotomies went on for seven and two minutes, respectively.

When the book manuscript was completed, Freeman informed Moniz that the dedication page of the volume would read, "To Egas Moniz, who first conceived and executed a valid operation for mental disorder." Moved by the dedication, Moniz called it "a mark of your kindness and friendship. I consider it a great proof of your generosity." Freeman also paid special attention to certain aspects of the book's design. Sick in bed with a cold, he cooked up an idea for a colophon, a small emblem that would appear somewhere in the book. His sketch for it depicted a skull bored with lobotomy holes from which a group of black butterflies was emerging and flying about. The image had been suggested by Marjorie, probably intimate by now with depressed feelings about her job and marriage, who told Freeman that a French expression for "the blues" was *"J'ai des papillons noirs tous les jours"* ("I have black butterflies every day"). Freeman sent his sketch for the colophon to Thomas, who liked it. In the end, the skull and butterflies, printed in gold and black, appeared on the dust jacket and the title page.

The cover, full of suggestive imagery and hyperbolic language, was also largely Freeman's inspiration. A pattern of intriguing blue swirls used as a cover illustration was painted by Mary Lawrence, a lobotomy patient working under the guidance of Ruth Faison Shaw, a proponent of the use of finger painting to reveal changes in personality. (Later questioned whether a finger painting was suitable for the cover of a book on psychosurgery, Watts answered, "Of course it's appropriate. Sells the book, don't it?") Freeman probably wrote the cover text, which urged, "Read the last chapter to find out how those treasured frontal lobes, supposed to be man's most precious possession, can bring him to psychosis and suicide!" Elsewhere the cover made a jarring claim: "This work reveals how personality can be cut to measure, sounding a note of hope for those who are afflicted with insanity." The "cut to measure" phrase is pure Freeman, a selection of words intended to produce shock and impress the reader with the pun, all while remaining factually accurate to a degree.

*Psychosurgery* came out in 1942, although five hundred copies bound for Europe went down with a torpedoed ship. Singlehandedly it nourished interest in the procedure during years when actual psychosurgeries remained relatively rare. As expected, pro-psychosurgery readers cheered it: In England, a supportive writer called it "an exuberant and brilliant monograph" that "brought some order to a confused and chaotic subject." Another English physician, the neurosurgeon Eric Dax, observed, "The book by Freeman and Watts hit us like a bomb. With over 300 pages and about the same number of references, we felt that the thoroughness of their enquiries and the amount of work they had done was so great that there was hardly anything to say." John Fulton asked if he could have the original manuscript of the book for the Yale University library. It changed the minds of few readers who had already formed low opinions of the effectiveness and legitimacy of lobotomy. Some critiques, mostly favorable, appeared in the popular press. A writer for the *New York Times* called it "more exciting than most novels. And why not? Probing into the brain, breaking up fixed pathways and compelling the thalamus and the prefrontal lobes to find new ones, watching uncontrolled minds slowly return to more normal ways of thinking—no novelist ever had a more thrilling subject."

Few medical journals reviewed the book. The volume—surely the most influential of Freeman and Watts's fifty jointly written articles and books—did enlarge, however, the public discussion of lobotomy's merits. "While those of us, with a reasonable control of our emotions and able to withstand a fair amount of mental or emotional pain may shudder at the thought of losing our sense of responsibility," wrote a correspondent for the *Spokane Chronicle*, "nevertheless if we should be suffering as these patients are suffering we would, for the sake of ourselves and our families, likely undergo operation for the relief of severe mental pain."

Even though Charles Thomas eventually sold all of the copies in the initial printing of *Psychosurgery* at $6 each, Freeman and Watts made no money as the book's authors. As was common in academic publishing, they fronted a substantial sum, $3,250, to cover printing costs. "However, Jim and I were making a good income from our partnership in lobotomy," Freeman wrote, "and we could afford it." The book, they believed, proved their seriousness as medical researchers, served as a calling card in the United States and overseas, and functioned as an outline of their defense of lobotomy.

Freeman went to great lengths to convey the book to European friends during the war. When in 1943 he received an invitation from the Nobel Committee in Sweden to submit a nomination for the award in medicine, Freeman sent a letter nominating Moniz by diplomatic pouch. He included a copy of *Psychosurgery* with a note asking Nobel committee members to forward the book to a Swedish friend when they were through with it. He and Watts felt their work validated when the book won a somewhat lesser honor, the Horsley Prize, given annually to an outstanding volume on surgery by a graduate of the medical school at the University of Virginia.

By the time the book was published, Freeman and Watts had treated more than two hundred patients with lobotomy. Although most patients lived in metropolitan Washington, some were drawn from other parts of the country by news accounts or by referrals from their own doctors. Freeman maintained "a sort of scrapbook of all these patients," which included a card on which he marked follow-up data; noted whether they were working, staying at home, or confined to a hospital; and attached before-and-after photos, X-ray images, and pictures of brain specimens extracted from patients who had died. "Fortunately we don't have many pictures of specimens," he noted. The follow-up information came from Freeman's tireless efforts to maintain contact with all of his lobotomy patients by mail, telegram, or phone. Of this large group of patients, 63 percent had improved, 23 percent had not changed, and 14 percent were in poorer condition than before the operation, Freeman and Watts asserted.

Because patients with affective disorders fared the best, Freeman started drawing conclusions about the procedure's curative powers. "I have been reviewing all the cases submitted to prefrontal lobotomy, after an additional year since the manuscript for the book was finished," Freeman wrote at the time, "and am glad to find that the patients with involutional depression have continued to improve almost without exception. Since some of these patients had been sick from five to fifteen years before operation and have remained well now from two to five years, I think we can say with some assurance that we have found a safe and permanent cure for this dreadful malady."

Patients with functional illnesses such as schizophrenia fared poorly by comparison. Although the "precision method" had reduced the incidence of debilitating side effects, many lobotomized patients went on to suffer convulsions, lack of initiative, incontinence, substantial weight gain, hemorrhage, and a condition that became known as frontal lobe syndrome, which was characterized by a sometimes shocking lack of inhibitions and absence of consideration toward others that, most puzzlingly to their friends and family, included no hints of meanness or intentional discourtesy.

The popular press not only approved of the book; it also uncritically accepted the theories of Freeman and Watts. An Associated Press report published in the *Houston Post* in 1941 called lobotomy a "personality rejuvenator" that worked by cutting "the 'worry nerves' of the brain running back from the forehead to the central and rear parts of it." The surgery held little danger, the article maintained. "Actually, Dr. Watts and other brain surgeons consider it only a little more dangerous than an operation to remove an infected tooth." Another Associated Press story noted, "Having your brain cut hurts less than having a corn removed from your little toe." Although this startling assertion is technically true because the brain has no nerves receptive to pain, readers received the false impression that lobotomy was not a serious medical procedure.

One such article sparked a storm that temporarily soaked the reputation of Freeman and Watts. Along with a 1942 article about psychosurgery titled "Turning the Mind Inside Out" by the *New York Times* science writer Waldemar Kaempffert, the *Saturday Evening Post* published photographs of Freeman and Watts operating on a lobotomy patient. The images outraged some of their medical colleagues, who considered the publication of such photos a form of advertising, then considered an unethical practice. At an executive meeting of the American Neurological Association, a member displayed the *Post* article and demanded an immediate investigation. Only John Fulton's intervention prevented the association from penalizing and possibly expelling both Freeman and

Watts. Kaempffert's article was later condensed without photographs and republished in *Hygeia*, a magazine of the AMA, and that condensation was further abridged into an article for *Reader's Digest*.

One of Freeman's early lobotomy patients, Harry A. Dannecker, produced a more beneficial account of psychosurgery with an article titled "Psychosurgery Cured Me," published in October 1942 in *Coronet* magazine. Although he began his story by describing lobotomy as "probably the most dangerous operation known to modern surgery," he went on to recount his own recovery in the brightest possible terms. An Indiana tool designer in his mid-fifties, Dannecker had suffered for years from depression and anxiety. He attributed many of his problems to his childhood experiences with a violent father and scenes of cruelty to animals he witnessed while young. He developed facial tics that he believed "were caused during those years by my trying to resist my tears of embarrassment or vexation." When he lost his job during the Depression, the tics increased in intensity. He had nightmares, suspicions that others were talking about him behind his back, and fears about performing common household chores. "It was about this time that I made up my mind that I would kill myself as soon as I found an easy way," he wrote.

In early 1937, however, Dannecker's wife, Hazel, showed him a newspaper article about Freeman and Watts's early lobotomies. Skeptical that anyone could help him, Dannecker took no action. But Hazel wrote a letter to the Washington doctors describing her husband's wretched life, and they replied with an offer to operate on him free of charge. Dannecker refused to go "until it occurred to me that the operation would be a simple way of dying without violating any religious injunctions against suicide." Once in Washington, he did not share his anticipation of death with Freeman, who explained the mechanics of the surgery by first invoking Moniz's theories on "fixed patterns of brain cells." Then Freeman added that Dannecker's thalamus exerted too much control over his frontal lobes. The leucotome would break the supremacy of emotion over reason. "You won't become a coldly thinking automaton," Freeman assured him, "but instead there will be a more even balance between your reason and emotions—as there is in all normal people."

As Dannecker reflected on the pros and cons of surgery, he looked out Freeman's office window and noticed passing pedestrians "whose worst obsessions were tuna fish on whole wheat three times a week or avoiding the cracks in sidewalks." Hazel told Freeman on his behalf: "Doctor, we have nothing to lose. Life is not worth living for either of us as it is. If there were no more than one chance in a million for him to recover, I would still want him to have that chance. Please, please go ahead with the operation." Dannecker agreed and underwent a prefrontal lobotomy on the morning of May 11, 1937. As he succumbed to the anesthesia, the "world dissolved in a blur and only sub-consciously was there any realization that it was my brain into which a drill was descending to prepare the way for the knife that would change my personality—or kill me." During surgery he dreamed of driving a car through the beautiful countryside and of confidently building a new house.

When he awoke, he quickly realized that the operation had not killed him. He was able to obey all the requests of his doctors and nurses to grasp things or to open his mouth. He regained the power of speech in a few days. He smiled "vacuously for no particular reason," slept well, and lost track of time. He read books and magazines with understanding, and his wife found him relaxed, without anxiety. Freeman and Watts discharged him from the hospital after nine days. He discovered that he could drive competently, but he was struck by the degree to which "the sparrows on the lawn seemed awfully funny to me, and I laughed merrily at their antics. Otherwise I was apathetic and slept much, waking only when roused."

As time passed, Dannecker's inertia dissipated, and he got back to the household jobs that had formerly intimidated him. An important stage of his recovery arrived when his mind seized upon a solution to a problem that had plagued him since before his operation: the design of a workable wheel-grinding device that could be used in the automotive industry. He patented it, found a job in a war materials plant, and began working sixty-five hours a week without ill effects. He considered his recovery a remarkable success story, and so did Freeman and Watts, who invited him to speak at medical meetings. "A chance to be normal again, to be friendly, to be good humored, to be able to emerge from that terrible underworld of the sick mind—all that, I feel, is worth any risk," Dannecker concluded. His story, though, ended tragically. Ten years after his lobotomy, Dannecker experienced a return of many of his old symptoms. A second lobotomy "resulted in profound inertia with permanent confusion and rather rapid decline to death a month later, March 1, 1947," reported his case history.

Dannecker was not the only grateful one among Freeman and Watts's early subjects. Their sixty-eighth lobotomy patient set down this poem after the operation:

Gentle, clever your surgeon's hands
God marks for you many golden bands
They cut so sure they serve so well
They save our souls from Eternal Hell
An artist's hands, a musician's too
Give us beauty of color and tune so true
But yours are far the most beautiful to me
They saved my mind and set my spirit free.

Freeman and Watts believed they could, nearly at sight, categorize the patients who would respond best to lobotomy. "Some patients come to operation at the end of a long and exasperating series of medical treatments, hospital treatments, shock treatments, including endocrines and vitamins mixed with the physiotherapy and psychotherapy. They are still desperate, and will go to any length to get rid of their distress," they wrote. "Other patients can't be dragged into the hospital and have to be held down on a bed in a hotel room until sufficient shock treatment can be given to render them as manageable. We like both of these types. It is the fishy-handed, droopy-faced individual who grunts an uhhuh and goes along with the family when they take him to the hospital that causes us to shake our heads and wonder just how far we will get."

The patient's strong feelings either for or against the surgery seemed to point to recovery after lobotomy better than apathy. Unconsciously, Freeman and Watts also targeted women as good candidates for lobotomy. Seventeen of their first twenty psychosurgery patients were women, and from then on women made up a solid majority of their cases. The preponderance of women who suffered from affective disorders may be partly responsible, but agitated and boisterous behavior in women was probably less acceptable to doctors of Freeman's and Watts's generation than such behavior in men. "Women with involutional depression feel hopeless about everything, including the possibility of benefit from operation," they wrote, "and require a great deal of persuasion to get them to enter the hospital."

After more than a decade of operations, Freeman believed that lobotomy benefited about one-third of those who underwent it and thus left 67 percent just as badly off or worse than before surgery. Death, which directly arose from about 2 percent of their operations, usually resulted from hemorrhage, seizures, or infection. (Freeman and Watts once expressed the opinion that the mortality rate was boosted by patients, like Dannecker, who hoped to die in surgery.) In one case, a patient died from meningitis when her compulsive hair-pulling produced an infection at the site of the entry hole to her skull. Another early patient died from "extreme inertia," an operative by-product not improved by a subsequent reoperation. Their 193rd patient died from dehydration and shock, the consequence of his refusal to receive liquids before operation. "We debated the safety of operation but decided to go ahead because of the danger of unrelieved catatonic stupor of such intensity," Freeman and Watts wrote, although it is not clear how a patient in a catatonic stupor could refuse nourishment. In May 1945 Freeman wrote to his brother Norman about two deaths more or less attributable to psychosurgery, "one from [a] convulsive state induced by operation, the other from a fractured hip suffered when [the patient] lost her balance after kicking at one of the attendants!"

Death, of course, was not the only harmful complication of lobotomy. In their first case of 1937 when they were still using the Moniz technique, imprecise placement of the coring leucotome gave the patient paralysis of the left arm and an unsettling explosive laugh. Ten years later, the laugh had not gone away. During one surgery conducted under local anesthesia, the doctors accidentally severed the anterior cerebral artery, a major blood vessel, while making the upward cut on the left frontal lobe. "This disaster was followed by an exclamation—'Oh, my!'—and [the patient] vomited and lapsed into unconsciousness," Freeman and Watts wrote. Then her pulse plummeted, respiration increased, blood pressure soared, blood spurted around the cannula, and the patient's brain bulged. After managing to staunch the bleeding and administering two blood transfusions, the physicians completed the lobotomy. "Recovery was slow and incomplete," they reported, "and death in status epilepticus occurred four months later." In 1943 Freeman told Norman about a reoperation that left a patient overcome by inertia and afflicted with a strange "bulldog" reflex in which "she gets something between her teeth [and] clamps down and won't let go." She could communicate only through movements of her feet. "I'm afraid she is a gone goose as far as useful life hereafter is concerned," Freeman concluded.

Several times Freeman and Watts found themselves the targets of planned attacks by lobotomy patients. A lobotomized patient named Michael May once telephoned from his home in Pennsylvania to inform Freeman that the doctor "was a crook, had never operated on him, and that he was coming to Washington to shoot me. He telephoned from

Union Station [in Washington], repeating the statement." The doctors notified the police, who refused to make an arrest until the patient actually appeared at their office door. "When he arrived the stage was set and almost before he had time to sit down a squad car answered the appeal and five officers walked in ready for action," Freeman wrote. Finding the patient unarmed, the police interrogated May in the office. When May confirmed that he had murder in mind, the police brought him to a nearby hospital for psychiatric examination. Freeman also wrote that twice "during my professional career, I have disarmed women [patients]; but I am not sure whether they intended to use the pistols on their husbands or themselves—or me. On one occasion, I told my small boys that I had a pistol in my bag. They later told their mother that I had a police dog in my bag." As a precaution against attack, Freeman usually arranged his offices so that he worked facing the entry door, and in an emergency he planned to tip over his desk to scare away any attackers.

Certainly their most notorious failure was the case of Rosemary Kennedy, the sister of President John F. Kennedy. Rosemary grew up an oddity among the nine children of Rose and Joe Kennedy, as she mixed poorly with her siblings in the family's strenuous physical and intellectual competitions. The oldest Kennedy daughter, she was known in social circles as the quietest and most beautiful of her sisters. Something—perhaps birth trauma, learning disabilities, or an inborn brain defect—slowed her mental development, although not enough to keep her from traveling overseas without a chaperone, participating in sailboat races, or learning to read and write. By the summer of 1941 the twenty-three-year-old Rosemary was living at a convent school in Washington, D.C. There, according to the Kennedy family biographer Laurence Leamer, "she had begun to suffer from terrible mood swings. She had uncontrollable outbursts, her arms flailing and her voice rising to a pitch of anger. . . . She sneaked out at night and returned in the early morning hours, her clothes bedraggled. The nuns feared that she was picking up men and might become pregnant or diseased."

"Joe liked to cut away at a problem," Leamer wrote, "and then move on," perhaps to explain why Joe Kennedy seems to have eagerly seized upon lobotomy as a treatment for Rosemary's behavior and as a solution for the problems she presented to her family. He first took Rosemary to a Boston physician, who refused to recommend psychosurgery. Next Kennedy conveyed Rosemary to Freeman, whose office was near the convent school. Freeman did not consider intellectual deficit or mental retardation to be conditions treatable by lobotomy, but debilitating tension and anxiety were. Whether Freeman observed strong emotional tension in Rosemary or the persuasive Kennedy convinced him that the tension was present, Freeman agreed that psychosurgery was worth a try. (Watts, in fact, diagnosed Rosemary with "agitated depression," one of the strongest indications they had for psychosurgery.) Kennedy authorized the surgery without the knowledge of his wife, who was traveling abroad.

The lobotomy left Rosemary inert and unable to speak more than a few words. Freeman would have anticipated this condition to result temporarily from a radical procedure. Joe Kennedy, however, was not a model for the forgiving and tolerant parenting that many lobotomy patients required during the slow healing in the weeks and months after surgery. Rosemary eventually regained the ability to walk—several times she wandered off on her own during visits with the family and prompted frantic searches—but permanently lost the initiative and mental capabilities she needed to live with even partial independence. In 1949 she was sent to live at St. Coletta's School for Exceptional Children in Jefferson, Wisconsin. Rose Kennedy did not see her daughter for twenty years, and no family member seems to have visited her there until 1958, when John F. Kennedy secretly paid a call while campaigning in Wisconsin.

A postscript to this sad case came in 1998. Apparently by chance, in a Nashville pancake restaurant Leamer encountered a retired nurse who said she had assisted in Rosemary's operation. "The nurse was so horrified by what she saw happening that she left nursing and never returned to the profession," Leamer wrote. "There were other operations, so many others, but it was Rosemary she mostly remembered." He concluded that Rosemary's lobotomy was "the emotional divide in the history of the Kennedy family, an event of transcendent psychological importance. . . . Unlike all the subsequent deaths and accidents, no mark of patriotism, heroism, daring or even dread circumstance could be attached to this act." The Kennedy family attempted to suppress public knowledge of Rosemary's medical condition, and Freeman's correspondence and private writings are silent on the question of her surgery and its outcome.

Despite such failures, Freeman's most controversial lobotomy patients were the children he treated—nearly all of them with a diagnosis of schizophrenia—and the controversy dogged him throughout his career. The first child patient, given a lobotomy in 1939, was a nine-year-old whose brutal tantrums and symptoms of schizophrenia did not improve and who returned after the surgery to a psychiatric hospital. Four years later, Freeman and Watts operated on a four-year-old boy, and they followed that the next year with psychosurgery on a six-year-old girl named Linda. Before surgery she had "stopped talking, tore her clothes, smashed her dolls, used toys as weapons and would often run away. . . . This child was exceptionally active and agile, had a withdrawn, abstracted look on her face, showed no interest in the various objects in [Freeman's] office, sat by herself rocking to and fro and uttering a sing-song, 'Ma-Me, Ma-Me,'" Freeman wrote. "In response to attempts to gain her attention, she struggled violently." Despite her home physician's diagnosis of encephalitis, Freeman and Watts thought childhood schizophrenia more likely, and they performed a lobotomy on her on August 24, 1944. A relapse soon followed. Eight months after her lobotomy, Linda again appeared in Freeman's examining room. She was chewing her clothing and fingers, incontinent, and "would often sit alone gazing into space, sometimes breaking into a smile and rocking back and forth, showing no affection for anybody," he observed.

Linda received a second lobotomy almost immediately after this visit. Her temper tantrums ceased and she stopped tearing her clothes. "She seems perfectly happy all the time and laughs a lot," her mother wrote. But Linda still smashed her toys and could not sit still. Freeman last visited Linda in 1949, when she was eleven. "She talked very little and was making no progress in helping in the home," he wrote. He believed, however, that a series of photo portraits he had taken of her throughout her treatment and recovery showed a brightening of her spirits and "are very revealing of her emotional state."

Another child, a twelve-year-old boy, presented Freeman and Watts with one of their most puzzling failures:

> There were no complications during a radical prefrontal lobotomy performed on November 15, 1945, and his condition was satisfactory throughout the procedure. His reaction the following morning was most unusual. He was apparently seeing things. At times there would be a look of stark terror on his face; then he might kiss the nurse on the arm or cheek, bite his finger nails, and kick his feet very rapidly. He was incontinent and vomited several times. He had a temper tantrum, then soon afterward broke into a smile. On the night after operation, as the nurse prepared to take his temperature, the patient became cyanotic and died within a few minutes. Autopsy permission was refused.

In the cases of all of these children, Freeman and Watts hoped to salvage the patients by attacking as early as possible the means by which their schizophrenic fantasies overwhelmed their everyday thinking. Most received radical prefrontal lobotomies, made necessary, Freeman believed, by the young age at which the illness had appeared. While these young patients usually remained utterly dependent on their families, "most of the patients have shown gradual improvement, have resumed some measure of helpfulness about the house, have regained some interest in the outside world and have definitely abandoned their fantasies, as no longer attractive," Freeman wrote. To Freeman, the gains outweighed the losses.

Freeman and Watts performed these lobotomies on children in a psychiatric environment undergoing deep wartime changes. Psychiatric evaluators had rejected 1.8 million American military volunteers and recruits for reasons related to mental fitness. For the men and women who passed their entrance examinations, military psychiatrists primarily concerned themselves with helping them adapt to the fields of battle. A stressed, tense, or mentally ill soldier was not a useful soldier if he could not function. By comparison, healing patients, even restoring them to comfort, dimmed in importance. During a two-week stint training physicians at the Army School of Neuropsychiatry at Lawson General Hospital in Atlanta, Freeman had noticed that "the army is interested in disposition of unfit cases rather than treatment of its [neuropsychiatric] casualties. 'If you can't cure 'em quick, get 'em out, but don't use any of these newfangled quick cures on 'em. It might make trouble later.'" As the war ground to an end and military psychiatrists once again donned civilian clothes, they carried with them these military imperatives. "Within this new framework, patients thus were not cured of a particular disease so much as they were restored as functioning citizens," wrote Jack Pressman. Because cures were rare, any psychiatrist who helped patients simply get on with their lives was doing a good job. Lobotomy moved a substantial number of patients out of hospitals—an excellent treatment for the psychiatry of a new era.

By the end of the war in 1945, the Freeman family's center of gravity had shifted to California. Walter, a navy recruit, was training in the use of military radar in Del Monte; Franklin was in training with the army and on the verge of departing for the Philippines; and Lorne and Paul were both studying in Palo Alto. That summer it made sense for the family to hold a reunion in Oakland as a prelude to Lorne's graduation from Stanford. Only Franklin was unable to attend. Afterward Marjorie returned home with Keen and Randy, and Freeman looked forward to a week of fishing alone in Yellowstone National Park. On the train to the park, Freeman heard the news that the Japanese had surrendered. "I was in the lounge car as the news spread, and presently an ice-skating group of girls started a snake dance up and down the aisle from one car to the next on through the train," he wrote. "I remember the calluses on their shins."

When he arrived at Yellowstone, the park was practically deserted. Freeman had all the solitude he wanted. He spent a day on the banks of a stream where he caught buckets of trout. Then the calm water beneath a nearby bridge tempted him, and he stripped down for a swim. He dove into the water eagerly without remembering to take off his glasses. The glasses slipped from his face and sank into the stream's depths. Freeman was helpless. Seriously myopic, he dove for the glasses and fumbled among the stones in the bed of the stream. He came up empty-handed. He dove again and again but to no success. Without his glasses, he would have a very difficult time returning to camp. Finally, after many dives, another fisherman came by and spotted the glasses in the water. "They had settled on top of a large flat white rock," Freeman recalled, "the most favorable spot in the whole river." After three more dives Freeman brought them up. Fortune, he thought, was treating him well. With persistence and some help from others, valued possessions lost in the cold, moving water of a stream could be retrieved.

# Chapter 9

# Waterfall

In the months immediately after World War II ended in the summer of 1945, Freeman felt a lethargy settle into everyone around him. He described it as an unbearable letdown, a sudden reduction of tension that hit Washington almost like a mass lobotomy. His colleagues at GWU, many of them just discharged from the military, lacked enthusiasm. Several young neurologists left their specialty and became psychoanalytically oriented psychiatrists. Freeman rankled under this cloud of apathy, this atmosphere in which drive and fervor seemed out of place. He had lost none of his zeal over the beneficial potential of lobotomy. In the last half of 1945, Freeman began spending odd hours in the autopsy room in the District of Columbia's Gallinger Municipal Hospital. It was not his habitual preference for solitary work that led him there. Nor was it his growing dissatisfaction with his marriage. Another challenge that had long troubled him made him furtively visit Gallinger in possession of a decidedly untraditional medical instrument—an ice pick.

The challenge, as Freeman saw it, was to find a solution for the hundreds of thousands of people in the United States with psychiatric illnesses who lacked access to psychosurgery. All of Freeman's work to develop, promote, and defend lobotomy had made little difference in the increasingly crammed wards of the country's 180 state-run mental institutions, where the bulk of America's seriously mentally ill lived. In these hospitals, patients languished without hope that their doctors could cure, relieve, or adequately care for them. Meanwhile, their families despaired, if they had not already washed their hands of the patient. The situation was only growing worse. As World War II ended, government-run mental hospitals experienced an overwhelming influx of patients: psychiatric cases filled more than half of the beds in public hospitals by the end of 1945, and by 1948 the American Psychiatric Association would estimate that state institutions were packed with 50 percent more patients than they could adequately accommodate. To further worsen the problem, psychiatric patients required periods of hospitalization four times longer on average than other patients.

Despite the need for some new treatment to help these large numbers of people in psychiatric institutions, American hospitals and physicians reported fewer than a thousand psychosurgeries between 1940 and 1945. To Freeman, three roadblocks kept psychosurgery from the many patients in need of help. The first was the inadequacy of the Freeman–Watts prefrontal lobotomy techniques, which he and his partner had championed for years. He and Watts had seen many patients emerge from the operating room with dreadful impairment and personality changes. "It took a number of dramatic instances of severe personality downgrading in well-preserved patients to make me realize that the standard or radical prefrontal lobotomy was too damaging," Freeman admitted in his memoirs. In its present state, he concluded, lobotomy was simply too harmful to use on anything other than the most desperate cases. What further refinements would make lobotomy less destructive? How could more patients benefit from psychosurgery?

Second, Freeman smoldered under lobotomy's dependence on the availability and approval of neurosurgeons. As a major surgical procedure requiring good hospital facilities, anesthesia, and a doctor trained in surgery, the existing operation was far too expensive and impractical for state institutions, many of them isolated in rural areas, where money and staff resources were scarce while operating rooms and the other requirements of surgery did not exist at all. Watts and Freeman worked well together and liked one another, but the neurosurgeon, more conservative than his partner, sometimes refused to operate on patients he believed were not ill or hopeless enough to warrant psychosurgery when Freeman contrarily thought a lobotomy was justified.

Third and even worse to Freeman, however, was his inability to help people still in the early stages of psychiatric illness, before their disease produced deep deterioration and permanent damage. Freeman was undergoing a revolution in his thinking on the circumstances in which lobotomy was appropriate. Waiting until the patient had exhausted all other courses of treatment, he suspected, was too late. For years he and Watts had repeatedly stated that lobotomy was the patient's last resort, a treatment to apply only when everything else had failed. The time would soon come when Freeman would declare, "Lobotomy, instead of being the last resort in therapy, is often the starting point in effective therapy."

Freeman coveted the role as the bearer of a new, safe, and more widely available psychosurgical procedure. He wanted to be the primary instrument of change, the person who brought the surgical treatment of psychiatric illness to the masses of people in such desperate need. Ever since his days as a medical student with the Army Medical Corps at Fort Dix, where he witnessed the rapid devastation that influenza brought upon its victims and the waste of human potential when young or talented patients died, Freeman followed a practical—but by the standards of laboratory researchers, reckless—mind-set that insisted that lab work and research were useless endeavors unless promising techniques were made available to patients quickly. Although talented in the lab and conscientious as a researcher, Freeman always focused on his responsibility to patients. He had no interest in a career like John Fulton's, with a carefully controlled laboratory, a secure spot in lofty academe, and legions of students and assistants eager to advance scientific theory another few inches. Freeman's job, as he saw it, was to *apply* medicine: to hear his patients' troubles, devise a treatment that left the patients better off, and execute that plan. If the treatment worked, Freeman was a success. If it failed, he would try something else.

In the state hospitals, where researchers working in distant universities nihilistically saw uncountable patients requiring custodial care until a sure and safe new medical procedure came along, Freeman recognized people in need of an immediate treatment plan. "There is an enormous inertia in these hospitals, and not all of it is due to lack of appropriations," he wrote in 1946. "When mental hospitals can show some results they are sure to get more money, but at present a lot of the money for research is frittered away on inconclusive and philosophical investigations on personality backgrounds and such like, with interesting formulations but no tangible results." To a limited extent, prefrontal lobotomy had been tried in the state hospital setting. The neurosurgeon Harold Buchstein lobotomized forty-six patients in seventeen months starting in January 1941 at Willmar State Hospital in Minnesota. Soon afterward, St. Louis City Hospital began a three-year-long lobotomy program that treated 101 patients.

Elsewhere in Missouri, Paul Schrader, a physician working at the state psychiatric hospital in Farmington, performed psychosurgery on nearly two hundred patients during the early 1940s, and he reported that about one-third had been discharged as a result. Among Schrader's lobotomy patients was Rose Williams, the sister of the playwright Tennessee Williams. Freeman praised Schrader's work as an example of "how the whole character of the disturbed service in a state hospital could be vitally altered by large-scale performance of lobotomy." In addition, the neurosurgeon J. Grafton Love and the hospital director Magnus Petersen had launched a lobotomy program at Rochester State Hospital in Minnesota. There, during 1941 and 1942, forty-six patients were lobotomized. "Personally," Petersen wrote Freeman in 1956, "I did not receive as much verbal abuse as you did. . . . At the same time, I was told afterwards that I came near being cited before the ethics committee of the Minnesota State Medical Society, not so much for carrying out the procedure but for doing too many. That appeared to me slightly illogical. If the operation was good for five individuals there was no reason why there might not be ten who might benefit from it."

Freeman realized, however, that the hospitals in Minnesota and Missouri were exceptional cases, and that few other institutions would or could follow suit. Most practitioners of lobotomy were neurosurgeons, Freeman noted, which "limits the value of this type of operation, especially when one considers that there are some 600,000 patients in mental hospitals in this country, and that most of the neurological surgeons are fully occupied with their own problems of tumors and discs, pain syndromes, epilepsy and head injuries."

Freeman longed to be the physician who would make lobotomy practical for all the other psychiatric hospitals in the nation. He would devise a plan, deliver the treatment, and teach others how to help these patients return to a life that approached normality. His allies would be the people most familiar with the dire predicament of state hospital patients: the administrators and caregivers who worked in the institutions. How could he alter psychosurgery to make it available to patients in state hospitals? What would make the application of lobotomy simpler, easier, and cheaper? What would work best in the combined opinions of the patients, their families, and their doctors? At this stage of Freeman's career, these questions counted most.

And those questions drew Freeman to the autopsy room at Gallinger. Of course he was no stranger to exploratory work on cadavers. He knew that his colleagues would not approve—at least initially—of what he had in mind. With his recall of practically the entire literature of psychosurgery research, Freeman had dusted off the work of Amarro Fiamberti, director of a psychiatric hospital in Varese, Italy, who shared something of Freeman's own philosophy of medical action. "In the present state of affairs," wrote Fiamberti, whom Freeman described as "terrier-like" in his intensity, "if some are critical about lack of caution in therapy, it is, on the other hand, deplorable and inexcusable to remain apathetic, with folded hands, content with learned lucubrations upon symptomatologic minutiae or upon psychopathic curiosities, or even worse, not even doing that." Fiamberti's contribution came in 1937, when he explored a seldom-used surgical path to the frontal lobes, which had earlier been pioneered by his countryman, A. M. Dogliotti, in the administration of ventricular puncture, and by Maurice Ducoste as a means of delivering malaria inoculations directly to the brain.

Rather than boring the top or the sides of the skull to gain access to the brain, Fiamberti took advantage of a preexisting opening in the skull: the eye socket. He penetrated the thin plate of bone separating the top of the eye socket from the brain cavity. The operative path steered clear of the frontal sinus, the optic nerve, and other fragile regions. Fiamberti lobotomized about one hundred patients by using this route to inject alcohol or formalin into the brain tissue, and in a few cases he inserted the Moniz leucotome and rotated its cutting blade. Although he felt he had demonstrated the safety of this technically simple procedure and documented only such minor side effects as headaches and short-lasting fevers, the outbreak of World War II and Fiamberti's own wandering interest kept him from further investigating the effectiveness of the leucotomies he performed by using this new approach. As a result, few such leucotomies took place outside Italy.

Freeman judged Fiamberti's results as poor, "probably because [his] patients needed something more," but he believed that this new approach held promise as an inexpensive alternative to the standard lobotomy, an alternative that could be performed in state hospitals without the expertise of neurosurgeons. At Gallinger, Freeman set off on an uncertain expedition. He experimentally journeyed through the eye sockets and up to the transorbital plates of his cadavers in his endeavor to break through to the brain cavity just as Fiamberti had. Once past the transorbital plate, Freeman hoped to revolutionize psychosurgery by directly and deeply cutting into the neural pathways that he thought relayed harmful signals between the frontal lobes and the thalamus.

For several years, Freeman had suspected that most of the effectiveness of the prefrontal lobotomy resulted not from injury to neural pathways in the frontal lobes but from atrophy produced in the thalamus when its neural connections to the frontal lobes were severed. Much of the therapeutic effect of lobotomy, in other words, came from altering the thalamus, not the frontal lobes. Freeman hoped that his new procedure, which he called transorbital lobotomy, would produce the atrophy of the thalamus while cutting only a minimum of neural connections in the frontal lobes. Unlike the neurosurgeons performing prefrontal lobotomies, he would sever nerves serving the thalamus "fairly far anteriorly without disturbing the cortex except to a minimal degree." Nerves feeding other areas, especially the gyrus rectus of the brain, would be left intact. He predicted that patients would recover without any of the changes in personality and deficits of initiative and energy that had plagued many subjects of the prefrontal operation.

Freeman was working in familiar territory. Several years earlier, as laboratories director at St. Elizabeths, he had tried to inject ventriculographic dye into the brains of cadavers through the orbital plate, but the best tools at his disposal, the spinal puncture needle and the paracentesis cannula, either cracked the bone or broke into pieces themselves. The Gallinger cadavers immediately gave Freeman similar troubles. As much as he tried, he could not penetrate the transorbital plate with the spinal puncture needle. They all bent or broke when he pressed them against the bony wall. Stymied, he searched for a tool that would work. He needed "some instrument, slender, sharp and tough," he wrote. In his kitchen drawer at home he found the solution: a tool with a strong shaft and sturdy wooden handle. It was a standard ice pick imprinted with the name of the Uline Ice Company.

With the ice pick now a part of Freeman's surgical toolbox, his experiments in transorbital lobotomy quickly produced results. He found that when he inserted his surgical ice pick beneath the upper eyelid and kept the shaft of the instrument parallel to the bony ridge of the nose and aimed slightly away from the center of the head, the point of the pick situated itself against the vault of the orbit, the portion of the skull that protects the eye and separates it from the frontal lobes of the brain. Fortuitously, this is usually the thinnest part of the skull, and, Freeman noted, "a light tap with a hammer is usually all that is needed to drive the point through the orbital plate." Because there were only small blood vessels in the path of the instrument if Freeman kept the point away from the midline of the skull, he felt confident that the risk of hemorrhage was small when he pushed the ice pick up to 5 centimeters deep.

Most important to Freeman, his new kind of lobotomy was quick and easily accomplished outside of a large medical center. "The transorbital method brings the possibilities of psychosurgery right into the psychiatric hospitals, where it is obviously of greatest value," Freeman noted. A method so simple, he believed, could be performed by physicians lacking surgical experience or certification—institutional psychiatrists, for instance—in an office without sterile draping.

In his 1950 instructional film, *Transorbital Lobotomy*, he demonstrates the procedure on a cadaver and follows that with a dissection of the cadaver's brain tinted with dye to show exactly which sections of the frontal lobes he had cut. Freeman begins the operation by lifting the upper eyelid of the corpse and inserting his tool into the tear duct. Two sharp strikes with a hammer breach the transorbital bone. Freeman then pushes the instrument 5 centimeters into the opening, and the real cutting begins.

He describes the rest of the procedure in detail:

> I pull the handle of the instrument as far laterally as the rim of the orbit will permit in order to sever fibers at the base of the frontal lobe. I then return the instrument half way to its previous position and drive it further to a depth of 7 cm. from the margin of the upper eyelid. . . . Then comes the ticklish part. Arteries are within reach. Keeping the instrument in the frontal plane, I move it 15° to 20° medially and about 30° laterally, return it to the mid-position, and withdraw it by a twisting movement, at the same time exercising considerable pressure of the eyelids to prevent hemorrhage. I then proceed with the opposite side.

At times, such as when writing to his son Paul, Freeman could describe the procedure in plainer language. Transorbital lobotomy "consists of knocking them out with a shock and while they are under the 'anesthetic' thrusting an ice pick up between the eyeball and the eyelid through the roof of the orbit [,] actually into the frontal lobe of the brain[,] and making the lateral cut by swinging the thing from side to side," he noted. He acknowledged to his son that the procedure was "definitely a disagreeable thing to watch." Many times Freeman paused in his descriptions of transorbital lobotomy in order to relate the distinctive sensation of the orbital bone giving way, which was often accompanied by "an audible crack."

Freeman had mapped out a treatment that in just seven minutes could accomplish as much as the complex Freeman–Watts standard lobotomy, which required hours in the operating room and weeks of recovery. As Freeman began identifying his first patients to undergo the new transorbital procedure, he felt a twinge of anxiety. "Just as in the case of Moniz, it needed some time to summon up courage to attack the brain through the orbital plate," he wrote. His apprehension certainly did not come from breaching the boundary between neurology and neurosurgery. During his earlier operations with Watts, Freeman had frequently done the surgical cutting on one side of the patient's brain, despite disapproval from some of his colleagues. Nor, of course, did risk alone intimidate him. Throughout his entire career, from his use of the "jiffy spinal tap" to his introduction of lobotomy and promotion of curare to ease the muscle contractions of electroshock therapy, Freeman never shrank from taking risks if he believed the results had a chance of helping patients in dire straits.

Cadavers had not complained about having an ice pick poked into their eye sockets; for living patients, of course, Freeman had to administer anesthesia. Traditional anesthesia could always be used, but only rarely would it be available in a state-run psychiatric institution. Freeman preferred to produce unconsciousness, or what he sometimes called "a state of coma," by using electroconvulsive therapy (ECT) machines to give patients a series of electric shocks at two- to three-minute intervals. These devices abounded in state institutions during the 1940s and 1950s, and they were an important part of that era's psychiatric regimen. A lucky benefit of electroconvulsive anesthesia was that patients often lost their memory of the minutes immediately preceding the lobotomy—moments when some struggled and felt great fear and anxiety. "Physically healthy patients might require three applications of the electroshock," Freeman observed in his instructional film, "the elderly and infirm just one."

Later Freeman would say that electroconvulsive anesthesia benefited patients in other ways. "Electro-shock appears to have a general disrupting effect on cortical activity, temporarily abolishing the psychotic manifestations and bringing the patient into a brief period of increased adaptability," he wrote. He also believed ECT aided in the clotting of blood and that the rapidity with which patients regained consciousness allowed him to promptly recognize any medical complications after the operation. But lobotomy patients with tuberculosis, he cautioned, should not undergo ECT because convulsions could hasten the spread of the disease.

The use of electric shock as an anesthetic was unusual but not new. (Nikola Tesla claimed to have experimented with electrical anesthesia as early as 1898.) But to many people who watched Freeman's demonstrations and operations, his use of electroconvulsive shock to anesthetize patients was one of the most troubling parts of the procedure. Strapped to a table with protective padding placed between their jaws, patients went into violent convulsions when the electricity passed through their brain. "I was there to hold the person's legs down," says Freeman's son Franklin, who assisted with a transorbital lobotomy in 1952. "We all went for a ride when he threw the [ECT] switch."

Freeman ruled out conducting the first transorbital lobotomies in a hospital because of the questions these experimental surgeries would raise, not to mention Freeman's lack of surgical privileges. He also felt irritated by the aseptic safeguards—pioneered in the United States by his grandfather, W. W. Keen—that he would have to follow in a hospital. Because the tear duct and the transorbital pathway were already sterile "98 percent of the time," Freeman thought that covering the patient with a sterile drape, wearing surgical gloves, and masking his own face pointlessly added time and complexity to the transorbital procedure. The only concessions he made to performing a sterile procedure were switching to a freshly sterilized leucotome when he began working on the second transorbital penetration and applying towels over the mouth and nose after the conclusion of the electroshock convulsions to avoid contamination of the surgical field. His later partner, the neurosurgeon Jonathan Williams, recalled Freeman's exhortations against "all that germ crap" and his annoyance with sterile procedures. During a surgery they were working on together, Williams once observed Freeman's characteristic lack of interest in aseptic technique and barked, "Walter, if you don't mind, would you kindly just step back and let me finish this procedure?" Freeman's impatience with sterile technique was yet another example of his intolerance of procedural or theoretical concerns that stood between him and the speedy application of the treatment.

Even as he gained confidence in his proficiency at transorbital lobotomy, Freeman believed no Washington hospital would allow him to undertake the procedure unaided in its operating room. "It was out of the question to carry it out except in private patients who were willing and whose families were willing to take a chance," Freeman noted. As a result, the first ten patients underwent this historic procedure in the informal setting of the office suite Freeman shared with James Watts.

For his first transorbital lobotomy patients, Freeman wrote that he chose people who "were not doing well under shock therapy but were so well preserved from the personality standpoint that the [prefrontal] operation was considered too severe." No one was totally disabled by psychiatric disease; no one was a hospital inmate. He told each patient that this new treatment was untried but would be far less grueling than the old Freeman–Watts operation. His first subject, in January 1946, was a patient named Ellen Ionesco. For a long time, the twenty-nine-year-old Ionesco had suffered from periods of manic behavior, bouts of depression that left her bedridden, and spasms of violence in which she sometimes struck her young daughter and required physical restraint. When she eventually grew suicidal—she once tried to jump out of a window—her full-time nurses began to keep knives hidden from her. In desperation her husband, a jeweler, had located Freeman and brought in Ionesco for an examination.

To Freeman, Ionesco's affective symptoms and suicidal actions added up to a good rationale for transorbital lobotomy. "I explained to her and her husband what I proposed to do and how Fiamberti and his colleagues in Italy had been performing this operation for nine years with very few accidents," he recorded in his autobiography. Ionesco consented to the surgery. She underwent the electroconvulsive anesthesia, and Freeman made his first entry into a living patient's orbital cavity with an ice pick. "With what may have been an excess of caution I operated on one side only and had her come back a week later for the second side," he wrote. "All went well."

Decades later, her daughter, Angelene Forester, retained a vivid memory of the day of the operation. While Ionesco and her husband joined Freeman in an examination room, Forester, then four years old, remained in a waiting room. She heard no sounds emerge. "I must have fallen asleep," Forester said. "When I woke up, [Ionesco] wasn't well; she couldn't walk by herself. They supported her over to the elevator [and we] took a taxi. She stayed in bed for a few days. It was a scary day."

The lobotomy, Forester remembered, left her mother with eyes that "were swollen and all black and blue. I asked my aunt who beat her up. She explained that they went through the eyes and that's why there weren't any cuts." Despite these visible injuries sustained at Freeman's hands, the young Forester had formed a positive impression of her mother's physician. His presence made Forester feel "very warm, very safe.... His eyes, his whole body was such that there was no tension, [only] gentleness and kindness. His eyes were very warm."

After the lobotomy, Ionesco's crazed outbursts, which had so frightened Forester, were gone forever. The violent and suicidal behavior "stopped immediately," Forester said. "[There] was peace, peace.... It was like, thank God it's over." Yet Ionesco's recovery was difficult. "I don't think you could call Mama a vegetable in any sense," Forester recalled. "Right after the lobotomy, she wasn't there—she was healing." A live-in nurse cared for her. Eventually Ionesco resumed her housekeeping chores and again took up gardening. "She came back," her daughter said. "She raised me, she became a [licensed practical nurse], she worked with Daddy in the jewelry store when he was alive [and] worked as a nanny for a lot of people. . . . She functioned on a level [of] earning money, being a productive person. She was a member of our church, a great cook." Four years after Ionesco's lobotomy, Freeman observed her progress and reported that she "is now enjoying good health."

Freeman and Ionesco corresponded for years, and their relationship may have crossed the boundaries of propriety. Ionesco saved up her earnings as a nurse to buy Freeman an expensive wristwatch for his birthday, and he accepted the gift. An aunt informed Forester that Ionesco was having an affair with Freeman. "There was a heck of a lot between them for a long, long time. . . . She loved him," Forester said. Meanwhile, the subject of the lobotomy was never raised in the Ionesco household. "She never ever mentioned it to me or her sisters." As Ionesco aged, her mental state declined; she talked to herself, suffered from delusions of persecution, experienced hallucinations, and sometimes exposed her underwear to strangers. Yet she was alive, not suicidal, and functioning at a respectable level of responsibility.

After operating transorbitally on several other patients, Freeman concluded it was safe for him to cut both sides of the brain on the same day. In fact, as he wrote to Moniz, he eventually preferred grasping a leucotome in each hand and making both cuts simultaneously. "This manoeuver not only makes for greater symmetry in the placement of the lesions, but also prevents the frontal lobes from moving laterally with the moving of the instruments. It also hastens the operation," Freeman observed. Most of the rest of these early transorbital lobotomies ended only with such minor complications as black eyes and headaches. "Some of the 'black eyes' are beauties, however, and I usually ask the family to provide the patient with sun glasses rather than explanations," Freeman wrote.

The results of the third operation encouraged Freeman. "It remains to be seen how these cases hold up, but so far they have shown considerable relief of their symptoms, and only some of the minor behavior difficulties that follow [prefrontal] lobotomy," Freeman wrote. "They can even get up and go home within an hour or so. If this works out it will be a great advance for people who are too bad for shock but not bad enough for [prefrontal] surgery."

Freeman sent most of his early patients home with a family member or in a taxi. But not the fourth patient, a man named David Berman, who required additional medical care because of "a hemorrhage and [who] was laid up for several days so I never did the other side. He recovered fairly well," Freeman wrote in his memoirs. Berman's fate was actually far worse. Freeman punctured a blood vessel in Berman's brain during the lobotomy, a wound that resulted in seizures and an emergency trip to the hospital. The patient was left permanently disabled and later could only support himself by selling newspapers on a street corner.

Not discouraged, Freeman proceeded. The operation held too much potential, Freeman believed, to allow mishaps to retard its advance. He learned from his mistakes and moved on. But when he advanced to his tenth transorbital lobotomy, he could not predict how this particular operation would forever alter his relationship with James Watts. The two men told different stories about this surgery, but both agreed that it marked the beginning of the dissolution of their thirteen-year partnership.

Freeman, according to his own version of the story, had informed Watts of his transorbital work on cadavers and had told him of his intentions to begin operating on patients. Watts refused to take part in this work. Later Freeman invited Watts up to his office to observe the tenth transorbital operation. Because this was what Freeman considered an urgent case, he operated on both sides of the brain at the same session. Watts did not like what he saw. "In fact he didn't like it so much that he told me that unless I stopped it he would look around for other offices," Freeman wrote. "He said that if I were to be sued he would stand up for me but thereafter he wouldn't back me up and would have to resign from the department. Under these circumstances I refrained."

Watts told a darker tale. Earlier he had asked Freeman not to give lobotomies to patients in the offices that Watts and Freeman shared. An office assistant, he said, informed him of Freeman's experimental surgeries. (In addition, he must have known about the complications in the treatment of David Berman, who was rushed to the hospital from their shared offices.) On the occasion of the tenth transorbital lobotomy, Watts walked upstairs and—astounded to see Freeman standing over an unconscious patient who had an ice pick protruding from his face—interrupted the operation in progress. Without showing any surprise at this intrusion, Freeman gamely asked Watts to hold the ice pick while he snapped a photograph. Watts refused and angrily protested that brain surgery should not be an office procedure. Later Watts informed his partner that he would fight any attempt by Freeman to perform transorbital lobotomies at George Washington University Hospital.

Watts must have made this threat with great difficulty. For years he had been charmed by Freeman's tireless and vivacious personality, and he deeply respected his partner's abilities. Watts once explained the reluctance of many of Freeman's residents to follow him into neurology by observing, "You almost could say that nobody could do what Walter Freeman could do." And Watts also felt a close bond with Freeman. In 1979 he told an interviewer, "I nearly always have to refer to *We*, that is Freeman and myself, because we did everything together."

Freeman, put off by what he called Watts's "dog-in-manger attitude, summed up as: 'I won't and you shan't,'" took his act elsewhere. Two lesser hospitals in the area, Washington Sanatorium in Takoma Park and Casualty Hospital near the Capitol in Washington, gave him permission to use electroshock anesthesia and to perform transorbital operations in their facilities. He still tried to engage Watts in a discussion of transorbital lobotomy and presented him with data from the operations on the first ten patients. Although Freeman and Watts jointly moved their offices from the LaSalle Building in Washington to new quarters at 2014 R Street—a row house near many other physicians' offices, with barely enough space for their practice—their rift widened in the spring of 1947, when Freeman made plans to tell the audience at a meeting of the Southern Medical Association that he wanted to teach psychiatrists, most of whom lacked certification as surgeons, how to give transorbital lobotomies. Watts vacated the offices on R Street after only a year. Freeman took on a new neurosurgical associate, Jonathan Williams, then resumed performing the surgeries in his office. Williams offered a contrast to Watts's conservatism and restrained personality. "His lively temperament fitted well with mine," Freeman noted. Meanwhile, Freeman and Watts kept their conflict a secret; they continued to behave cordially around each other and amicably cochaired GWU's department of neurology and neurosurgery.

Watts's early objections to Freeman's transorbital lobotomies echoed the protests that Freeman would hear for years from other surgeons. Watts believed brain surgery to be the province of neurosurgeons. "It is my opinion that any procedure involving cutting of brain tissue is a major surgical operation, no matter how quickly or atraumatically one enters the intracranial cavity," he wrote. "Therefore, it follows, logically, that only those who have been schooled in neurosurgical technics and can handle complications which may arise should perform the operation." Watts feared the prospect of hastily trained state hospital psychiatrists—most of them far less gifted and knowledgeable in brain anatomy than Walter Freeman—performing lobotomies in their offices or on the wards. "Well, he knew more anatomy of the brain than most of the neurosurgeons, but I seriously objected to having the psychiatrist do it, who probably hadn't seen a brain since he was in medical school," Watts later declared. He thought that an operating room in a hospital, complete with sterile draping and emergency medical equipment, offered the safest setting for any lobotomy. Watts acknowledged that sterile draping was not absolutely necessary "because all you're going to do is pull up the eyelid and use a leucotome, a modified ice pick. I said, 'Walter you're going to kill the damn thing if you're going to use an ice pick. And you should use a gilded hammer.'"

Freeman ignored these admonitions and occasionally performed transorbital lobotomies in unusual settings. In 1950, after a patient had three times not appeared for office appointments to receive a transorbital lobotomy, Freeman learned that police were holding the man after a disturbance in a motel room near Bethesda, Maryland. Frustrated by the missed appointments, Freeman loaded his car with medical gear and a portable electroshock machine and drove toward Bethesda. When he arrived at the motel, Freeman informed the patient's brother that the lobotomy would happen immediately or not at all. The brother consented, and the police wrestled the patient to the floor of the motel room. Freeman applied the terminals of the electroshock machine to the patient's head, knocked him out, and performed the lobotomy in less than ten minutes. Jonathan Williams recalled that Freeman submitted a medical insurance claim for this lobotomy, and when Blue Cross turned it down—on the conditions that Freeman was not a surgeon and had performed the operation outside a hospital—Freeman persuasively recounted the emergency circumstances of the lobotomy and won his payment.

Watts recoiled at the idea of electroconvulsive shock as surgical anesthesia. To this, Freeman had a ready, although unpersuasive, response: "The surgeon is offended by the spectacle of a convulsion on the operating table. The surgeon should recall, however, that the sensibilities of the psychiatrist may be equally wounded by the sight of blood. In any event the psychiatrist is thoroughly at home with the electrically induced convulsion, and the technic of transorbital lobotomy is well within his capacity." As for Freeman's release of patients just hours after their lobotomies, "Dr. Watts thought that was insane," recalls Freeman's son Franklin.

Finally, Watts flat out disagreed with Freeman's growing conviction that the operation should be tried on patients in the early stages of psychiatric illness. Freeman argued that an early transorbital lobotomy could prevent psychiatric disease from worsening and disabling patients. He advocated the use of lobotomies on patients who had not improved after just six months in a psychiatric hospital, "before the personality of the patient has undergone severe changes," as he wrote to a colleague. Freeman increasingly wanted to use lobotomy as an early intervention to prevent the worsening of moderate psychotic symptoms into major ones, "no question about it," Watts declared. "He felt more strongly about that than I did . . . about the suffering patient." Watts took the opposite view. "If it is a major operation, disability constitutes the indication for transorbital lobotomy," he stated. In other words, brain surgery should have been the treatment of last resort for mentally ill people whose disease was overwhelming and otherwise untreatable. Watts would not consider or recommend transorbital lobotomy until other more time-consuming treatments, including ECT, had been tried. By the fall of 1947, the partners had not mended their differences. In discussions with Watts, Freeman "stressed the fact that now I had tasted blood and was out for more," he wrote. "He still thinks the operation is a major one and should be done only in a hospital by a trained neurosurgeon, while I am equally insistent that it is a minor operation since it involves no cutting or sewing, and that it should be performed by psychiatrists."

Freeman believed that the results of the transorbital lobotomies he performed were enough to refute Watts's position. Examining his first ten patients two years after their operations, Freeman found that "six of them were fully restored and only one was in hospital." (Returning to their cases after another three years, Freeman reported similar outcomes.) Compared with those who underwent prefrontal lobotomy, transorbital patients fared far better in the short term. "Recovery from transorbital lobotomy is spectacular," Freeman proclaimed. "Within an hour the patient may be sitting up or even able to walk," he observed. "There is little headache or pain in the orbit. Vomiting and incontinence are unusual. Swelling of the orbital tissues is marked in some cases, absent in others." Infections were rare, Freeman told Moniz, "even though the patients were allowed to rub their eyelids."

Watts opposed transorbital lobotomy mainly in the settings and under the circumstances that Freeman wished to perform it. "He believes that each case should be studied in detail and watched for developments before the next case is undertaken," Freeman wrote. Watts was not opposed to it under other conditions. In 1948, not having seen Freeman operate transorbitally for about a year, Watts asked for a demonstration of Freeman's newest technique. Freeman invited him to observe a transorbital lobotomy he was performing in a Maryland nursing home for two other surgeons. "The patient was wheeled into a kind of attic," Watts recalled. "Freeman took off his coat. He kept on his vest and—I'm sure this was for the benefit of the Baltimore surgeons and me—instead of using a silver-coated mallet, he used a carpenter's hammer and drove [the leucotome] through and did the operation on one side, put the hammer on the patient's abdomen, on his nightshirt, and then did the other side." Although Watts did not believe Freeman's showy application of a carpenter's mallet and his use of the patient as a table were unsafe, he found them to be unwise and aesthetically offensive. "Freeman liked to do it in what you might call a more dramatic or exceptional way," Watts concluded. "Those were the things that we used to pull and tug about."

Nevertheless, Watts performed twenty-eight transorbital lobotomies between 1949 and 1956 to treat patients diagnosed with schizophrenia, psychoneurosis, obsessive tension, anxiety tension state, and agitated and involutional depression. In one case, he demonstrated the procedure at the University of Virginia, where he advised the university's staff, "As you probably know, it is my opinion that transorbital lobotomy should not be done unless the patient is disabled. . . . I'm quite sure that we could agree upon a suitable case for operation, one that is not too bad but not too good, either." In another instance, Watts transorbitally lobotomized a patient at St. Elizabeths Hospital when he feared that the schizophrenic patient, who had tuberculosis, could not survive the prefrontal operation.

Freeman and Watts, in fact, agreed that transorbital lobotomy was worth trying on terminally ill patients, many of them with cancer, who did not have psychiatric problems but were experiencing pain without relief. Burdened by memories of his father's death, Freeman still regarded cancer with a personal animosity, as a disease whose victims must be relieved of suffering. Both doctors had previously observed that patients seemed less attentive to pain after prefrontal lobotomy: "I watched two lobotomized women during delivery," Freeman wrote. "They were almost flippant about their pains until the babies' heads were on the perineum. Then they hollered bloody murder and had to be anesthetized with all haste. As soon as it was over they were cheerful and said there was nothing to it." Therefore the two doctors reasoned that the transorbital operation might offer the same reprieve from the suffering of chronic pain with fewer complications and a faster recovery from the surgery than the earlier operation. The partners eventually patched their rift. Watts did not object when Freeman began performing transorbital lobotomies in an examining room at GWU Hospital, and he accompanied Freeman to the annual conference of the American Psychiatric Association in 1949.

"This procedure can be carried out, apparently without danger, even in patients who are almost in extremis, and it does away with the necessity of morphine and its derivatives and brings about a certain euphoria and resignation that are very gratifying from the standpoint of the nurses and the doctors who have to care for the patients, and also for the relatives who, too, are relieved of their distress in seeing the suffering," Freeman wrote. He wondered whether a patient near death could withstand the convulsions produced by electroshock anesthesia, but he found that wasted muscles responded with weak convulsions and saw no bone fractures in such cases.

The results often astonished patients in pain and their families alike. One patient, whose terminal illness had left one leg disabled, "no longer complained of pain when being moved and was even able to get out of bed and walk, although she complained that the affected leg was shorter than the other," Freeman reported. "When I suggested that the other foot could be cut off to make the two sides equal, she laughed heartily at the witticism. The patient having spent six months in bed previously, with great distress, the family was amazed at the change that had come over her. I believe the operation has actually prolonged her life."

Watts, of course, performed his pain-limiting transorbital operations in a hospital with standard anesthesia and with Freeman sometimes assisting. "Jim made a fancy operation out of it," Freeman complained. "He had two or three men giving the anesthetic, an assistant and two nurses, and he draped the patient until I protested that I couldn't see anything to sight the instrument. He used an expensive surgical instrument with various attachments and from the surgical standpoint did a very neat job." Watts's patients experienced mixed results. One of the earliest, a woman with throat cancer, arrived with burning pain in her throat, severe discomfort in her ear, difficulty swallowing, and a fear of suffocation and starvation. She had been bedridden for a month. Watts thought a transorbital lobotomy was warranted because of the degree of her suffering and the shortness of her life expectancy. He performed the operation on May 9, 1947.

Within five days, the patient was at home, on her feet. She refused narcotics and showed no mental impairment. Two months after the operation, she told Watts how she felt. "She described in a factual way the difficulty she was having in breathing and swallowing and did not mention pain until she was asked about it," he noted. "When questioned, she said, 'No, I don't have much pain, there is a burning on the side of my head and a fullness in my ear. I do not sleep well at night and often have to sit up in a chair.'" She died seventeen days after that interview. As she had feared before her lobotomy, she suffocated to death.

Watts rated this transorbital lobotomy a success, but as a treatment for four subsequent patients with chronic pain, he scored the procedure a failure. Two had cancer, one had tabes dorsalis (a neurological degeneration resulting from an untreated syphilis infection), and another was an amputee afflicted with phantom limb syndrome. Watts ultimately opposed the use of transorbital lobotomy to treat patients in pain, except "in the relief of suffering associated with carcinoma." But he conceded that the new procedure left patients with less inertia than the Freeman–Watts operation. Freeman, in contrast, gave a reporter at a 1948 medical conference in Cleveland a far different impression. "It's all over within ten minutes, and the patient usually goes home the next day," Freeman said. "The pain clears up like magic."

Complications from transorbital lobotomy were infrequent. In his first year of performing the surgeries, Freeman operated on a Washington, D.C.–area police officer whose illness (which Freeman never specified) had not improved with electroshock therapy. The officer, in fact, had suffered a fractured vertebra when Freeman had administered the ECT. Freeman selected him for transorbital surgery, with tragic results. "The patient had hemorrhage into both frontal lobes, I believe, and lay for a number of days in an unconscious state before making a gradual physical recovery, but with such severe brain damage that he was never able to do more than the simplest tasks around the house," Freeman wrote. "I would rather have taken a surgical fatality than such a distressing wreck."

Of the first four hundred transorbital operations he performed, Freeman reported that one patient died from hemorrhage within a day, two died later from bleeding in the brain, and one experienced serious hemorrhage but "survived in a state of bland indifference considerably resembling that observed after radical lobotomy." Six others suffered convulsions, inertia, weakness on one side of the body, or other complications. Fewer transorbital patients died or were permanently disabled than those who underwent prefrontal lobotomy. Freeman documented the cases of many patients who bounced back after transorbital surgery to resume their careers. One was a physician who established a ten-member medical clinic and obtained his pilot's license, another was a symphony orchestra violinist, and yet another was a successful psychiatrist. These showcase patients, however, were not typical. Freeman eventually determined that 85 percent of his "private" transorbital patients—those not originally confined to state hospitals—were able to live at home, and two-thirds were "usefully occupied." Of his transorbital patients in state hospitals, 45 percent returned home. In 1948 Freeman reported that three patients "have been married since operation and are reported to be carrying on normally."

The emotional and mental condition of transorbital lobotomy patients also exceeded the state of earlier ones. Freeman told caregivers to immediately expect to see the patients "cheerful to the point of elation" for a few days to a couple of weeks after surgery. Inertia and disorientation might occur at the same time, or occasionally "a certain indolence and tactlessness. For the most part the patients are quite themselves, the only thing lacking being a certain subtlety and a modicum of insight into their own mental mechanisms," or as he later described it, "a certain lack of subtlety in their relationships with other people." Later he seized upon a striking analogy for their behavior; he declared that they exhibited "the Boy Scout virtues in reverse" and showed a lack of trustworthiness, loyalty, helpfulness, friendliness, courtesy, kindness, obedience, cheerfulness, thriftiness, bravery, cleanliness, and reverence.

Freeman warned that patients might exhibit a temporary return, sometimes for several months, to some of their old patterns of behavior. A slow and gradual improvement would follow. "Relapses are fairly common but can sometimes be effectively treated by shock therapy," to which they might now be more responsive, Freeman reported. As a sop (and a sly wag of the finger) to his peers who disavowed biologically based treatments, he noted that some of his patients had emerged from transorbital lobotomy amenable to psychotherapy "for the first time." "My parting words are usually: 'No restrictions,'" Freeman noted.

Initially, Freeman regarded transorbital lobotomy as most effective for patients suffering from acute anxiety, hypochondria and obsessive tension states, as well as for "well-preserved patients with involutional depressions, even though the disease conditions may be of long duration." He later concluded that his new technique could help people suffering from a wider range of psychiatric conditions. "Anxiety and emotional tension, for instance, are often overcome; depressive ideas, phobias, and obsessional thoughts usually improve," he wrote. "Fixed delusions and hallucinations are rather resistant, and motor compulsions or tics may continue relatively unchanged. If [the] operation is performed before the psychosis or neurosis becomes fixed, however, even the most ominous symptoms may subside within a few days." He did not recommend transorbital lobotomies for deteriorated or long-suffering schizophrenics—he noted that "severe emotional dilapidation is the outstanding contraindication to [lobotomy] of any type"—but believed that patients with less severe schizophrenia benefited from the operation.

By the end of 1948, Freeman had devised a refinement of the procedure—what he termed the "deep frontal cut"—for patients whose disorders had not improved after the standard transorbital lobotomy. This more radical procedure, in which he levered the leucotome against the orbital plate for access to the base of the frontal lobe, sometimes cracked bone, but Freeman found it preferable to referring relapsed patients to Watts and others for prefrontal lobotomy. Meanwhile, Freeman's enthusiasm for the transorbital procedure became so well known that rumors circulated among physicians that some patients who came to Freeman for electroconvulsive therapy had left his office with the tell-tale black eyes of an ice pick operation.

This new variety of transorbital lobotomy exceeded the capability of the standard transorbital leucotome and spurred Freeman to develop new hardware. The new instrument built from tool steel by Freeman's faithful machinist, Henry Ator, "was tested by inserting the point through the keyhole of a door and lifting with a force of 25 kilograms on the handle, without bending or breaking the instrument." Freeman christened it "the orbitoclast," and at last "I felt safe in using maximum power to produce the deep frontal cut."

Only patients seemed underwhelmed by transorbital lobotomy. Many denied ever having received the operation. This curious amnesia was more attributable to the effects of the electroshock, which commonly produces memory problems, than to Freeman's actual cutting of brain tissue. With pride, Freeman told Moniz of one patient who "sat up [after the operation] with a broad smile of relief on her face and said I must have pulled an eyelash from her because she felt so much better."

He found that the degree to which patients retained their personalities after the transorbital operation, as well as how well they recovered from the surgery and their illness, depended on such factors as individual brain anatomy, the support of families, the permanent deterioration their diseases had already caused, and their strength of character. "The general impression I had of the few patients I knew was that they were remarkably indifferent, with a 'Who cares?' and 'So what?' attitude," recalled Freeman's son Franklin. "They didn't seem to participate in life. There didn't seem to be a loss of intelligence, just a curious indifference." He recalled one of his father's transorbital patients, the son of a California cattleman, who came down with schizophrenia while a student at Harvard. "He spent 18 years in the back wards of Stockton [State Hospital]," Frank said. "Dad gave him a lobotomy, and it muted his symptoms. After he went home, he was nicely dressed . . . but he was still hallucinating, rolling his eyes and mumbling." Freeman took a personal interest in this patient by helping him regain his driver's license and getting him a job as a golf course groundskeeper. "[The patient] later took up the game and became a very popular addition to the course. But he had no ambition. Later he married a nurse, to more or less look after him," Franklin said.

Another of Freeman's favorite transorbital patients liked to bicycle over to Freeman's office for social visits after her operation. "Her appearance and language were pretty wild, but she was married for a short time," he recalled. To Freeman, the fact that these patients were out of the hospital and a part of human society at all was a major victory. The quality or status of the life outside of confinement was of less importance.

In making claims of the relative safety of transorbital lobotomy, Freeman assumed that his patients would have continued to deteriorate without the operation, never to recover. This assumption was reasonable had he been speaking only of schizophrenic patients. Freeman, however, was claiming credit for the restoration of all the patients who received transorbital lobotomies, including those with affective, or emotional, disorders. Even before the advent of the shock therapies and lobotomy in the 1930s, however, people hospitalized with affective illnesses frequently did recover enough to leave the hospital. Elliot Valenstein reported that during the 1930s nearly 70 percent of patients with affective illnesses earned discharges from state hospitals after being cured or being found to be improved enough to leave.

Even before Watts published his opposition to transorbital lobotomy except as a treatment for chronic pain and for psychiatric patients as a last resort, John Fulton, the former director of the primate lab at Yale University, chimed in. Transorbital lobotomy presented Fulton with a troubling conflict. He and Freeman were friends with a warm professional relationship. As Freeman was experimenting on cadavers and treating his first transorbital patients, Fulton had begun to issue public warnings about the reckless pace at which lobotomy was being practiced. While Fulton acknowledged that Freeman's new approach "is much sounder physiologically" than the Freeman–Watts operation, he also found it "unconscionable from a surgical standpoint."

In the fall of 1947, before Freeman had reported in print on any of his transorbital operations, Fulton chastised his friend. "What are these terrible things I hear about you doing lobotomies in your office with an ice pick?" he wrote. "I have just been to California and Minnesota and heard about it in both places. Why not use a shot gun? It would be quicker!" Although Freeman joshed back that transorbital lobotomies were "much less traumatizing than a shotgun and almost as quick," the rest of his response lacked honesty. Freeman denied that he considered the operation suitable as an office procedure, and he implied that Watts approved of it—impressions that Freeman would have to erase in the coming months. Fulton later tried to dissuade Freeman from demonstrating transorbital lobotomy on his home turf in Connecticut.

Fulton and Freeman continued their friendly relations, and they eventually reached an impasse in which each realized he could never convert the other, but their dispute over the appropriateness of transorbital lobotomy laid bare their differences in personality, approach to medicine, and attitude toward the opinions of others. Fulton, a researcher and experimenter who did not treat patients, was the director of a prestigious university laboratory whose list of grants, fellowships, and honors filled several pages. As admirable as his skills as a scientist were, his talents lay in negotiating the mores and rules of academe—and at this, beginning with his early days as a Rhodes scholar at Oxford, he was a stunning success. He had no children and regarded his extensive network of colleagues, peers, assistants, and students as an extended family that buoyed him with its respect and reverence. He belonged to sixty-three medical and scientific societies, had founded a journal of neurophysiology, and had been recognized as an honorary member of the British Society of Neurosurgeons.

In addition, Fulton believed that subjecting patients to new treatments without previously using animal research, carefully designing human trials, and thoroughly studying the trials was unscientific, reckless, and a waste of time. Only after experimentation and clinical trials, Fulton thought, could anyone possibly establish a system for evaluating and choosing the patients to be treated with a new technique. Physicians should use new techniques only after trials have proved them to be effective and applicable with sound therapeutic advantage. Doctors like Freeman who applied new treatments haphazardly and appraised the results anecdotally were to Fulton the antithesis of the scientific method. Fulton viewed each of Freeman's transorbital lobotomy operations as a tragic waste of data, a potentially dangerous misuse of technology that nobody was adequately studying case by case. None of these reservations, however, kept Fulton from recommending a "limited lobotomy" for a relative experiencing psychiatric troubles.

Freeman, however, headed a department at what was then a second-rate university, despite his Yale pedigree. He frequently complained that grant makers never took him seriously, and he felt most comfortable examining and handling patients in a clinical setting. He took an interest in animal research but—given the differences between apes and humans—

sometimes found it flawed as a guide to appraising new treatments. He dismissed small clinical trials as prone to unreliable results from overlooked environmental factors, the biases of investigators, and variables in the condition and family background of the patients. Laboratory researchers, he maintained, had contributed no useful treatments in the fight against psychiatric illnesses, and the only new beneficial techniques—insulin shock therapy, Metrazol therapy, electroconvulsive therapy, and lobotomy—had originated with clinicians like himself. Freeman often pointed toward the topectomy—Lawrence Pool's brain operation in which segments of the brain's gray matter were excised—as an example of a scientifically elegant but ultimately useless procedure. The clinical results of topectomy were no better than that of the plain, ugly ice pick lobotomy, and it was far too expensive and complicated to do much to reduce the number of mentally ill people crowding the nation's mental hospitals.

The battlefields of medicine—the hospitals, clinics, and doctors' offices—were to Freeman the only real testing grounds of a new treatment, and only in the field of battle could an investigator determine whether a treatment worked. Laboratory findings and the results of tests administered to small numbers of patients meant nothing compared to the opinions of patients and their families and caregivers on how well patients could function in the big, wide world. To withhold transorbital lobotomy from people who truly needed it to reclaim a place in society was itself a colossal waste of human potential.

Freeman's struggle to establish transorbital lobotomy as a legitimate psychiatric treatment took place alongside an event that shook his family and accelerated the deterioration of his marriage. During the summer of 1946 the annual meeting of the American Medical Association in San Francisco attracted Freeman's attention, and he decided to travel there, in typical fashion, in his trailer with children in tow. Most of his kids were now adults engaged in university study or military service, but the two youngest boys, Keen, age eleven, and Randy, age nine, had no commitments for the summer. Freeman's nephew Jef would accompany them, and another nephew, David, would join them in California. The road trip was uneventful on the way west, but Keen frequently displayed his scientific bent of mind. Near Ogallalla Lake in Nebraska, Keen busied himself by collecting animal bones and teeth and piling them on the trailer's dining table. He showed Freeman an unwashed kitchen knife and posed the question, "How many things can you make out on this knife, Daddy?"

Freeman identified mustard, peanut butter, cherries, gingerbread, butter, and the remnants of hot dogs.

This kind of inquisitiveness and spiritedness was characteristic of Keen, who excelled in science in school. "Keen . . . is doing well in school, has a rather intense nature, reads voraciously, and follows the exploits of Terry and the Pirates and other thrillers with increasing skepticism," Freeman had written the previous year. He had a strong streak of mischief that prompted him, for instance, to sometimes chew his cod liver oil capsules and breathe the potent fumes on his siblings. Named after the family line that included Freeman's illustrious grandfather, Keen had already told everyone that he wanted to be a surgeon. The boy was about to begin middle school and was taking more responsibility in the upkeep of the Freeman household. Randy, his much less intense and less competitive younger brother, happily orbited him.

On the road, Keen assumed leadership over the other boys and behaved far more responsibly than his older brothers did at the same age; he even took charge of the younger children for a day while Freeman attended to professional business in Omaha. "On the rest of the trip he was enthusiastic, going beyond the rest of us in his exploring," Freeman wrote. When a storm on the Great Plains soaked the floor of the trailer, Keen immediately attacked the mess with a mop. As the Freemans neared their first terminus in San Francisco, Freeman noticed that "Keen has a well-developed sense of seeing-what-ought-to-be-done, and during my absence he bossed the job of getting the trailer in apple pie order . . . and also getting his clothes in order." Keen displayed the family traits of diligence, intellectual curiosity, and impetuosity that Freeman most loved within himself.

Marjorie, traveling by rail, joined the family in San Francisco, where Freeman attended the AMA meeting. After the last session, Freeman urged Marjorie to join him and the boys, now augmented by son Paul, for ten days at Yosemite National Park. But Marjorie did not enjoy roughing it in Walter Freeman style. She preferred to return home, and she set off driving east with daughter Lorne.

Freeman had last visited Yosemite in the spring of 1944, when he admired the power of the park's waterfalls. He was especially taken by the way in which "the frozen spray would drop off and thunder down a thousand feet with the roar of an avalanche," he wrote at the time. This time, two years later, the water drew his attention again. "There had been plenty of snow, so the streams were full," he observed. He dropped the trailer at Camp Curry, and they all enjoyed good weather, despite overcrowding, overflowing garbage cans, and cramped public washrooms. On July 8 Freeman went out on a hike with Keen, Randy, and Jef, with the goal of climbing to the top of Vernal Falls, a beautiful 325-foot drop of the Merced River. It was a hot day, but Freeman forgot to fill their water canteen. While hiking along the river's edge near the crest of the falls, Keen felt thirsty and kneeled at the bank to fill the canteen. The events of the next few seconds Freeman later remembered in clear, horrifying detail.

Keen dropped the canteen in the water and lost his balance as he tried to retrieve it from the river. Freeman heard a shout, possibly from Keen, and turned around to face the water. Keen was moving rapidly with the river current toward Vernal Falls. "I was some distance back and was as if paralyzed," Freeman remembered. "I thought later that if I had vaulted the railing and extended my cane I might have saved him." Instead, a twenty-one-year-old man named Dale Loos, discharged from the navy five days earlier after two years of service, jumped over the rails and dove into the river. He reached Keen about fifteen feet from the brink of the falls and grabbed the boy but was unable to gain a foothold on the rocks in the swiftly moving water. Another man, never identified, climbed the railing and held onto it with one hand as he extended the other to Loos. The sailor could not reach him. He and Keen were carried to the precipice. "The last look I had," Freeman recalled, "was Keen's face as he went over the edge. A cry of horror rose from the small crowd at the top, and Randy and Jef and I clung together." Freeman jerked himself out of his paralysis of terror and ran down the trail. "Keen, Keen, you're dead, you're killed," he cried.

Freeman's intuitive fear was correct. Vernal Falls drops down a wall of rock higher than many office towers. At the bottom, the water crashes into a nest of boulders before resuming its course in foaming rapids. Keen and Loos vanished into the roaring plume. Park rangers soon arrived at the base of the falls and began to search for the bodies. "There is so much water going over Falls that whole area around pool is constantly deluged to proportions amounting to a hurricane and rainstorm combined," wired park superintendent Frank Kittridge to Loos's parents. "Below pool water flows through narrow gorge with turbulence over huge boulders. Is completely white stream of churning water. There is no possibility of seeing bottom or any object whatsoever under water's surface." While the recovery efforts proceeded, Freeman may have flashed back to an event of his youth in which he watched a child nearly drown in a canal. Another boy, paralyzed with fear, had refused to grasp the hand of the endangered child. Finally, a man performed a rescue with a boat hook. "The dry child was still groveling in fright on the towpath," Freeman, as a medical student, had written about the incident. "Damn fool kid, why didn't he pull him out?"

Ignorant of the tragedy, Marjorie and Lorne were somewhere in the Midwest driving home. At Freeman's request, state police were given a description of their auto and were asked to stop them. The police never found them. Instead, two days after the accident, the mother and daughter read about Keen's death in the *Chicago Tribune* while overnighting in Springfield, Ohio. Marjorie "came across the item on Page 2," Freeman recalled. "She said the names seemed to jump out of the paper but she couldn't read what it was about even though she had her glasses. She called Lorne and they read it together. Then she called me. It was a relief in this way, sharing the sorrow."

Marjorie and Freeman decided that when their son's body was found, he would be buried in California. "To bring him to Washington would be just too depressing," said Marjorie, who continued with Lorne to Washington to await news on the search for the body. "Somehow," Freeman told son Franklin, "although we have lived in Washington for a quarter of a century, and although Keen and all of the rest of you were born there, there's no particular joy in picking out a lot in the cemetery there. . . . We feel now that California definitely belongs to us. I don't know what the future will bring, but there's little doubt that the Yosemite will see us more than once. There is no bitterness about it, no blame or recrimination. There was no presentiment of disaster. It was just one of those things that happens."

Meanwhile, teams of rangers explored the river every day and inspected nets they had stretched across the water's path. For a week, the searchers found no trace of Keen or Loos. Then, on July 15, Keen's body appeared in plain sight between two rocks. Someone left this note for Freeman: "Mr. Kittridge telephoned to say that Keen was found this morning at 11:00 A.M., about 100 yards below the pool at the base of the falls. They had him out of the water by 11:00 P.M. They have called the undertaker according to your instructions, and Keen will be there by 5:00 P.M." Loos's body surfaced about a week later.

Freeman, who had examined countless bodies in his neuropathology lab at St. Elizabeths Hospital, was now called upon to identify the corpse of his son. "Of course being in the water a week or so, even though it was cold had led to some swelling of the tissues and body gases, and some peeling of the skin," he wrote to Franklin. Taking emotional refuge in his powers of medical observation, he continued, "I had to look twice to identify the body, but there was no mistaking it. Fortunately there was no disfigurement, and yet the back of the head was badly crushed, showing that death was immediate and painless. I imagine that after the slip over the falls there was a sort of 'this is it' sensation, and then nothing more."

Marjorie and Lorne had difficulty making it to California in time for the funeral. Freeman's son Walter received a military discharge on July 10 and headed straight for Yosemite. Franklin, stationed in Japan, was unable to obtain a leave of absence. Walter and his brother Paul, showing a Freeman trait that went back to their otolaryngologist grandfather, took comfort in the wildness and beauty of the outdoors and made a 2:00 A.M. hike to the top of Half Dome. Freeman did not have the heart to join them, but he too found solace in his surroundings. He bathed daily in the Merced River. "It is extraordinarily restful up here," he wrote. But the pain of Keen's death was too much for him to directly confront, and he turned his thoughts to his work. The things his mind could manage to consider had "mostly to do with lobotomy patients, past and future," he wrote on July 22. His youngest son, Randy, Keen's constant companion, also failed to realize the depth of his loss. "Randy is our mascot, and he doesn't seem to be depressed or upset or lonely," Freeman noted. "He seldom mentions Keen, and seems content to wear his clothes. He asked for the little hoard of money that Keen had. I think he wants to buy a present for somebody, so I'm giving it to him."

Keen was buried in Merced, not far from the river that took his life. After the funeral, Marjorie and Lorne went back east by train; during the trip of several days Marjorie spent much of her time seeking comfort from another passenger, a Catholic priest. Freeman returned home with Paul, Randy, and Jef in the trailer, their trip a dreary blur. The journey began with follow-up visits to lobotomy patients in Southern California. They stopped to check up on a prefrontal lobotomy patient from 1938 living in Sugar City, Idaho. "Her home is far from beautiful, and when I got there in the middle of the afternoon the beds were unmade and the ironing board was in the middle of the living room, and she was off visiting and gossiping with a neighbor," Freeman observed. "However, that was better than being locked up in a madhouse." Somewhere nearby, they stopped to watch a garter snake devour a large mouse. "Randy didn't know what to make of it, but it was Paul who came to the rescue: 'That's life for you.' Kill or be killed," Freeman wrote. "We allowed them to carry on."

In Yankton, South Dakota, Freeman performed several transorbital lobotomies at a state hospital—his first such operations outside of the Washington area. Freeman later met with Dale Loos's parents in Dayton, Ohio, and had a minor accident with the trailer in Virginia. Upon their return to Washington, he sold the trailer and traded in his car for a more luxurious Lincoln. Never again would he travel by trailer with his family, and none of the events of this final trip after Keen's death would leave a lasting imprint on his memory. "The rest of that trailer trip, the last of the series, is forgotten," Freeman wrote decades later. "I don't know what route we took back, or anything that happened."

## Chapter 10

# Fame

At home by mid-August, Freeman did not fail to notice the grief of his wife and children. Randy, whose sunny disposition had darkened, could hardly mention Keen's name at all. In a melancholy way, Marjorie, drinking more frequently, sometimes played Keen's favorite pieces on the piano. "Marjorie is still deeply affected by coming across the little belongings of Keen, his school things, his clothes and so on. She really broke down on meeting several of Keen's teachers a couple of days ago," Freeman wrote. He later observed, "Marjorie personifies Keen a great deal, talks of him as though he were an unseen visitor, now with one, now with another of the family." She went through a bout of severe hay fever and dental pain that resulted in the extraction of all her teeth, and she remained emotionally fragile through the end of 1946. "I would say that in some respects her life ended," son Walter remembered. Desperate for assurance that she would see Keen sometime again, she made a temporary return to the Catholic Church. Freeman saw her distress, but because he allowed himself only limited expression of his own sadness, he could not help comfort her. Keen, in fact, had been much like Freeman in spirit and personality—and was a devastating loss to the father.

The month Keen died, Freeman reflected enough on his grief to write a poem that included these verses:

> The children grow from birth by easy stage
> And each one leads the next by heritage
> Nor is there strangeness in the family grout
> Unless death takes one at a tender age. . . .

Take pride in building with incessant toil
A home no bitter enemy can spoil
But think not that your roots are firmly set
Until you've dug a grave upon that soil.

That was Freeman's immediate response. His deeper reaction to Keen's death—the months-long shock that Marjorie, unlike him, did probe and gradually came to accept through her handling of Keen's possessions and her contacts with her son's teachers—remained unexplored. Upon his return to Washington in the summer of 1946, Freeman buried himself in two months of accumulated correspondence. He then occupied himself with preparing for a new academic year at GWU. "There seems to be nothing but work, work and more work, but Gosh! How I love it," he wrote in early 1947. By later that year, Freeman had taken up a grueling daily schedule in which he awoke between 4:00 and 5:00 A.M. to read and work for several hours; spent a full day at his office, the hospital, and the medical library; and after dinner at home, resumed working for a few more hours. "I knocked myself out with Nembutal every night because the work was so exciting, and slept the sleep of exhaustion," he observed. "I saw little of the family and my disposition became more and more testy. I'd cook a little supper for myself at home rather than go out for dinner with the family, and Marjorie thought I didn't love her any more."

Marjorie's thoughts eventually explored another possibility—that her husband was having extramarital affairs. The older children harbored similar notions as well. "I suspected it was just one of those things he was subject to, a one-time fling," Franklin remembered. Freeman may have become involved with a nurse, a secretary, and a former colleague's widow—all women whose proximity and familiarity might have made them convenient diversions from the emptiness and guilt that Keen's death produced. "He was a vigorous man, and there was never any commitment in these things," Franklin added.

Freeman effectively lopped Keen and his family's grief out of his life by instead embracing his work. "Whatever work I have to do must be done while I have the life and the enthusiasm to carry it on," he wrote five months after Keen's death. "Maybe it is this in part that makes me tick. . . . So, I don't waste time in futile questioning and hoping. Keen's death I might compare to an amputation. One can get along, although with a difference, and bemoaning the fact will not alter the situation." All Freeman would allow himself was simple regret, which his unsentimental realism abruptly terminated. "Keen was marked for a surgeon, a bright mind, an independent spirit, with a touch of ruthlessness," he wrote in the 1960s. "But we lost him."

Neurosurgery was among the newest of medical specialties, but in a few short decades neurosurgeons had successfully laid claim to all types of physical intervention in the brain. Not only had Freeman threatened that medical monopoly, but he had done so by introducing a procedure that seemed appallingly primitive and unaesthetic. As the psychosurgery historian Elliot S. Valenstein pointed out, Freeman's electroshock anesthesia produced seizures that neurosurgeons normally considered it their job to prevent. They disliked the bare hands and unmasked face Freeman brought to his operations, and many decried the practice of operating "blind" on concealed regions of the brain.

Although Watts ultimately would not do so, some neurosurgeons took Freeman's background into account in judging whether it was proper and in a patient's best interests for Freeman to be conducting brain operations. Freeman was unusual for a psychiatrist—he had a vast knowledge of brain anatomy and had frequently performed half of the earlier prefrontal lobotomy surgeries. Among his psychiatric colleagues, Freeman stood almost alone in his command of brain structure. But when Freeman proposed to train psychiatrists all over the country to perform transorbital lobotomies, the neurosurgical profession responded with outrage. How, its members asked, would a psychiatrist lacking surgical experience or certification respond to hemorrhage, seizures, brain damage, or any unexpected outcome of a transorbital operation? Would a psychiatrist know how to properly sterilize instruments and guard against infection? Freeman breezily dismissed these concerns. He believed that neurosurgeons were simply trying to stake out the brain as their exclusive property, to the detriment of tens of thousands of needy patients. "The operation can be performed by the psychiatrist as soon as he has familiarized himself with the landmarks and practiced on the cadaver," Freeman told readers of *The Lancet* in 1948. "It requires little more preparation of the patient than is demanded for electroshock therapy; the operation is brief, safe and simple." After a few hours of instruction, any doctor—psychiatrist, podiatrist, or medical resident—could learn it.

To prove his point, Freeman frequently demonstrated transorbital lobotomy before audiences of psychiatrists, surgeons, and other physicians—friend or foe, it did not matter to Freeman. He believed that the simplicity and safety of the procedure spoke for itself, although he relished the ways in which transorbital lobotomy transfixed anyone who observed it. It was the greatest showstopper in his repertoire, a medical treatment that to many looked like nothing more than a medieval form of torture. "There was a certain amount of horror and fascination," Freeman acknowledged. Edwin Zabriskie, a seventy-four-year-old clinical neurologist who had seen bloody combat during World War I and was a professor emeritus at the Columbia University School of Medicine, watched one demonstration at New Jersey State Hospital and fainted dead away, an event Freeman called "a climax" of his advocacy of the procedure.

Freeman even demonstrated transorbital surgery before the unblinking lenses of television cameras. During one dramatic demonstration in 1952 at Herrick Memorial Hospital in Berkeley, California, before a group that included sons Franklin and Paul, Freeman encountered severe hemorrhaging after the lobotomy and had to react quickly. He inserted a cannula through the operative opening, removed the clotted blood, and irrigated the area with saline until the bleeding ended. The patient survived. For years Freeman pointed out that Fiamberti, the originator of the transorbital approach, was a psychiatrist, and that a high percentage of transorbital lobotomies—most of them performed by psychiatrists—went off without complications. "The psychiatrist is in a position to know the needs of his patients, to give the necessary care before and after operations and personally to confer with the families concerning the rehabilitation of patients after discharge," Freeman argued.

Most neurosurgeons were not convinced. One protested that transorbital lobotomy "is being used in psychiatrists' offices without provision to deal with complications which must occasionally result. Such a blind procedure is unjustifiable, and it is to be feared that its simplicity will cause it to be used too freely and bring disrepute on a potentially very useful measure." Another noted, "You can't treat the brain like it's a block of ice," while yet another neurosurgeon called Freeman's transorbital leucotome a "stiletto" and indirectly branded Freeman a criminal.

An especially forceful warning against all forms of psychosurgery came from Gösta Rylander, a Swedish psychiatrist, surgeon, and friend of Freeman who had been one of the earliest advocates of lobotomy. In December 1947 Rylander told members of the Association for Research in Nervous and Mental Disease that psychosurgery patients frequently experienced devastating personality deterioration. "He cited as examples eight of his cases, all persons with high intelligence quotients, but with various obsessions," wrote a reporter who attended Rylander's presentation. "After frontal lobotomies, he said, the obsessions were cured, but the I.Q. of each fell." Rylander's patients had lost their grasp of the value of money, disavowed their earlier fervor for political radicalism, and, according to relatives, suffered the amputation of their souls. Freeman, who was present at the meeting, challenged Rylander's assertions from the floor. He asserted that lobotomy might prove to be "a great leveler" that slightly reduces the I.Q.s of intelligent patients but increases the scores of those with low I.Q.s.

Physicians in other specialties, of course, protested as well. Nolan Lewis, a psychiatrist who directed the New York State Psychiatric Institute, questioned whether lobotomy most benefited caregivers or patients. "Is the quieting of the patient a cure?" he asked. "Perhaps all it accomplishes is to make things more convenient for the people who have to nurse them. . . . I think it should be stopped before we dement too large a section of the population."

Freeman enjoyed chipping away at the ethical and medical high ground that neurosurgeons and others assumed in criticizing him. He compared himself to his grandfather, W. W. Keen, who had been called a liar when he tried to promote antisepsis in the United States. Freeman toyed with his neurosurgical opponents like a cat with mice, partly because he enjoyed intellectual combat but also because he knew he could never bring them to accept his vision of how transorbital lobotomy should be administered. Only when patients wore head-to-toe surgical draping and underwent transorbital operations in a hospital under traditional anesthesia, with a certified neurosurgeon in attendance, would these critics be satisfied. So Freeman focused his attention on training psychiatrists to perform the lobotomies his way: as an out-of-hospital treatment, inducing coma by jolts of electricity.

As part of this outreach to psychiatrists and other candidates who wanted to learn the transorbital technique, Freeman oversaw the production of a new lobotomy instructional film in 1949. He frequently showed this movie "with good effect," but he seems to have considered it unimpressive because it included no "live" sound, only narration added after the filming. Freeman wanted a less stilted movie, so he made another attempt at preserving a transorbital operation on film with natural sound. He commissioned the photographer James Ankers to transform a studio into a replica of an operating room, with microphones strategically set. Freeman judged a reluctant and quick-tempered patient, Gladys Cartier, suitable for the filming because she was "a good example of an intelligent obsessional," thus someone with a high likelihood of benefiting from the transorbital operation.

When doctor, nurse, and filmmaker were ready to begin, however, everything suddenly fell apart. Cartier panicked, and Freeman had to hurriedly administer the electroshock anesthesia. Freeman, who hoped to capture the patient's recovery on film, was disappointed to find that Cartier retained her bad temper after the operation and insisted on wearing her sunglasses long after her bruises disappeared. Freeman performed a second transorbital lobotomy two months later, free of charge, but the patient was again unimproved. Cartier tried to join the Carmelite order of nuns, was rejected by the mother superior, and asked Freeman to write her prescriptions for tranquilizers. He refused, declined to perform another lobotomy, and instead suggested increased physical activity to combat her anxiety. "About eight months after the operation, I received a letter from her mother in Florida saying that her darling daughter had died of a heart attack," Freeman wrote. "I wrote to the bureau of vital statistics in Miami and found out the truth of the matter. Gladys had died from an overdose of Phenobarbital. The film has never been shown and never will be." Freeman called the effort "a monument to my misguided enthusiasm."

Freeman remained dissatisfied with transorbital lobotomy's lack of success in treating patients with symptoms of schizophrenia, especially hallucinations. He found inspiration for a possible new approach to treatment in the case of one of his longtime neurological patients, a woman named Bella Harrison who had undergone brain surgery as a teen to remove a tumor of the cerebellum. The operation saved her life but left her blind. She bounced back by learning Braille and starting a career in the Braille division of the Library of Congress. Two decades after her brain surgery, however, Harrison became schizophrenic; she began hearing voices and unexpectedly seeing hallucinations that appeared in patterns of Braille. To Freeman, Harrison's case strongly suggested that the visual hallucinations of schizophrenia originated in the uncinate fasciculus, a stream of neurofibers that curves over the fissure dividing the frontal and temporal lobes. Because the middle cerebral artery passes through these fibers, any attempt to cut them is extremely hazardous. Instead,

Freeman proposed to disable the fibers by soaking them with a syringeful of the patient's blood, which he would inject into the brain through the transorbital route.

"I focused my attention on this area after being presented with some patients for reoperation after transorbital lobotomy had failed," Freeman noted. "I did so by repeating the original operation and then using a syringeful of blood drawn from the patient's arm and injected into the lower part of the incisions on each side. I used blood because of its viscosity, its availability, and the tolerance of the brain to spontaneous hemorrhage," although he acknowledged that hot water might work even better. Nobody else seemed convinced by this approach to treating hallucinations, though. "I have tried in vain to get permission to operate upon chronically hallucinated patients, except in a handful of isolated patients," Freeman complained. The psychiatric historian Elliot Valenstein reported that Freeman and the neurosurgeon Jonathan Williams eventually did operate on Bella Harrison. But instead of injecting blood into her brain, they gave her an amygdalectomy, which cut portions of the temporal lobe. The outcome is unknown.

In the eleven years after Freeman first plunged a leucotome through the orbital cavity and into the brain of Ellen Ionesco, he operated on 2,400 more patients. By 1950 he had not performed any prefrontal lobotomies with Watts or anyone else for several years, and he took the opportunity at that year's annual meeting of the American Psychiatric Association in Detroit to make a momentous announcement. Eight years earlier he had called prefrontal lobotomy "a safe and permanent cure" for affective disorders. But in Detroit he told his APA colleagues that he would never again perform one. "I believe that transorbital lobotomy can replace prefrontal lobotomy in all cases except those showing extreme and prolonged violence, and in psychotic children," he later wrote. "In both of these types maximal operation is the only effective procedure and necessarily leads to a rather vegetative condition." Transorbital lobotomy was far safer and much less prone to complications. Fifteen percent of the prefrontal patients suffered seizures (the figure rose to 47 percent after two such operations), compared with 0.5 percent of the transorbital lobotomy patients. The Freeman–Watts series of prefrontal lobotomies at last concluded in 1948 with a total of 624 patients who had undergone 702 psychosurgeries. Watts continued giving prefrontal lobotomies to patients for another five years.

For fifteen years Walter Freeman had devoted his career to advancing prefrontal lobotomy as a beneficial treatment for the mentally ill. That era, he said, had passed. He had discovered something better not only in the sense that patients fared more favorably than after prefrontal lobotomy but also because transorbital lobotomy was a procedure Freeman could promote without Watts—without the necessity of any surgeon; without the specters of Moniz, Fulton, and others hovering above and blocking some of the spotlight; without the restrictions that normally bound neurologists and psychiatrists; and without the confinement of the operating room that usually shielded its events from the eyes of spectators. Above all, Freeman was free to perform, refine, and champion transorbital lobotomy as a solo operator. He liked his own company best. Even more fervently than he had promoted prefrontal lobotomy, Freeman began a campaign for transorbital lobotomy. He would take that campaign out of the pages of textbooks and medical journals and into the psychosurgical battlefield.

Meanwhile, as the crossfire flew over the pros and cons of transorbital lobotomy, a distinguished cadre of neurosurgeons, psychiatrists, and other neurological investigators was joining Freeman on the field of battle. Psychosurgery, despite the loud condemnation of critics, was spreading. Of the forty thousand to fifty thousand cases of psychiatric surgery performed in the United States during the forty years after 1936, Freeman was involved in less than 10 percent of the total. By the late 1940s, lobotomy had moved from the edge of psychiatric practice to the mainstream. Its practitioners included some of the most prominent names in neuroscience during the mid-twentieth century: Walter Dandy, W. Jason Mixter, Wilder Penfield, James Poppen, William Scoville, William Peyton, J. Grafton Love, and Lawrence Pool. Lobotomies were carried out in the surgical rooms of such well-known institutions as Columbia, Johns Hopkins, Harvard, and Yale universities, the Institute of Living in Connecticut, McLean Hospital in Massachusetts, and the Mayo Clinic.

As the psychosurgery historian Jack Pressman pointed out, lobotomy flourished not because its practitioners were bad doctors or ethically flawed but because of the position and role of medicine in the culture at the middle of the twentieth century. Throughout history, Pressman wrote, the perceived effectiveness of physicians is "not wholly dependent upon the actual physiological status of the ailment under treatment but is linked to the perception that the problem has been resolved: thus medicine itself is embedded within the larger social system." He concluded that "the fact that psychosurgeons seem so bizarre to us today is an indication that even in the recent past the world itself has changed substantially. The multiple 'locks' that psychosurgery once were meant to open have already faded away and the particular frames with which participants viewed the procedure's results have since been discarded."

During the 1940s and the first half of the 1950s, a growing number of American physicians advanced lobotomy as the solution to a tangle of issues in the treatment of the severely mentally ill: overcrowding in psychiatric hospitals, the shortage of other effective treatments, a swelling sense that warehousing patients without treating them was shameful, and the conviction—especially strong among psychiatrists—that physicians were obligated to try to return disabled patients to productive functioning in society as much as to cure individuals of disease. These issues resonated with the concerns of patients. By the spring of 1945 Freeman was receiving letters "from all over the country, sometimes a half dozen inquiries in a day," from people who were convinced that they or their relative would benefit from a lobotomy. And unlike those engaged in theoretical or academic psychiatric research with foundation support, he noted, "we can report results in terms of discharges from hospitals and restoration of economic competence." Lobotomy, he believed, saved time, wear and tear on hospitals, and money—as well as the overall mental health of patients.

Freeman battled for transorbital lobotomy at a time when his academic responsibilities, practice, and professional activities demanded a greater share of his attention. After the war, GWU made plans to build the new hospital that opened on campus in 1948. Freeman helped plan the hospital's neurological clinic and was pleased that it had "fifty beds of our own and many additional facilities." But he did not approve of the doors to the rooms in this ward, as they lacked shatterproof glass. After the hospital administration ignored his request for safer glass, Freeman arrived one day and asked a nurse to remove her shoe. Using the heel as a hammer, he smashed some panes. Freeman got his shatterproof glass. "Walter believed in the frontal attack," Watts noted of the incident.

When administrators wanted to divide neurology and neurosurgery into different hospital divisions, Freeman and Watts successfully fought the initiative, and they continued to work as joint heads of a combined department despite the dissolution of their partnership in private practice. In addition, Freeman accepted a consulting position at Walter Reed Army Hospital. He served as president of the Philadelphia Neurological Society, the American Association of Neuropathologists, and the Medical Society of the District of Columbia. In the latter organization, he championed the admission of African American members, who previously had been excluded. Members of the women's auxiliary of the society, which Marjorie headed during her husband's term as president, especially fought Freeman's proposal to end membership discrimination because they "did not want the black doctors' wives at its social events," David Shutts wrote. In the end, Freeman pushed through his membership changes, and he regarded this breaking of Jim Crow traditions as his best achievement as president. "If this is fame, it doesn't connote sauntering," Freeman observed.

His third term as secretary of the American Board of Psychiatry and Neurology had ended in 1946 after a total of twelve years. The removal of his responsibilities pained him, but he acknowledged that someone known primarily as a psychiatrist, not as a neurologist, should have a turn. Still, after a year in a diminished capacity as an associate examiner with the board, Freeman returned to its leadership for another twelve-month term as president.

Although Freeman may have viewed himself as a neurologist, many others did not. His renown as a neurologist was diminishing in the United States due to the public's identification of him as a lobotomist and the increasingly psychiatric nature of his practice. He worked in the psychiatric division of the GWU hospital, for instance, when he administered electroshock treatment. He began treating some patients with a form of psychotherapy that bore little similarity to the psychoanalysis of Freud and his followers: Freeman listened to patients' physical complaints—although he refused to let them discuss their emotional distresses, sexual problems, or impressions of their inner lives—and advised them to exercise daily. He told one patient, a woman suffering from hypochondriacal pains, to imagine as she walked that the passing rocks and plants were her enemies and to beat them with a stick. She followed her exercise regimen with a vengeance and eventually became an accomplished weight lifter.

Freeman, however, also took a strong interest in the inner lives of his lobotomy patients, particularly their sex lives. He recorded some instances in which patients greatly increased their frequency of sexual intercourse after the operation. In one case, a middle-aged bachelor "with religious obsessions had denied himself intercourse for more than twenty years preceding prefrontal lobotomy. Six months later he commented upon the renewed pleasure he experienced, complaining only that the girls cost him more money than he could afford," Freeman wrote. Another patient had sex with his wife four to six times daily during his first week home from the hospital. Freeman also treated a homosexual patient, suicidal because of his sexual orientation, whose lobotomy completely extinguished his sexual drive. Overall, Freeman and Watts believed that the inertia produced by lobotomy often made patients less interested in sex.

One of the strongest signs of the lobotomy's entry into the mainstream of psychiatric care was the degree to which the U.S. Veterans Administration embraced it in the years after World War II for use in the growing psychiatric wards of its hospitals. VA Hospital staff performed only seventy-one lobotomies in all of 1946. The following year saw the government agency codify the process by which hospital committees would consider and select applications for prefrontal lobotomy. By 1949 the VA was performing an average of forty-eight lobotomies per month. That year the agency showed its interest in investigating further use of the procedure by commissioning a comparative study of the various forms of psychosurgery involving more than two hundred patients.

Several other American physicians had used psychosurgery before the VA. The earliest American disciple of Freeman and Watts was James G. Lyerly, a Jacksonville, Florida, neurosurgeon who was a friend of John Fulton. Lyerly disliked "blind" approaches to lobotomy that left the surgeon unable to view what he was cutting, and in 1937 he began investigating psychosurgical techniques that let him see the fibers he cut. Three years later he developed a procedure that allowed him with the aid of a lighted speculum to view the nerve fibers he was targeting after opening a large hole in the patient's skullcap and lifting the cortex of the brain out of the way. "In the grisly finale to this difficult method," wrote David Shutts, "the nerve fibers were coagulated with an electrical device, which crackled as it seared the tissue along the plane of the coronal suture." Freeman and Watts could find no advantage in the "open" operations of Lyerly and others. "The surgeon sees what he cuts but does not know what he sees," Watts commented. Freeman observed, "It is easy to get lost in the frontal lobes, and sometimes with results ranging from the negligible to the disastrous."

Francis Grant in Philadelphia and W. Jason Mixter in Boston were other early adopters of psychosurgery. In New York City, the physician who most actively performed psychosurgery was J. Lawrence Pool, a researcher affiliated with Columbia University Medical Center who received patient referrals from no less than John Fulton. In 1946 Pool pioneered a technique called topectomy, in which specific areas of the gray matter of the frontal cortex—not the connective nerve fibers—were excised with a scalpel. The procedure was similar in method to Burckhardt's operations in the late nineteenth century. Pool believed that the Freeman–Watts method of severing white fibers inevitably incapacitated brain functions not associated with emotional tension and thus affected intelligence.

Topectomy went head to head against Freeman's techniques of lobotomy during two comparative studies called the Columbia–Greystone Projects. The first project, begun in 1947, was "an effort to determine which part of the frontal lobe was sacrificed with most benefit and least harm," Freeman wrote. While Pool and others performed topectomies on forty-eight patients, Freeman was invited to operate transorbitally on eighteen patients at New Jersey State Hospital in Greystone Park, but "there was not a single one that I would have chosen from my own practice. The results were as bad as I anticipated." In one of these deteriorated Greystone patients, Freeman broke the tip of his leucotome in the base of the brain. "Fortunately, there were no unfavorable effects, but the embarrassment was mine."

The results of the first Columbia–Greystone Project compiled by Alfred Kinsey were inconclusive but suggested that topectomy was beneficial. Freeman maintained that the benefits of topectomy mainly derived from the postoperative care of social workers, not from the operation itself. In the second Columbia–Greystone project, Freeman treated only nine patients with transorbital lobotomy. The outcomes of those operations, however, impressed Pool, who noted, "I think Dr. Freeman has demonstrated that the transorbital lobotomy is an effective, useful tool, and should be used more." He thus became one of the earliest endorsers of the procedure outside of psychiatric hospitals, although he qualified his enthusiasm by declaring "that we, as neurosurgeons, should handle the operation from start to finish, and not the psychiatrists—with the exception of Dr. Freeman."

Despite the sophistication of the surgical technique, topectomy could claim few benefits for patients over other forms of psychosurgery; the operative risks were high, and patients developed seizures and other complications at a rate no less than with other methods. Pool eventually acknowledged to Freeman that he could not prove the superiority of topectomy over prefrontal or transorbital lobotomy, and he even admitted that the main advantage of topectomy was the opportunity it afforded for the experimental application of electrical stimulation during surgery.

Another active psychosurgeon—and eventually a fervent foe of Freeman—was James Poppen, who practiced at Boston's Leahy Clinic. This neurosurgeon favored an approach to psychiatric surgery in which targeted portions of the brain were aspirated or sucked away. After Poppen went on record as a disparager of transorbital lobotomy, Freeman called Poppen's technique "a rather gross method" and commented that it reminded him of running a "vacuum cleaner over a tub of spaghetti." To ice the cake, Freeman observed, "These brains, at least the one or two that I have seen following [Poppen's operations], are horrifying."

Other lobotomists were scattered across the North American landscape, and they devised some thirty different ways to produce psychosurgical lesions in the brain. In 1947 E. A. Spiegel and H. T. Wycis adapted a technique first performed half a century earlier in medical experiments on cats—stereotaxic surgery—in which they placed over the skull an elaborately designed metal frame fitted with electrode-tipped needles to produce small brain lesions. In Kentucky in 1948 William Scoville began using a technique of cortical undercutting, which severed the fibers feeding the orbital cortex. That same year the Canadian neurosurgeon Wilder Penfield first tried gyrectomy, the removal of parts of the frontal gyri structure of the brain; William Peyton at the University of Minnesota performed frontal lobectomy, in which sections of the frontal lobes were amputated; and Ted Scarff tried lobotomy on only one side of the frontal lobe to minimize personality damage.

Other neurosurgeons treated psychiatric disorders by freezing, burning, and radioactively and ultrasonically exposing parts of the brain. These doctors used a dizzying array of surgical tools to accomplish their aims, including nasal septum elevators, paper knives, cutters with rotating blades, knives fitted with arc-measuring protractors, leucotomes with sharp, curved, and blunt edges, and perforated spatulas. "I don't think the various techniques at the time made much difference," Watts later said. "We claimed ours was the best, but it's hard to know."

Lobotomy was gaining wider acceptance overseas as well. In England the champion leucotomist was Wylie McKissock of London. McKissock tallied his five hundredth psychosurgery in April 1946—earlier than Freeman and Watts reached that same milestone—and by 1950 had performed more than thirteen hundred leucotomies, which was more than Freeman or anyone else at that point. McKissock, who used a plain cannula for his brain cutting, noted the primitiveness of the tools at his disposal in comparison with the elegantly designed instruments available to the average dental surgeon.

Just as Samuel Johnson had James Boswell to study and appreciate his work, McKissock had the intrepid Maurice Partridge, who shared Freeman's passion for the chase and his fondness of telling ribald stories about patients—yarns that did not conform to the usual format of medical and scientific reporting. For three years Partridge, a psychiatrist with St. George's Hospital in London, followed up on three hundred of McKissock's leucotomy patients. Traveling more than sixty thousand miles to nursing homes, asylums, group homes, and private residences, he wrote engagingly of his experiences and affectionately of the patients he encountered. Partridge's examination of complications, death rates, and the lack of correlation between the results and the techniques of psychosurgery mostly mirrored the studies of Freeman and Watts. At times he was deeply impressed by the recovery of patients after psychosurgery, including one man who after seven years of motionlessness punctuated by "bouts of being violent with double incontinence" was within a year of operation able to achieve a score of 50 at cricket. "Bizarre illnesses may require bizarre treatment," Partridge noted after completing his study, "and in psychiatry they often get it. They show so often a stubbornness and resistiveness to treatment, they expose so clearly the ignorance of their pathology and aetiology, that they arouse aggressive reactions in the baffled and frustrated therapist."

Partridge's talents as a storyteller matched his skills as a medical investigator. He wryly wrote of the "minor martyrdoms" he endured to bag his interviews with scattered leucotomy patients. "It has been impossible to avoid talking on all sorts of topics quite devoid of interest to the interviewer: it has been necessary to help with the washing up or the preparation of meals, to do some weeding, play with children, take them out in the car, bring the washing in and out of the rain, etc.," he observed. Partridge conducted his examinations in hotel lobbies and public parks as well as in private rooms—though he did not always focus his sights on the right person. "On one occasion in a district of Swansea, where all the streets looked the same and everyone was called Jones, the writer was misdirected; the lady who opened the door of the house seemed quite unfamiliar, but offered a cordial welcome, saying that she well remembered being visited in the hospital and she was delighted at the reunion," he wrote. "Following the first principle, though with some misgiving, of getting the patient to talk spontaneously, there was ten minutes of talk on the rather intimate topic of vulvo-vaginitis before it was possible to persuade the patient that we had never met before." He also told a story illustrating the unreliability of relatives as judges of a patient's recovery: "One mother, after a private exordium on her son's 'cure,' cried, 'It is miraculous,' as she dramatically threw open the door to reveal a sagging, hunched, disheveled, unshaven, remote figure who readily proved to be the subject of affective incongruity with multiple delusions and ideas of reference, while much preoccupied with sadistic fantasy."

Like Freeman, Partridge took interest in the sexual lives of patients. Recording details not normally found in follow-up studies, he once observed, "One woman, in general somewhat unrestrained, when alone with her husband would raise her skirts and dance in Oriental style, after which she would say, 'Well, Freddy Weddy, what about beddy weddy?' And if he demurred, she would leave amorous notes in the kitchen, which he would find on going to fill his hot water bottle, and which were designed to whet his appetite by a statement of what awaited him in the boudoir."

McKissock and his chronicler were by no means the only overseas physicians inspired by Freeman's and Watts's psychosurgical activities in the States. In Japan, where lobotomies began in 1939, the procedure became quite fashionable, according to Sadao Hirose, who taught at the Nippon Medical School in Tokyo. In June 1947 Hirose began a series of prefrontal lobotomies using the Freeman–Watts methods and transorbital lobotomy, with a total of 477 cases by 1956. Freeman kept track of other active lobotomy programs in Czechoslovakia, New Zealand, Canada, Brazil, and Venezuela. In the latter nation, Humberto Fernandez-Moran, a former GWU medical student, began performing transorbital lobotomies as early as July 1946.

In the eyes of his international colleagues, Freeman remained both a neurologist and a psychiatrist—a physician unusually active in promoting a form of psychiatric treatment that still puzzled many doctors in Europe. When the Royal Society of Medicine sponsored a conference on psychosurgery in April 1946, he accepted an invitation to speak. With the British still suffering under postwar shortages of luxury items, he carried a suitcase full of presents with him on his first transatlantic airplane flight. "I brought some nylons with me, unfortunately too small for most of the English ladies," he wrote, "as well as some lipsticks and compacts for the girls. These were confiscated by the mammas, I'm afraid. Cigars for the men were welcome, however." During several days visiting colleagues in Scotland and Bristol, he showed the lobotomy movie that he and Watts had produced six years earlier. The filmed operation, Freeman recalled, "was a bloody one," and he unwisely allowed a group of high school students to be his audience in Bristol. The unfortunate youngsters were not prepared to see moving images of a skull being drilled and a leucotome being inserted into the holes. During the showing, "five of them fainted and had to be dragged out." Freeman called this event "a particular high point in my evangelistic career. . . . I said afterwards that Frank Sinatri [sic] could hardly do better."

Freeman then proceeded to the conference in London. He was mildly critical of the reports he heard from British psychosurgeons. He found their fatality rates too high, their criteria for improvement too low, their use of the procedure too late, and their reluctance to reoperate too cautious. He also criticized their practice of requiring patients to recover in the hospital rather than at home. One finding, though, cheered his heart. He noticed that "there was general agreement that prefrontal leucotomy was very helpful in certain types of cases[,] particularly in the obsessive neuroses and the involutional depressions," he wrote to Moniz, who was unable to attend the conference because of a flare-up of his gout. As for the efficacy of psychosurgery against schizophrenia, the British physicians were less certain.

Freeman made his next overseas trip two years later to attend an affair that began brewing in his mind, and in Watts's, during the London conference. With close to two thousand lobotomy cases completed worldwide, they believed the time was ripe for an ambitious international congress that focused on the exchange of information among psychosurgeons. In 1946 Freeman requested Moniz's assistance in organizing such an event and suggested Portugal—the birthplace of psychosurgery—as the best site. When Moniz earned honorary memberships in the New York Neurological Society and the American Neurological Association in 1948, the Portuguese government decided to exploit its native son's renown by financing the congress. After gaining further support from practitioners of leucotomy and lobotomy around the world that year, Freeman brought the First International Congress on Psychosurgery into existence. Moniz would serve as its honorary president.

The congress was scheduled for early August 1948, but Freeman packed up his family—including all five of his children, ranging in age from twelve to twenty-three, although Franklin came a few weeks late due to his school schedule—for a two-month European excursion that began on July 15 and consumed several thousand dollars of savings. The Freemans would not undertake a sea voyage. "We flew the ocean in three different planes on the theory that it was a bad thing to put all the eggs in one basket; or, stated differently, that there ought to be somebody to enjoy the insurance." Upon arrival in Portugal, the family took up residence in a seaside hotel in Estoril. There they swam, saw the sights, and took day trips into Lisbon. Shortly before the start of the congress, Freeman gave a talk before the Lisbon Academy of Sciences and was awarded honorary membership. The tokens of this honor were a diploma and a heavy gilded star that hung on a chain. Moniz, after years of official Portuguese indifference to his medical achievements, hung this medal around Freeman's neck and whispered, "This is my revenge." Freeman wrote, "I found out later that the only other American so honored was Albert Einstein—me and Al!" (Einstein had received his membership in the academy before obtaining U.S. citizenship.)

The conference gathered one hundred participants from India, Iran, France, Spain, the Netherlands, Argentina, Brazil, Mexico, Panama, and twelve other nations. Events began on August 3 with a speech by Moniz, whose gout did not interrupt his enjoyment of the proceedings. There were roundtable discussions, banquets, parties, the presentation of papers, and Freeman's demonstration of two transorbital lobotomies. Wearing the only gray clothes in a sea of black suits, Freeman posed for photographs with his colleagues. "Altogether we had reports on some 8,000 cases. Practically all of the new work came from the U.S.," which sent a dozen delegates, Freeman reported. James Watts was conspicuously absent, possibly a sign of his growing disaffection with his old partner. While Freeman prepared for his own presentation of a paper, a review of the role of the frontal lobes in mental disorders and a tribute to the scientific prescience of Moniz, his children tore through the nightclubs, beaches, and sailing lanes of Lisbon.

Freeman left the congress elated with its success. He and his family remained an additional week in Europe, and Freeman traveled to the German city of Freiburg. Two years earlier Freeman had noted that "the dark continent of Africa seems to be the only place where psychosurgery has not extended, omitting, of course, the dark continent that was Germany." Germany's physicians remained wary of lobotomy partly because of the international condemnation of Nazi medical experimentation before and during World War II. In Freiburg, Freeman encountered strong opposition to the procedure. Freiburg neuropsychiatrist Karl Kleist, who had examined the effects of brain injuries during both world wars, put up the fiercest resistance.

At the congress, Kleist "was emphatic when discussing lobotomy," Freeman remembered in summing up Kleist's position. "No patient with bilaterial injuries to the frontal lobes could ever earn a living; this was mutilating surgery that would change a functional mental disorder into an organic one for which there was no treatment, and Germans should avoid world criticism for experimental procedures." Freeman tried responding using "execrable German, that I had learned from a series of governesses," and made the point that just as a gangrenous leg demanded amputation, the same "should be done to the man who had a gangrenous Oedipus complex." Not surprisingly, Kleist was not moved by this weak and—uttered in flawed German—most likely incomprehensible attempt at humor. Freeman nevertheless demonstrated two transorbital lobotomies in Freiburg. Kleist came to watch the demonstration and did not hide his disapproval: "He shook his head in doubt and wonderment," and "I almost had to push him out of the way in order to perform the operation," Freeman recalled. But Freeman succeeded in winning over some physicians who were present, including "one of the young assistants who reported on the success of one of my lobotomy cases [and] lost his position." Freeman called the doctor's dismissal "guilt by association!"

By the standards of the 1947 Nuremberg Code of Medical Ethics created by an American military tribunal that judged the cases of Nazi doctors, Kleist stood on firm ground. The code held that experimental medical procedures—a category that unquestionably included transorbital lobotomy—had to be based on either animal trials or definitive knowledge of the illness, must be performed in facilities that could treat the subject in the event of complications or injury, and required the informed consent of the subjects, who must be competent to give such consent. Transorbital lobotomy, as practiced by Freeman, often failed to meet one or more of these conditions. A Soviet physician cited these ethical conflicts, as well as the transformation of the patient into "an intellectual invalid . . . an insane person changed into an idiot," when his nation banned psychosurgery soon afterward.

Freeman made it plain that he found such ethical complaints a waste of time. "While it is essential to have an assessment of the moral, ethical and legal aspects of such a therapeutic procedure," he wrote, "the prolongation of such discussions accomplishes little. Meanwhile, thousands of patients are added annually to the rolls of state hospitals. This, added to the personal distress, anguish and disability of the patient as well as the hardship to his family should stimulate us to look at the problem from the viewpoint of the state mental hospital." Freeman rarely replied to ethical concerns in any more detail. Ethics were his blind spot—he considered them, like rigorous antisepsis during surgery, annoying obstacles to treatment. He believed opponents brought them up when they could not comprehend the urgent distress of sick people or the wasted potential of debilitated psychiatric patients. He lumped ethical argumentation along with Freudian analysis as forms of intellectual dithering—activities that in no way benefited the lives of seriously ill patients.

Freeman traveled from Germany to France, where he performed that country's first transorbital lobotomy at a psychiatric hospital in suburban Paris; the patient was the daughter of the U.S. consul to Peking. The girl's mother "was more than pleased. . . . The doctors were very much impressed, and Dr. Tison, the superintendent, said he was going to have other patients operated on as soon as the surgeons got back to town," Freeman wrote his relatives. Moving on to England, he returned to Bristol for a transorbital lobotomy demonstration at the Burden Neurological Institute. "The British observers reacted to the transorbital technique with both horror and fascination," David Shutts wrote. "However, in Bristol, there was no guilt associated with radical surgery and no attempt made to physically prevent Freeman from operating." Freeman also took the opportunity to deliver a manuscript on the procedure to the editors of the influential British medical journal *The Lancet*. The paper was published two months later.

Dosed with Nembutal for the flight back to America, Freeman returned with his family to Washington in the middle of August. He would make many other trips abroad during the next several years. At the Fourth International Neurological Congress in Paris in 1949, he demonstrated a transorbital lobotomy in which he used the deep frontal cut and compared brain treatments for mental illness to the techniques being used to repair the early computing devices then in production. He had heard that "one could overload the circuits, equivalent to electroshock, and if the machine was still misbehaving, one bank of tubes could be removed, and this was like a leucotomy," he said. Around 1950 Freeman met Jean Delay, a Sorbonne professor and the director of the Institute of Psychology. Freeman had no idea that within a few years Delay would play an important role in the development of chlorpromazine, a tranquilizing drug that to most psychiatrists would make lobotomy obsolete as a psychiatric treatment. Freeman also traveled extensively in Latin America to meet with his former residents and fellows from GWU.

None of these trips would later resonate in Freeman's memory as much as the excursion to Lisbon in 1948. The accomplishments of the First International Congress on Psychosurgery and the honors that had fallen upon both Freeman and Moniz had the effect of drawing the two men closer together. Moniz felt deeply moved by the accolades he received, and he dedicated more than one hundred pages of his memoirs to an account of the congress. They had already exchanged inscribed photographic portraits. Soon Freeman would send letters to Moniz bearing the salutation, "Cher Maître," and would confess, "When I think of your magnificent contributions in the face of difficulties, I take pride in following in your footsteps." Freeman would see the completion of the ultimate tribute to his mentor, a campaign to award Moniz the world's highest honor in medicine.

Moniz had raised the issue two years earlier in a letter to Freeman dated February 4, 1946. In a paragraph headed "Confidential," he wrote, "I have hesitated to beg of you a special favour because I have no authority to do so. But as you made references in your [last] letter to 'false modesty,' I take courage to explain my wish to you." He went on, clearly embarrassed, to ask Freeman to nominate him for a Nobel Prize in medicine and physiology. "If this request is not reasonable, forget it, and I shall continue to be on good terms with you and this explanation will be as non existent," Moniz said in closing. "My audacity is very inconvenient, but this prize would be a kind and useful conclusion for my life."

There was no need for Moniz to fear he was making an impudent request. As early as 1943, Freeman had already nominated Moniz for the Nobel Prize, and Moniz's Portuguese colleagues had submitted nominations twice before that. Freeman promptly replied to the anxious Moniz, "I hope, indeed, that before many more years have passed the Nobel Committee will recognize the fundamental contribution that you have made. . . . When [psychosurgery] is adopted in the most conservative institutions you may be sure that it will be more generally appreciated." Moniz, in turn, appreciated that the nomination had come "without any indication from the candidate," meaning himself. "It is, coming from you, as valuable as the prize itself."

Freeman may have hoped that he would share the international recognition if Moniz won a Nobel Prize. Franklin remembered an argument between his parents over Freeman's long working hours and frequent road trips. "My father said something to the effect that [he was] going for a Nobel Prize. To which my mother said, 'Walter, you're too controversial.'"

The Nobel honors did come but not for Freeman. Just after the International Psychosurgery Conference in Lisbon in 1948, a group of Brazilian and Portuguese physicians made a convincing appeal to the Nobel selection committee on Moniz's behalf. In October 1949 the Nobel committee announced that Moniz had won the prize in medicine and physiology, along with Walter Rudolf Hess. The committee did not mention the development of cerebral angiography at all in citing Moniz's accomplishments but specifically lauded "his invention of a surgical treatment for mental illness" and "his elaboration of the psychophysiological concepts that made it possible." It was only the second time since 1901 that the Nobel committee had given the award to a physician who had developed a treatment for mental disease. (Julius Wagner-Jauregg had won the prize in 1927 for his work on the therapeutic use of malaria to cure neurosyphilis.)

A smashing blow against the critics of psychosurgery, the committee's recognition of Moniz helped launch a worldwide wave of lobotomies. In the month that Moniz received the award, Freeman estimated that five thousand psychosurgeries had taken place around the world in the previous fourteen years. But the pace at which American physicians performed psychosurgeries doubled in the last four months of 1949 over that of the first eight months of the year. Twenty thousand people in the United States underwent lobotomies in the four years after the Nobel announcement—and one-third of the total were transorbital operations. Psychosurgery became a form of treatment used at more than half the public psychiatric institutions in the United States.

If Freeman felt disappointed that the Nobel committee had excluded him, he did not show it. He continued his diligence in following up on the postoperative lives of his lobotomy patients. "The follow-up was my hobby, my compulsion, or my expiation, depending upon how one looked at lobotomy," he wrote. "This has, to my mind, been the most gratifying work of my life. To have been the means of relieving suffering and restoring a goodly number of patients to their homes has outweighed the disappointments and failures." He went to great lengths to locate patients who had relocated or vanished from sight. He racked up enormous long-distance telephone bills, contacted distant relatives of patients, and succeeded in tracking down patients who had resettled in Venezuela and Australia.

This great effort to ascertain the fate of his hundreds of lobotomy patients satisfied his wishes to refine the technique of psychosurgery, determine the operation's effectiveness, and bring each case in his portfolio of psychosurgical investigation closer to completion. Freeman also, however, wanted to use this large mass of follow-up data in a revision of the book on psychosurgery that he and Watts had published in 1942. "The first [edition] was based on a mere 80 cases, with a limited follow-up," he wrote in 1947. "Now we had experience with some 500 cases as well as several autopsies in which the retrograde degeneration of the thalamus was worked out." There were new psychosurgical procedures to detail, including transorbital lobotomy. In addition, doctors with greater frequency were using lobotomy to treat chronic pain, "and as I said to Dr. Watts on a recent occasion this is the [use] that will really make prefrontal lobotomy a respectable operation," Freeman wrote. "It presages wide acceptance by neurosurgeons who have hesitated to become involved with their psychiatric confreres." With these overly optimistic goals in mind, Freeman convened with Watts to discuss the new content of a revised book in May 1947. Freeman had already prepared for the start of a revision by reading every lobotomy study he could find and corresponding with psychosurgeons around the world.

With Freeman and Watts each working on separate chapters, the first draft of the book was completed the following summer. Freeman dogged his partner mercilessly to finish his assigned sections. One weekend in August, he trailed Watts to the Farmington Country Club near Charlottesville, Virginia, where Watts had gone with his family on vacation. "I was impatient of delay since I could foresee the opening of the school year with a full schedule," Freeman remembered. "On walking into the dining room I got a black look from his wife, Julia." She clearly resented his presence, but Freeman was determined to hasten Watts's slow pace on his work. The next day Freeman brought a pile of papers out to the veranda, set them on a table, and put Watts to work. "I was full of the subject and pressed Jim for his part. We worked all morning," Freeman wrote. "Sometimes he dictated while I wrote and sometimes vice versa. When his hand tired I started reading to him one of the new chapters. Presently his head drooped and he was off in a doze. I was proud of the phrasing so I was irritated and must have showed it. It was a sultry Sunday afternoon so we knocked off work and had a swim before I started back along the Skyline Drive, keyed up to the point where I sang. We were accomplishing things." Hard work nearly always cheered Walter Freeman.

This work came a year and a half after Freeman's initial experimentation with ice picks through the orbital cavity, and somehow their book had to reflect their differing opinions on the procedure. "The authors regret to announce that they have been unable to reach an agreement on the subject of transorbital lobotomy," they wrote in the book. "Freeman believes that he has proved the method to be simple, quick, effective and safe to entrust to the psychiatrist. Watts believes that any procedure involving cutting of brain tissue is a major operation and should remain in the hands of the neurological surgeon." Then followed essays by each man detailing his position.

One of their lobotomy patients completed the typing of the manuscript, and Freeman and Watts met the deadline set by publisher Charles Thomas, who had accepted $5,000 from the authors to produce the book. In selecting photographs, Freeman clashed with his publisher over an image he had captured in 1947, possibly at St. Elizabeths Hospital, showing a tormented, nude patient struggling against attendants before her radical lobotomy. He captioned the photo "Other Patients Came in Chains." While Freeman believed the photo illustrated the agony of hospitalized mental patients, Thomas thought it exemplified sensationalism and bad taste. Freeman lost this battle and replaced the photo with another he had taken.

Thomas also had the unenviable job of dealing with Freeman as he assembled the volume's bibliography. Freeman explained to Thomas his obsession with accuracy in his listed sources by relating that "my reputation in the Republic of Argentina, among the neuropsychiatrists and neurosurgeons there, is rather on the low side because of a glaring error that I perpetrated some 15 years ago. They still haven't gotten over it. It was almost as though I had insulted an Irishman by calling him a Greek, just because he happened to live next door." The publisher's revenge came in long delays during the proofing, printing, and binding process. When the book at last appeared in print in 1950—with the pronouncement in the new preface that "psychosurgery has come of age"—three other new volumes on psychosurgery had preceded it.

Like the first edition of their book, this new volume—retitled *Psychosurgery in the Treatment of Mental Disorders and Intractable Pain*— has a literary sensibility that separated it from its drab competitors on the medical bookshelf. Freeman and Watts mixed lessons in brain anatomy and descriptions of psychosurgical technique with detailed case histories of their lobotomy patients. They used vivid language throughout—in one instance referring to a group of patients as "scrawny 'fraidcats." The case histories began before treatment and often ended years after the operations. From the stories of these patients, the authors drew conclusions about why their lobotomies succeeded or failed. Freeman and Watts reported that of the 711 lobotomies they had together performed when the book was written, 45 percent yielded good results, 33 percent produced fair results, and 19 percent left the patient unimproved or worse off. (They also presented evidence that African American women and Jewish patients benefited in higher percentages, the latter group probably because of "the greater family solidarity manifested by these people.") Three percent of the patients died directly or indirectly from their lobotomies.

The book also included the results of psychological studies on lobotomy patients conducted by Mary Frances Robinson, whom Freeman and Watts hired for a year with $5,000 of their own money when all of their applications for foundation support failed. Robinson had previously tested lobotomy patients in Missouri. "She is enthusiastic, says she wonders when she will wake up and find it's all a dream," Freeman wrote soon after she arrived. "When we had a number of patients in for a follow up study she was like a little girl in a toy shop, running from one end to the other and not knowing which one to study first." Robinson concluded that although psychosurgery did not significantly affect the intelligence of patients, as measured by testing, it did limit their abstract thinking and attention span, curbed their self-awareness, and harmed their ability to plan for the future.

As with the first edition, the narrative emphasis of the book alienated many medical publications. Some delayed their reviews or declined to review the volume at all. "It is rather disappointing to know that the book has been out for a year and still there have been no reviews from the *American Journal of Psychiatry* and *The Archives of Neurology and Psychiatry*," Freeman grumbled. The reviews that did appear frequently mentioned the authors' distinctive style of presentation. "The nonsurgeon will perhaps be impressed by the theatrical description. . . . The characteristic Freeman drama is quite noticeable in discussion and description," wrote David Cleveland, an opponent of transorbital lobotomy, in *Surgery, Gynecology and Obstetrics*. "The surgeon will shudder, and rightly so, at the thought of cerebral surgery becoming an office procedure in the hands of the usually, very unsurgical psychiatrist."

The *Journal of Abnormal and Social Psychology* characterized the book as "far from being a scholarly production," and a reviewer for the *British Medical Journal* concluded that the volume was "neither a textbook on leucotomy nor a source of material for the research worker." Not all the reviews were dismissive. The perhaps biased *Medical Annals of the District of Columbia* noted, "It is indeed a pleasure to read a medical book where the dose of one chapter is not equivalent to three grains of Seconal. The vivid descriptions and personal expressions of the authors in their text should be studied by all who anticipate writing a medical book that they expect to have read."

Nevertheless, the completion of the book—which remained in print until 1956 and sold 1,728 of 3,000 copies published—gave Freeman an exhilarating feeling of accomplishment. "As I approach my 52nd birthday I have a feeling of competence and assurance that is almost grandiose," he acknowledged. "Maybe it comes from superb health, and maybe from the fruition of dreams that have proved within my grasp. But anyhow, I'm sitting on top of the world." He bought a handsome green Lincoln four-door automobile and cultivated the appearance and bearing of a distinguished scientist of the mind. People who met him frequently commented on his piercing gaze. "I don't like to talk to Dr. Freeman, because he knows what I'm thinking," one observed. As he aged, his external features, subtly wielded pipe, and other accessories confirmed to many that he was a brilliant physician. While Watts, who had a lesser need for attention and acclaim, was often ignored by the public, Freeman "was internationally known, a VIP and, in some circles, one might say he was a celebrity," Watts remembered.

Freeman settled into a pleased acceptance of his growing fame. During the mid-1940s he commissioned the Belgian artist Alfred Jonniaux to paint his portrait. Jonniaux had deserted his native country for Washington, San Francisco, and Latin America, where politicians, military leaders, and celebrities hired him to paint their likeness. In Washington he worked from a studio in the basement of the Corcoran Gallery. "A slender rather intense man with short gray hair (what was left of it) and a graceful manner, he started out on his mission by a visit to our home to look over my wardrobe," Freeman wrote. "He decided that a sport jacket and slacks would best express my personality." It was a doubtful judgment, but Freeman agreed with it and posed in six sittings for Jonniaux at the Corcoran. The two negotiated over the placement of Freeman's pipe in the picture and the treatment of his eyes. "Those eyes, they are always the most difficult," Jonniaux told him. "I want them kind but not soft."

As the portrait took shape, Freeman grew increasingly delighted. "I don't know what better investment I could make at the present time in terms of personal satisfaction and gratification of the ego," he told his daughter, Lorne. After its completion, he observed that "the portrait shows me at the peak of my professional career." The painting was subsequently exhibited in the foyer of the Natural History Building of the U.S. National Museum, along with several dozen other works by Jonniaux. In a clipping of a review of the exhibition from the *Washington Star*, Freeman underlined two phrases: "all of the present group are attractive" and "his masculine sitters are all 'men of distinction.'"

Although many people saw the painting on exhibit, millions more gained impressions of Freeman through articles about him and his techniques of psychosurgery in the popular press. From the end of World War II through the early 1950s, these magazine and newspaper stories appeared frequently, sometimes twice in a single week. Psychiatric medicine was a hot topic fueled by movies and novels that included Alfred Hitchcock's *Spellbound* and Mary Jane Ware's *The Snake Pit*. Although Freeman acknowledged that the articles were exaggerated and often lurid—one article in the *Boston Traveller* included the statement, "Two prominent Washington physicians are convinced some people think better with less brain"—he boasted to his son Paul that they were "certainly no more spectacular than the facts."

If he was quoted accurately, Freeman seems to have frequently lost control of his tongue during interviews with the press. In 1948 a reporter for the *Hartford Times* attended a Freeman lecture at the Institute of Living and came away with this paraphrased quote: "If you are clumsy with your hands, can't get along with people, have no imagination and no spiritual awareness, chances are that the front portion of your brain is either underdeveloped or missing." In addition to fan letters and requests for signed photographs, these stories usually generated an avalanche of letters and phone calls from people who wanted lobotomies for themselves or for relatives. Freeman recalled that the first such request they fielded during the late 1930s "was a man who complained of asthma and wanted to know if a brain operation would relieve that."

Freeman and Watts culled many patients from these pleas for help, although Freeman noted in 1946 that "when the fan mail is all sorted out the pickings are a bit thin." One batch of correspondence that year included a plea for surgery from a woman obsessed by guilt from her placement of a sheet of flypaper that was full of insects into an open fire. "I might suggest that she wanted to set fire to her children, but I'll restrain myself," Freeman noted. There was also a request from a man who considered lobotomy a way to make his racing greyhound less excitable on the track. "However, if we waited for psychiatrists to send patients to us we'd still be on our first hundred cases instead of our fifth hundred," Freeman observed.

Many of Freeman's medical colleagues condemned his willingness to promote lobotomy in the popular press. "In recent years much has been written about different surgical approaches to the treatment of insanity," wrote neurosurgeon James Poppen, whose competing method of psychosurgery Freeman had insultingly dismissed. "I am certain that there will be more to follow. I do hope that in the future we will not be informed initially through the weekly popular magazines. Any procedure which is instituted for such a serious condition should be thoroughly tried and proved to a certain degree before it is advised. Premature information through weekly magazines (not always accurate) has a tendency to give patients or relatives false hopes or impressions."

Freeman's garrulous personality and his refusal to take criticism personally probably also contributed to the antagonism of Poppen and other vocal critics. In the spring of 1948 at a cocktail party during the annual meeting of the American Psychiatric Association in Washington, Freeman approached psychoanalytically oriented opponent Henry Stack Sullivan with the greeting, "How goes it, Harry?" Sullivan's response shocked Freeman. "He raised his fists overhead, contorted his face, thoroughly enraged, saying, 'Why do you *persist* in annoying me?'" Another psychiatrist who overheard the exchange led Sullivan away.

Articles about Freeman appeared in such nationally circulated magazines as *Life*, *Time*, and *Newsweek*. But the magazine story that most gripped his attention was one written for the *Saturday Evening Post* in 1951 by Irving Wallace, who would later gain fame as the author of *The Man*, *The Seven Minutes*, *The Word*, and other best-selling novels. In the article, titled "The Operation of Last Resort," Wallace deviated from Walter Kaempffert's adulatory approach in the same magazine a decade before. Wallace chronicled in lengthy detail the poor outcome of the lobotomy of a brilliant Freeman–Watts patient named Herbert Kaufman who emerged from surgery a more contented but undeniably less intellectually capable man. (The *Post* had retitled the story, which Wallace had originally headlined "They Cut Away His Conscience.")

When Wallace submitted the article to the physicians for their approval before publication, Freeman's response was volcanic. Although he congratulated Wallace on his craftsmanship as a writer, Freeman declared that the article was too unfavorable and liable to open him to censure from colleagues in the medical profession. Wallace had characterized, for instance, the view of lobotomy opponents in this way: "On the other hand, there is the school of thought that can prove, also with facts, that prefrontal lobotomy converts patients into docile, inert, often useless drones, stripping them of their old powers, giving them convulsive seizures, making them indifferent to social amenities, filling them with aggressive misbehavior and impairing their foresight and insight. There are those who feel the operation tampers with God's substance, who feel that if it cuts out a man's cares, it also cuts out his soul and his conscience."

"I am not overly sensitive to censure, but I am not going out of my way to earn it," Freeman wrote. "Therefore, I must decline to assist you in the correction of a number of errors unless you accept also the elimination of the personal references to myself and Doctor Watts in the body of the paper." He concluded his letter by reaching back into his childhood to paraphrase some lines from "If," a poem by Rudyard Kipling: "If you can bear to hear the truth you've spoken / Twisted by knaves to make a trap for fools." Two months later Watts turned down the *Saturday Evening Post*'s request for photos of a lobotomy operation and repeated Freeman's demand that their names be removed from Wallace's article.

Writing about the experience years later, Wallace clearly felt he had come out on top. After the article's publication and high praise from readers, Freeman "did a complete about-face. Unashamedly, he wrote his knave asking for assistance in placing for publication a popular medical article that he had written." Not inclined to show charity to the doctor who had insulted him, Wallace recorded that he "dropped his literary request into the wastebasket and did not reply to him."

Despite its complications, fame gratified Freeman, but it was not enough to make him professionally complacent. Keen's burial in Merced had loosened his roots in Washington. As early as 1948, sick of the sticky summers and mindful of the diminished role of neurological study at GWU, he began considering relocating. He had started to fill out his application for a medical license in California when members of the Medical Society of the District of Columbia asked him to serve a term as president. Around the same time, GWU's announcement that it would erect a new hospital had compelled Freeman to play a role in the planning of its neurology and neurosurgery facilities. The passing of those obligations freed Freeman to take his work to the road and to reconsider moving. No one could imagine how great a temptress he would find life outside Washington.

## Chapter 11

# Road Warrior

Since the early 1940s the masses of mentally ill patients in the nation's government-run psychiatric hospitals had drawn Freeman with a strong but unfulfilled longing. On extremely limited budgets, these hospitals offered few treatments to the hundreds of thousands of inmates they housed. Freeman viewed the efforts of Paul Schrader in Missouri and J. Grafton Love in Minnesota to perform prefrontal lobotomy on large numbers of hospital patients as a hopeful sign but not as a strategy that many other institutions lacking available or willing neurosurgeons could copy. Many hospital administrators who were unable to develop lobotomy programs had eagerly embraced the new shock treatments using Metrazol, insulin, and electricity that appeared in the 1930s. But within a few years, only electroshock had retained any widespread acceptance or held much hope as a therapy that could benefit a large number of patients. This should not have been surprising: even today, most new strategies for combating illnesses ultimately fail.

In the mental hospitals, though, the news of the difficulty in successfully applying the shock therapies greatly troubled staff members who had little else to offer to the desperately sick people in their charge. Medical directors in these overcrowded facilities lost the optimism that the shock treatments had introduced. To make matters worse, patients diagnosed with schizophrenia, then a singularly untreatable illness, were filling the wards at a fast pace—especially in the United States, where the disorder was identified far more often than in Britain and other countries. Training and cultural perspective accounted for some of the difference, but the incidence of schizophrenia was indisputably on the rise. Increasing numbers of people, mostly young, were discovered to be suffering from the telltale symptoms: inability to form healthy human relationships, hallucinations and delusions, disorganized thoughts, and poor adaptability to normal stresses. State-operated psychiatric hospitals bore the burden of admitting and attempting to treat these multiplying cases.

The transorbital lobotomy treatment that Freeman held out to hospital superintendents was not just a possible method of restoring sick patients to full citizenship but a way to disarm the potentially disastrous institutional time bombs of overcrowding, poor morale, medical stagnation, and political invisibility. "To the hardpressed hospital superintendents, an offer by Freeman to visit and demonstrate transorbital lobotomy seemed a godsend," wrote Elliot Valenstein. "In addition to the possible benefits to patients, a visit by a man of his prominence was a welcome relief from the dreary routine of most state hospitals. Furthermore, an enterprising superintendent could publicize Freeman's visit and possibly parlay it into increased appropriations for his hospital."

From his years at St. Elizabeths Hospital and his familiarity with psychiatric hospitals and hospital administrators around the world, Freeman understood the seriousness of the despondency that hung over the wards of state-run hospitals. He had seen mentally ill people confined for three, four, and five decades, and had once photographed John Washington, a ninety-year-old patient at St. Elizabeths who had been hospitalized for seventy-one years, as well as Otto Poole, a patient at the state hospital in Cambridge, Maryland who had not left a mental institution for the previous seventy-six of his one hundred three years. "Suppose it should happen to one of my children, or one of their friends!" he wrote.

In a sense, these old-timers were the lucky ones. "The extinction curve of schizophrenic patients, that is, of patients dying in the hospital, crosses the halfway line at 28 years," Freeman declared. He believed that lobotomy provided patients and hospital staffs an escape from a spiral of hopelessness, and eventually he convinced scores of people responsible for the administration of mental hospitals that he was right, or at least on the right track. He admitted that through the mid-1940s, prefrontal lobotomy could claim only a mixed rate of success—a rate that dropped significantly for patients with schizophrenia. Given the spotty early success that transorbital lobotomy had achieved in making schizophrenic people well enough to live outside of hospitals and take on work or family responsibilities, Freeman could have abandoned it and devoted his staggering energy to the investigation of other promising treatments. Predictably, Freeman and many other psychosurgeons refused to follow this course.

"To have abandoned the enterprise at this stage, however, would have been for them an abdication of their responsibilities as medical scientists," wrote the medical historian Jack Pressman. "Indeed, it is precisely those treatments that yield satisfying results only some of the time which afford the best opportunities for further research; this is what science often does best—improve upon its mistakes."

Freeman annonced his intention to intensify his efforts to bring lobotomy into psychiatric treatment programs by declaring, "Transorbital lobotomy is a simple, effective method of treatment. It offers hope of returning a relatively high percentage of 'incurable' psychotics to their communities." And by adding that "it seems not only possible, but obligatory, to extend the program of psychosurgery into the state mental hospitals in an effort to relieve misery," he indicated that he would shoulder the task personally, one hospital at a time, although he readily acknowledged that transorbital lobotomy had no lasting value if only he could obtain satisfactory results. Freeman saw himself as a man of action, someone who could make things happen. "The definition of genius that I like best is the ability to put the cart before the horse—and make them both run," he later wrote.

Freeman's first attempt to operate transorbitally in the setting of a state psychiatric hospital came in the summer of 1946, less than eight months after his first ice pick surgeries. "I may have some opportunity to carry it on a large scale this summer in the South Dakota State Hospital," he wrote to his son Franklin in May 1946, "since the superintendent is definitely interested and may be able to swing it." Frank Haas, the director of the hospital in Yankton, did manage to arrange it after he and Freeman had met and discussed transorbital lobotomy at an American Psychiatric Association meeting in Chicago. Freeman considered Yankton an especially fitting institution in which to test the new procedure. "What I wanted was a large group of cases to work on that would under no conditions have the services of a surgeon available and I thought to find them in an isolated location like South Dakota," he wrote.

Freeman hoped that Haas would provide a dozen hospital inmates for whom electroshock therapy had not brought relief of recently developed psychotic symptoms. When Freeman arrived in Yankton on June 16—he had driven there on a detour during the cross-country road trip that would later result in Keen's death—he was dismayed at the condition of the patients awaiting him. "Dr. Haas was a bit timid I guess and submitted only a [group] of old crocks that he had monkeyed around with for ten years or so without getting any place," he reported. They were long-term, chronic, and wrecked schizophrenics. Nonetheless, Freeman went ahead with his plans to operate on some of them. He "did five transorbitals in about an hour," he wrote to Watts later that month. "I turned down five other patients on the basis of their long-continued psychosis, but I have heard since then that the five [lobotomized patients] are getting along all right, four of them improved already. It remains to be seen how much and how long." In a concession to Watts's concerns over proper surgical procedure, Freeman made a point of mentioning that he had performed the lobotomies in an operating room.

On the return trip from California, still dazed from Keen's drowning, he stopped once again at Yankton for a day and a half to examine the recovering patients and to supervise operations on three new ones, with Haas's assistant performing the actual lobotomies. "The patients I worked on the way out had had no bad effects although I was not particularly pleased with the results," he wrote. "The cases were [too] far gone I think to offer much chance of improvement, but it was a useful demonstration and I think I have started off these folks on a program of effort." But Haas did not permit further operations in his hospital after the three patients in the second round of surgery failed to improve. "Frustrated again," Freeman wrote of the episode a year later. He subsequently confessed, "I felt many a time as though I were out on a big game hunt armed with a pea shooter when I saw these tempestuous catatonics who had been in a disturbed condition for years and who were supposed to be relieved by a ten minute operation."

A frustrated Freeman, however, was not the same as a discouraged Freeman. He tapped an old friendship with William N. Keller, the superintendent of Western State Hospital in Fort Steilacoom, Washington, to obtain an invitation to perform transorbital lobotomies in another remote hospital setting. Back in 1941 Freeman had been Keller's guest for a two-week-long stay in which he taught at a postgraduate seminar and lectured on prefrontal lobotomy. This time, Freeman arrived with an attitude reshaped by his disappointment in South Dakota. "Whoever undertakes to perform psychosurgery in a mental hospital will have to make the choice of operating upon chronic schizophrenics or not operating at all," he wrote. Keller provided several cadavers for demonstrations and "lined up some 75 cases on which he had gained permission, and I selected 13 after hearing the history and interviewing the patients." Most of the selected patients were schizophrenics, but overall the duration of their illness was not as long as the patients in South Dakota. "Then on the next day, Aug. 19, [1947,] we got to work." During an intense three hours, Freeman performed the first couple of operations, then handed the leucotome to hospital staff members and supervised the remainder of the surgeries. The Western State doctors evidently felt nervous about undertaking this new treatment. "One of the men had such shaky hands that I was afraid he would touch the skin of the face before he got the point of the instrument into the conjunctival sac (which is normally sterile) but he managed finally," Freeman observed.

Keller, like many mental hospital superintendents of his generation, was not a psychiatrist. But he knew that public attention could increase the level of care his hospital provided. Hoping to generate publicity and more funding, Keller invited several newspaper reporters to cover Freeman's visit. Calling the surgeries "super-orbital" lobotomies, a writer for the Seattle *Post Intelligencer* noted that Freeman had donated his services and reported Keller's prediction that "at least 10 of the 13 should show definite signs of improvement." The publicity turned Freeman into a medical celebrity. When Freeman flew to San Francisco the evening after the operations, an Associated Press reporter awaited him at the airport and requested an interview. The following year, Freeman reviewed the progress of the Western State patients. He found that "six of them were usefully occupied outside of the hospital."

When Freeman returned to supervise five more transorbital lobotomies at Western State in October 1948, Keller again notified reporters. "Dr. Freeman . . . explained to a staff meeting that he would accept only patients 'facing disability or suicide,'" the *Post Intelligencer* noted, which was an apparent contradiction of Freeman's repeated declarations that the operation should no longer be regarded as a treatment of last resort. Several relatives of patients seeking treatment had to be turned away. In some patients he applied his newly developed "deep frontal cut," which severed neural pathways further back in the frontal lobes. "In ordinary language, Dr. Freeman described his operation as severing nerves which deliver emotional power to ideas. He said that the operation resulted in a reduction of the imaginative powers of the patient, but he added: 'That is what we want to do. They are sick in their imagination,'" the newspaper reported. Published photos showed Freeman talking with patients before their lobotomies, operating, and sitting on the bed of a recovering patient. "Less than an hour after being operated on each patient was able to sit up in bed, behave rationally and report feeling 'fine.' Dr. Freeman predicted that for three out of the five the lobotomies would mean a return to normal living," the *Post Intelligencer* said. Eventually, in fact, Western State doctors reported that seventeen of their first forty-one patients lobotomized by hospital staff using the transorbital method were released.

A disturbing scene in the film *Frances*, a 1982 biopic about the life of the rebellious movie and stage actress Frances Farmer, shows a balding and goateed psychiatrist who closely resembles Freeman performing an ice pick lobotomy at Western State Hospital on the supine heroine. The movie relied on an account published in 1978 by William Arnold, a film critic for a Seattle newspaper, in his book *Shadowland*. Arnold maintained that his information on Farmer's lobotomy came from hospital nurses who recalled the operation when interviewed decades later. Interviewed by the psychosurgery historian David Shutts around the time that Arnold's book came out, Freeman's son Franklin declared his conviction that Farmer was once his father's patient. "It's her all right," he said, indicating a woman in a photograph taken during the transorbital lobotomy series at Western State. The photo shows a crowd of hospital staff and administrators watching as Freeman, his hairy arms bare to the shoulder, operates. The face of the patient, although identifiable as a woman's, is obscured.

Twenty years later, Franklin Freeman felt less certain and recalled only that he had heard the identification of Farmer in the photo secondhand. There is no evidence in Freeman's voluminous writings about his former patients that he ever met Farmer. Nor is there evidence that Farmer ever had a lobotomy. In her family memoirs, Edith Farmer Elliot, the actress's sister, wrote that doctors only contemplated giving Farmer a lobotomy. "I got there just in time to head them off from some danged experimental brain operation on her," their father wrote to Elliot. "I notified them one and all if they tried any of their guinea pig operations on her they would have a danged big law suit on their hands. The head man answered my letter agreeing that nothing would happen without my written consent." In their account of their lobotomy program published in *Northwest Medicine*, Western State physicians described nearly fifty of their psychosurgery patients. Only one matched Farmer's gender and age, and this patient was discharged at least two years before Farmer's release in 1950.

Given Farmer's personal accomplishments after her release from Western State—marrying, regularly hosting a television program in

Indianapolis, and appearing on *This Is Your Life*—Freeman would probably have mentioned her with pride had she been his patient. As it turned out, Arnold's book and the film it inspired turned Farmer into a well-known symbol of the excesses of psychosurgery: a patient supposedly selected for her nonconformist political opinions who was operated on with only the consent of a vindictive mother, and with her soul and spirit vanquished after surgery. That image, reinforced by the portrayal of lobotomy in Ken Kesey's *One Flew Over the Cuckoo's Nest* (also adapted into a widely seen film) and other literary works, still persists.

Freeman followed his return to Western State with his surgical debut at Herrick Memorial Hospital in Berkeley, a general hospital that would prominently figure in his career for years to come. He came to Berkeley at the invitation of the neurologist Abram Bennett, the exploiter of curare's use as an anticonvulsant agent during electroshock, who had moved to California from Nebraska. Because of the general hospital setting, Freeman substituted intravenous anesthesia and sedatives for his customary use of electroshock to immobilize his patients. An audience of fifty physicians witnessed his operation on two female patients at Herrick in November 1948, an event also covered by the local press. "An extremely revolutionary, spectacular and delicate brain operation was successfully performed on two mentally ill Berkeley women," wrote one reporter, who called Freeman "a distinguished American brain surgeon." The newspaper noted that Herrick officials reported "an 'excellent recovery' in the case of one woman and 'marked improvement' in the other."

Returning to Herrick four years later, Freeman provided a startling display of his chutzpah and resourcefulness in emergencies. Operating this time before a smaller group that included sons Paul and Franklin, he performed transorbital lobotomies on two women. One patient emerged from the operation without complications, but the other was paralyzed on her right side. "We got trouble," Freeman announced before proceeding to a waiting room, where he told the patient's husband that it would cost an additional $1,000 for him to perform a procedure to alleviate the hemorrhaging on the left side of the brain, which he suspected to be the cause of the paralysis. After obtaining the husband's consent, Freeman strode into the operating room and removed from his bag an instrument resembling a bicycle pump; it was connected to a metal tube. He filled the bulb of the pump with saline solution and placed the tube into the opening left by the transorbital leucotome in the left eye orbit.

Freeman's energetic pumping flushed the site of the lobotomy with water, and "then he extracted a dark, bloody mess, which he squirted into a basin," reported the lobotomy historian David Shutts. "Then, filling the bulb again, Freeman repeated the grisly procedure over and over, all the time keeping up a running patter with the nurses and doctors, who were unaccustomed to the procedure. At one point, he ordered an alarmed nurse to inject the patient with vitamin K to speed up the process of coagulation. In the meantime, Paul Freeman periodically ran a car key up the sole of the patient's [right] foot to test for a reflex response. Nothing happened. Freeman continued to pump the saline solution in and out of her brain until the liquid in the discharge basin changed from dark red to light pink. After a seemingly interminable period of time, the key finally elicited a response, which became stronger and stronger. Eventually, the patient recovered fully."

Freeman had found a role he savored. Mental hospitals around the country wanted him and increasingly sought him out. Operating on their patients, he could display his skills, pass his knowledge to others, and salvage people trapped in the worst and most hopeless medical facilities in America. He could act as innovator, teacher, savior, and commonsense man of action. He relished it so much that he once refused to let even a fractured arm keep him from demonstrating transorbital lobotomy. The long stretches of driving soothed his senses and calmed his worries between operations. Friends sometimes asked him whether his transorbital lobotomy trips qualified as vacations. "I tell them by all means yes, but that it was also a field trip and that I left a string of black eyes all the way from Washington to Seattle," he wrote.

"Freeman thrived on the 'horror and fascination' that accompanied such demonstrations," wrote Jack Pressman, "especially the publicity and notoriety they engendered. His concern was not neatness but impact." Freeman's travels kept him removed from his troubled marriage and his declining prestige in his medical school. He merrily called these excursions "head-hunting expeditions," a characteristic wordplay on the imagery of travelogue, cannibals, and "headshrinking" psychiatrists. (His chapter in his unpublished memoirs about this phase of his life, "Head-and-Shoulder Hunting in the Americas," seems to add the name of a popular dandruff shampoo to the mix of insinuations.) Over several years Freeman covered the Mid-Atlantic States, Appalachia, the Midwest and Great Plains, the South, and the Pacific Coast—one of the strangest series of journeys in medical history, a set of voyages into the darkest pits of psychiatric despair. Eventually, he left his mark on more than fifty-five hospitals in twenty-three states. Even when patients in these hospitals did not recover from their illnesses as a result of their lobotomies, Freeman believed the institutions still benefited. "The noise level of the ward went down, 'incidents' were fewer, cooperation improved, and the ward could be brightened when curtains and flowerpots were no longer in danger of being used as weapons," he observed.

The first of Freeman's major tours of state hospitals began in the nation's midsection in February 1949. With him he carried a typewriter, dictation equipment, a camera and floodlights, record cards, and his lobotomy tool case, which fit in his pocket. He first drove to Texas, where he demonstrated the transorbital technique at institutions in Galveston and Rusk. These visits bore immediate fruit. During the next eight months, physicians at the two Texas hospitals used Freeman's training to perform more than five hundred transorbital lobotomies. These doctors were the first of many around the country who nurtured the lobotomy seeds that Freeman scattered. Many other institutions, including New Jersey State Hospital in Trenton and Wernersville State Hospital in Pennsylvania, launched independent transorbital programs inspired by Freeman's example. "Although transorbital lobotomies comprised 30 percent of the psychosurgery performed during the peak years, in state hospitals they were more than half the total," wrote Elliot Valenstein. "Virtually all of the transorbital lobotomies in the United States can be traced to the influence of one man: Walter Freeman."

Freeman made his way from Texas to institutions in Little Rock, Arkansas; Lincoln, Nebraska; Rochester and Hastings, Minnesota; and Columbus, Ohio. It was in Hastings State Hospital, a half-century-old facility set on four hundred acres of landscaped campus and institutional farmland, that a young neurosurgeon named Lyle French observed Freeman operating on several patients. French, who performed more traditional psychosurgeries at the University of Minnesota, did not like Freeman's transorbital procedure. "I didn't think much of it," he said. "It seemed like it was not an aseptic and sterile technique, and you couldn't really determine what you were doing. . . . It seemed to me that it wasn't a very scientific approach."

Other observers had reacted with similar distaste to Freeman's transorbital lobotomy demonstrations. In 1948 Patricia Derian, a student nurse at the University of Virginia in Charlottesville, watched Freeman perform a transorbital lobotomy at a nearby state hospital. Freeman selected the patients for operation, she reported, by twisting their joints to determine their flexibility, not by reading or taking histories. After "a special lunch in honor of the occasion" of his visit, he occupied a conference room and had each patient shocked and photographed. "When all was ready, he would plunge [the leucotome] in," Derian noted. He wore no gown, mask, or gloves. "Afterwards he would sit the patients up and have them walked out of the room. He was very proud of the fact that the people walked in and walked out. None had to be carried, though one or two of them sagged badly on the way out," she remembered. After several operations, Freeman enlivened the demonstration by cutting nerve fibers on both sides of the brain simultaneously. "Then he looked up at us, smiling. I thought I was seeing a circus act. He moved both hands back and forth in unison, cutting the brain identically behind each eye. It astonished me that he was so gay, so high, so 'up.'" Derian recalled the sequence of events as a living nightmare, a deeply disturbing performance.

In 1950 Freeman made only scattered trips through the East and the South. A minor stroke that summer slowed him down, but even while hospitalized at George Washington University Hospital, he made patient rounds in his bathrobe. Still he made two stops at Milledgeville, Georgia, home of one of the nation's biggest state mental hospitals. (One of the Milledgeville staff psychiatrists who learned the transorbital technique later made the notorious quip, "You can change your mind, but not like I can change it.") The following year saw one of his most ambitious travel itineraries. Starting in June and traveling in a station wagon with sons Walter and Randy, Freeman set up shop in Arkansas, Texas, California, Washington State, South Dakota, Nebraska, Missouri, and Iowa. At Cherokee State Hospital in Iowa, three out of twenty-five patients he treated died. Freeman inadvertently caused one of those deaths when he stopped to photograph the position of the leucotome; the instrument sank deep into the patient's midbrain.

Freeman covered eleven thousand miles during the summer of 1951, including trips to demonstrate transorbital lobotomy in San Juan, Puerto Rico, and in Willemstad on the Caribbean island of Curaçao. The miles traveled could have been greater had a group of patients been made available to him around this time at Tuskegee Veterans Hospital in Alabama, a segregated facility for African Americans. (Several years earlier the hospital had notoriously participated in an experiment that deprived African

American men of syphilis treatment.) The English neurosurgeon William Sargant had recommended Freeman and his transorbital procedure to a psychiatrist working at Tuskegee. "Why not operate on fifty of the chronically agitated patients, whose personalities had not yet deteriorated too far, and compare the results with a group of fifty untreated patients who, if the experiment proved successful, would undergo the operation later on," Sargant argued.

Because no qualified Alabama physicians could be found who would agree to operate on African American patients, Freeman volunteered to perform the operations without charge. Abruptly and at the request of one of its neurosurgical consultants, however, the Veterans Administration banned the use of transorbital lobotomy at Tuskegee and, in Sargant's words, the "whole Negro-rescue plan had to be cancelled." Writing in 1967, Sargant was sure that the experiment had value and that "a few at least of these patients are still enduring mental hell, more than fifteen years later, in the back wards of Tuskegee Hospital, and still receiving no treatment beyond tranquillizers to muffle their groans for help."

The year 1952 marked more ambitious lobotomy trips for Freeman whose significance far exceeded simply the miles covered. That year, at the invitation of state officials, Freeman inaugurated large transorbital lobotomy programs in the state hospitals of Virginia and West Virginia. In Virginia he operated on 353 patients, the heavy majority of them schizophrenics, during the next several years at hospitals in Staunton, Williamsburg, Roanoke, Marion, and Petersburg. In these institutions, Freeman declared, 10 percent "of patients per year are candidates for lobotomy." One ugly incident marred his Virginia campaign: rushing to catch an airline flight, he left a psychiatrist whom he had just trained in transorbital lobotomy and who was in the middle of an operation. After Freeman's departure, the psychiatrist broke the metal tip of a leucotome in the patient's brain. The patient recovered after a surgeon removed the fragment, but a state board later investigated the accident. Freeman argued before the board that such accidents were not uncommon—one such mishap had left a Maryland patient blind in one eye—and he reviewed the benefits of the lobotomy program for Virginia patients. The board felt compelled to take some action, however, and mandated that only physicians experienced in psychosurgery could perform lobotomies in Virginia state hospitals.

In 1960 Freeman tracked down 316 of his Virginia patients. Sixty percent remained hospitalized and the rest had been paroled to home, with

23 percent of the total working or keeping house. The numbers were far better for the subset of Freeman's Virginia lobotomy patients suffering from depression and other affective disorders, 55 percent of whom lived at home, in most cases with employment. Freeman took pride in his Virginia successes. In a 1961 letter to the superintendent of Southwestern State Hospital in Marion, he specifically mentioned one schizophrenic man, "who, after 15 years of hospitalization and operative hemorrhage, was reported five years later as managing his own business," and another, "who, after 13 years of hospitalization, was reported as employed steadily" years later. To the superintendent of Eastern State Hospital in Williamsburg, Freeman similarly highlighted the achievements of a patient who was farming his own land in Mecklenburg County after transorbital lobotomy had relieved his symptoms of schizophrenia.

Edward F. Reaser, a hospital superintendent and a former colleague of Freeman at St. Elizabeths Hospital, helped bring Freeman and his transorbital leucotome in 1948 to the impoverished state hospitals of West Virginia. Reaser, who directed Huntington State Hospital, "apparently thought that transorbital lobotomy could accomplish miracles (although I had told him to the contrary), and supplied six old-timers who were pretty badly dilapidated and hardly suitable even for radical operation," Freeman wrote. Seven additional patients from West Virginia's Western State Hospital had been ill and hospitalized for considerably less time, though, and Freeman found with satisfaction that six had been discharged thirty months after lobotomy.

Freeman devoted an intensity and energy to his mission in West Virginia that he surpassed nowhere else. Over the next four years he frequently visited the state, with the result that its per capita rate of lobotomy was the highest in the nation. In 1950 Freeman operated on additional patients at West Virginia hospitals in Huntington, Lakin, and Spencer. In the spring of 1952 he presented the results of these operations to the West Virginia Board of Control. "Arrangements were made for me to devote three weeks to visiting the various hospitals and to perform transorbital lobotomy on suitable patients," Freeman wrote to Moniz. "The [hospital] superintendents thereupon wrote to relatives of nearly 500 patients in all, asking their consent to operation[s] upon the sick people." About half of the families gave their consent to the treatment. Freeman maintained there were few outright refusals and that many families simply could not be located. He contradicted this assertion, however, in an article he coauthored in the

*Journal of the American Medical Association* on the West Virginia Lobotomy Project: "The program did not meet with wholehearted acceptance by relatives of patients, so that a control group was available whose relatives refused permission for operation," the article stated. "This control group numbered 202 patients."

Between July 18 and August 7, 1952, Freeman operated on 228 West Virginia patients. "Actually only twelve days were devoted to operating," Freeman wrote, "and the others were taken up by trips into various parts of the state visiting patients that had previously been operated upon. I mention this matter to reveal how mass surgery can be carried out against a background of shortages of everything except patients." (Freeman repeated with pride his daughter Lorne's prediction that his skill as a mass lobotomist would earn him renown as the "Henry Ford of Psychiatry.") At the hospital in Spencer, Freeman operated on twenty-five female patients in a single day, a record perhaps never equaled. Watts later acknowledged that assembly-line treatment opened Freeman to harsh criticism, but he maintained that his former partner always knew the case histories of his patients, even when there were two dozen of them in one day. "I know he had abstracts on these patients," Watts said. "He knew how long they'd been in the hospital. He knew what the symptoms were. He knew what the medical officers in charge of them thought of them. [He performed] an examination of his own. And then, they were simply lined up, and they were operated on, and I think that did make a bad impression."

Charging only $20 to $25 per operation, Freeman did not make a financial killing in West Virginia. (He earned as much as $2,500 for transorbital lobotomies he performed on private patients; more commonly he made about $200 each.) But summing up the results in West Virginia a few months later, Freeman called the lobotomy campaign "a gratifying experience." He attributed the high rates of postoperative release at some hospitals to their administrators' direct manner of informing families what to expect. At the hospital in Spencer, for example, the superintendent "invited the families of patients to come to the hospital for a discussion of the problems they would have when the patients went home. . . . The patients operated upon were also in the auditorium, easily distinguishable because of their black eyes," Freeman wrote. At the hospital in Lakin, a segregated institution for African Americans, Freeman was pleased when he returned "a week or so after operating upon twenty very dangerous Negroes and found fifteen of them sitting under the trees with only one guard in sight. It was the first time that they had been out of the seclusion rooms for anywhere from six months to seven years." The Lakin superintendent soon discharged half of this group from the hospital.

A month after the end of his West Virginia sojourn, Freeman learned that 81 of the 228 transorbital patients—36 percent of the group—had already obtained their paroles from the state hospital system. (An additional 5 patients had gone home but almost immediately required readmission to the hospital.) He wrote that "the Board of Control is considering the possibility of appointing me to the post of Consultant, with arrangements to make the rounds of the hospitals twice a year and operate upon patients who would presumably benefit from the procedure." Eventually, Freeman hoped to assemble enough data on the lobotomized patients to enable him to present a report to the state legislature on how to make the West Virginia hospitals even less costly to operate than they already were. The savings could come in freed beds, food, laundry, nursing charges, less broken furniture and equipment, fewer injuries to patients and staff, and shorter hours of restraint and seclusion. "I have maintained that the timely application of psychosurgery to mental cases will eventually change the insane asylums to old peoples' homes," he wrote, "but evidently the problems of the hospitals cannot be met until the large backlog of the chronically disturbed patients is much more effectively handled than it is at the present time." Freeman's prediction came to pass, but new medications, not the large-scale application of lobotomy, brought about the change.

One aspect of the West Virginia Lobotomy Project that ultimately disappointed Freeman was his failure to train many local psychiatrists to undertake the transorbital procedure. Several unexpected circumstances made it difficult for Freeman to leave trained disciples in his wake. As in other state hospital systems, the most able West Virginia staff psychiatrists won promotions to administrative positions that removed them from the psychiatric wards. "Then there is the possibility that an accident or two occurring at a hospital [during surgery] would hurt the hospital in the eyes of the public," Freeman noted. "People would be apt to say that the accident was due to inexperience. Being myself an expert, I can afford an accident now and then, but a less expert individual could easily be criticized, and this could not be easily tolerated by the hospital." Freeman did make mistakes in West Virginia. There were four deaths among his patients, "two of these in all probability due to hemorrhage, but necropsy in the other two revealed clean incisions with little or no bleeding," Freeman wrote. "Death in these cases was attributed to heat prostration."

Freeman remained interested in the fate of his West Virginia patients for the rest of his career. In 1958 he reported on his attempts to locate the patients who had been released from the hospital during the summer of 1952, and he tracked down nearly all of them. "Sometimes it meant climbing a rutted road on foot to a dwelling perched precariously on the edge of a cliff," he wrote. "Sometimes it meant asking around at various homes to find out where the people had moved. . . . 'Over the mountain and down the hollow' was an oft-repeated phrase. I crossed streams on swaying cable bridges where coal-blacked water flowed a few feet below. I encountered the smelly aftermath of a flash flood where carpets and upholstered chairs were festering in the sun." He conducted one interview with a released patient who had to shout to Freeman from a second-story window because his family had locked him inside the house. Unfortunately, "this patient, who didn't know me from Adam but was glad to have anybody to talk to, was hardly an asset to the home."

At one remote farm, he discovered a former patient in a state unfit for interviewing: "I rounded the corner of the barn and found Hubert on a pile of straw, face down, shoeless, laughing and gesticulating as though at the funniest movie ever. I could not distract him long enough to tell me what work he was doing. Yet when I came by there again the following year he was farming intelligently and taking the produce to market, buying necessaries and handling his money efficiently." In 1960 Freeman followed up on a total of 787 West Virginia patients he treated with transorbital lobotomy between 1952 and 1955. Nearly ninety percent were schizophrenics. Eighty-four of the patients had died (eighteen as a direct result of their lobotomies), but of the remainder, 44 percent were no longer institutionalized. He concluded that the "personality changes brought about by lobotomy definitely increase the chances of discharge from the hospital even after confinement there of 2, 5, or even 10 years."

Nearly every state-run psychiatric hospital that Freeman visited during that summer of 1952 was crowded, overtaxed, and staffed by doctors, nurses, and caregivers unable to meet the demands of the patients in their overflowing wards. The sole exception was Wyoming State Hospital in Evanston, and Freeman noted with satisfaction that some of its beds were empty because its staff had performed nearly two hundred transorbital lobotomies, which equaled about one-fifth of its patients. The following summer, Freeman concentrated on a single institution in Ohio, Athens State Hospital. Here, Freeman encountered a grade of patient that, in his opinion, was better suited to transorbital lobotomy than those in most other hospitals he visited. "The results at Athens have been superior to those in the other state hospitals owing largely, I believe, to the choice of patients, specifically the small number of those hospitalized for over 10 years," he wrote during a 1961 follow-up. His survey of 168 Athens patients showed that 43 percent remained hospitalized, 37 percent were employed or keeping house, and 19 percent were being cared for at home. He found that five deaths were directly attributable to the surgery. "My only regret is the high proportion of patients who died as a result of operation," he wrote to the hospital's superintendent.

He returned home from Ohio and his other destinations with souvenirs that he enjoyed reviewing between trips: rolls of film, patient record cards, and wax cylinders from his dictation machine. During one trip, the wax cylinders had partly melted after too much exposure to the sun, and "the transcribing was a most difficult job." On that same expedition at a campground in Washington State, Freeman lost a notebook that contained the names of all his transorbital lobotomy patients to date. He did not realize his loss until he had driven all the way to South Dakota. "The damage was repaired but it took some time to identify the photographs," he wrote.

The film Freeman took home from the hospitals afforded him a chance to indulge the interest in photography that he had cultivated since childhood. "The many rolls of film I brought home from these excursions I had developed and then made the enlargements myself, in a closet under the staircase [in his home]," he wrote. "The sweat used to roll off my forehead and also down my back in the summertime, and the tins of squeegees were stood on end for the prints to drop off when dry. Trimming, mounting, checking and sending duplicates to the superintendents made the time pass quickly."

In addition, Freeman used the time between head-hunting journeys to take advantage of the latest technological aids to his record keeping. He acquired IBM card punching and sorting equipment that he found essential in compiling data, and he even invented a device that made the graphical expression of percentages easier to accomplish. It was a complex machine fabricated for him by the Armed Forces Medical Museum in Washington. "Forty-eight metal bars with varying cross hatching were introduced into grooves at the base of the machine and each one pushed a certain distance corresponding to the percentage calculated from the IBM cards. The bars were then clamped in place and the chart upended for photography."

The year 1954 was one of Freeman's busiest for lobotomy-related travel. In addition to stomping the familiar ground of Virginia, West Virginia, and Ohio, he began frequenting hospitals in California and made yet another visit to Herrick Memorial. He became a familiar figure at California state institutions in Fresno and Patton. As in previous years, Freeman recorded all of the travel details in what he called "my little transorbital lobotomy books," which gave a hint of the staggering amount of ground Freeman covered during the 1950s. "Since 1954 I have averaged nearly 100 miles a day in driving, not counting plane trips and rented cars," he later acknowledged, a formidable claim in the days before long-distance commuting was common. "I put 86,420 miles on my 1954 car, and turned it in August 1956. My Ford 1961 had gone over 200,000 miles in 6 years."

Freeman's lobotomy trips gave the procedure wide exposure, especially in the poorest and most geographically isolated regions of the country. He believed that he had succeeded in showing that his new method of psychosurgery led to better psychiatric treatment and earlier releases, and as a result, "thousands of chronically ill and violent patients have been rendered more amenable to management in psychiatric hospitals." Press accounts of the successes of transorbital lobotomy appeared all over the country. "At the week's end, all the patients excepting one had shown improvement and most were up and about," wrote one newspaper of transorbital lobotomy trials in Lincoln, Nebraska. "One patient began playing the piano as beautifully as she did prior to her illness. Another spoke coherently for the first time in many months and still another asked to be allowed to aid in the construction of a building." Modest improvements were more typical, Freeman acknowledged, but he told a reporter that even if transorbital lobotomy "enables the patient to sleep on a bed instead of under the bed, it is worthwhile." Although some hospital workers did not approve of Freeman's treatment, others thanked him. "Thank God for men such as you that have had the courage and perseverance to undertake new approaches to the problems of the mentally ill," wrote a staff member at Traverse City State Hospital in Michigan in 1958. "From you we gain strength."

Freeman reached the height of his head-hunting travel in 1954. He would continue to tour, very extensively in some years, but an important development in psychiatric medicine altered the focus of his trips. In March 1954 a new drug called chlorpromazine won FDA approval for use in the United States. Sold under the trade name Thorazine—a mark derived from the Norse god Thor—the drug soon began to assert its power. It was not the first pharmaceutical product to gain favor in the treatment of psychiatric illnesses: many others, including such sedatives and sleep-inducing drugs as apomorphine, bromine, hyoscyamine, scopolamine, chloral hydrate, and barbital, had enjoyed their days in the sun. Chlorpromazine, however, was different. It was a true antipsychotic, acting, according to its manufacturer, "like a chemical lobotomy" that calmed patients, distanced them from their emotional disturbances, and even eliminated hallucinations, delusions, and other symptoms of schizophrenia. Such a drug could have widespread application in the most hopeless wards of the nation's state hospital systems, the very places where Freeman had championed the effectiveness of transorbital lobotomy. Without surgery, without leucotomes, and at a low cost, hospitals could use chlorpromazine to end the screaming, raving, and violence of their most unmanageable patients. The state hospital system in New York became one of the first to adopt the new pharmaceuticals, and within a year it saw a drop in its patient population. Elsewhere the drug caught on quickly. By the end of its first year of use in the United States, chlorpromazine had been used to treat about 2 million mental patients. Starting in 1955, not coincidentally, the number of patients in government-run psychiatric hospitals began a steady decline.

Freeman was at first attracted to chlorpromazine, which he called "a drug that seemed to have a selective quieting action upon the central division of the autonomic nervous system." He prescribed it for his patients and initially believed it beneficial for emotional disturbances. But as he observed the drug's side effects of restlessness, a severely runny nose, diarrhea, urinary problems, tremors, and a peculiar muscular rigidity that gave some patients a loping gait, he changed his mind. He came to view chlorpromazine as a stopgap medication that obscured the symptoms of mental illness while failing to treat its roots—the same criticism ironically that many critics leveled against psychosurgery.

As chlorpromazine and other new psychiatric drugs appeared, lobotomy faded away. During the late 1950s and early 1960s, state hospitals abandoned their psychosurgery programs and stopped inviting Walter Freeman to come to demonstrate transorbital lobotomy. The number of papers published about psychosurgery quickly dropped from more than one hundred fifty annually to fewer than fifty. Neurosurgeons did not cease experimenting with new psychosurgical procedures, because a small number of patients failed to respond to the new pharmaceuticals. But the wholesale application of lobotomy in psychiatric hospitals—as well as the vestigial use of insulin shock therapy—came to an end.

Even without the arrival of new drugs, large lobotomy programs may have been marked for extinction in America's psychiatric hospitals. "Virtually every patient who might qualify for a lobotomy had already received one," wrote Alex Beam of the situation at McLean Hospital in Massachusetts in 1954. "Eighty-five percent of the lobotomized patients were discharged within eighteen months of having the operation." The remainder, the failures, occupied a small number of beds earmarked for incurables, while McLean made an effort to populate the wards with younger patients who had a greater chance of successful treatment by more conservative means. To most psychiatrists, lobotomy appeared to be in fast retreat. "The procedure faded away in the early 1950s almost as abruptly as it had risen up, a blip in the history of psychiatry, though an illuminating one as a study in medical hubris," as the psychiatry historian Edward Shorter summed it up.

To Freeman, however, lobotomy was not fading away. It had merely been temporarily obscured by something new. He believed that chlorpromazine might diminish the use of surgical psychiatric treatments like lobotomy, but it would never eliminate them. He pointed out that "the same phenomenon occurred in 1937 following the introduction of the shock methods of treating serious mental cases, so that I think in a couple of years that surgery will again find its place in the treatment of those patients." In the years that followed, Freeman unwaveringly maintained his conviction that psychosurgery offered a solution—and often the best solution—for patients entangled in the snare of mental illness. He held to that belief despite the undeniable successes of the new psychopharmacology, the noninvasive nature of drug treatments, and the eventual migration of most of his biologically oriented colleagues to the drugs.

Freeman refused to open his eyes to the possibilities that the new medications offered. He had staked his career on the efficacy of lobotomy, and he would not let go of his beliefs. The stubbornness, egotism, and ambition that directed his thinking were his greatest flaws. Freeman's assumptions proved false. Following chlorpromazine came other drugs even more effective in treating neuroses and psychiatric illnesses: reserpine, lithium, meprobamate, amitriptyline, chlordiazepoxide, and diazepam. Meanwhile, psychosurgery never again returned to center stage in the treatment of psychiatric disease.

Several popular novels seized upon the drama of the decline of psychosurgery. In Ken Kesey's *One Flew over the Cuckoo's Nest* (1962), lobotomy has evolved into the tool wielded by a mental hospital's administration to break the will of the story's antihero, Randle McMurphy. *Daybreak*, a novel that the physician and prolific author Frank G. Slaughter published in 1958, vividly captures the conflict between psychosurgery and the new pharmacology. The central character, a recently widowed neurosurgeon named Jim Corwin, has taken a position at a large psychiatric hospital in a southern state, "a backwater far from the great mainstream, a final haven for the doomed and damned." He has previously observed Anton "Ziggy" Ziegler, a proponent of prefrontal lobotomy modeled in part on Freeman, perform his psychosurgeries. "Once the connection between the thalamus and the frontal lobes is broken," Ziegler explains, "tensions have no chance to accumulate. Pathological emotional conditions are thus under permanent control. . . . At the same time it leaves enough of the frontal lobes to allow the patient capacity for productive labor—and enough judgment to make a practical adaptation to his environment."

In his new position, Corwin begins performing transorbital lobotomies, with the result that many patients emerge from close confinement in the locked wards. But he encounters opposition from other doctors. "I still maintain lobotomy is wrong, both morally and clinically," one character tells him. "You've no right to seal off a section of a man's brain, just because his emotions don't operate by normal rules. . . . You simply can't probe into a certain section of the brain and insist it's causing the illness you hope to isolate." Corwin soon meets Lynn, a catatonic schizophrenic patient whose beauty and artistic skills charm him, and he experimentally treats her with the new drug reserpine. Despite the machinations of the hospital's director to continue the use of lobotomy to build a vast labor pool of submissive patients, Corwin uses reserpine and chlorpromazine to save many from the disagreeable side effects of lobotomy. Lynn is among these lucky patients, and in the end he not only cures her but marries her.

In Frank Slaughter's fictitious world, psychiatrists discover new cures, set aside their old treatments, and find love and contentment. Walter Freeman, on the way to a new chapter in his life, enjoyed none of those storybook turns of fate. Without realizing it, he was slipping into a new chapter of his life in which all that he considered precious would tumble upside down. His greatest accomplishments and strongest convictions would turn into liabilities; the stability he fought to maintain in his personal life would dissolve.

# Chapter 12

# Leaving Home

The rugged beauty of America's West Coast had always exerted a strong pull on Freeman. Until 1929, when he traveled west for the first time, he had little experience of the ocean other than the warm and gentle beaches of the Atlantic around Cape May and no conception of valleys more dramatic than the Delaware Water Gap. In 1938 from a train he saw the Sierra Nevada for the first time, and was stupefied. "I think I fell in love with California on waking up and looking out of the window at the houses with the frosting like marshmallow around Colfax, and then the almond and cherry trees in bloom around Roseville," he wrote. A few months later he drove along the shore of Lake Tahoe, "that lovely lake rimmed by mountains, [which] gave me the feeling that this was the place I wanted to live in."

Keen's death and burial in California cemented Freeman's desire to remove himself from Washington's congestion and miserably humid summers. (He frequently called California "the land of the dry shirt.") In addition, Freeman increasingly found his work at George Washington University dissatisfying. He noticed a decline in the interest of students in clinical neurology. "No applications for graduate training in clinical neurology have been received for more than a year," he wrote in his department's midyear report for 1954. "The last Fellow stayed a week, and the prospective Fellow never came." After World War II the ascendance of psychoanalysis had lured away most of his teaching assistants. He and James Watts, though still amicable coleaders of GWU's department of neurology and neurosurgery, no longer performed lobotomies together.

Just as disheartening, his relationship with St. Elizabeths Hospital, the institution that had launched him into the practice of pathology and fueled his interest in the problems of the mentally ill, was deteriorating. In 1952 the *Washington Evening Star* incorrectly reported that Freeman had performed many of his lobotomies at the hospital, an error that the outraged St. Elizabeths superintendent, Winfred Overholser, blamed on Freeman. Overholser, in fact, had earlier prohibited Freeman from performing transorbital surgeries at the hospital. After twenty years Freeman found himself bounced out of his appointment as the hospital's neurological consultant, a job that added nothing to his bank account but paid handsomely in honors and emotional satisfaction. Upset, Freeman obtained a correction from the newspaper, but Overholser would not budge and, furthermore, told a reporter that he intended all discharged St. Elizabeths patients to return home "with intact brains."

Overholser indeed wanted patients' families to feel assured that Walter Freeman had no connection with St. Elizabeths. "A great deal of antagonism had built up in the hospital staff against lobotomy, as it had wherever poorly chosen chronic patients were operated upon and failed to improve," Freeman later recalled. "Since the hospital doctors saw only the failures their ideas of the value of lobotomy were quite distorted. This episode, however, was another of the disagreeable events of my final year[s] in Washington." He would later give to St. Elizabeths some bound editions of his most important medical journal articles on lobotomy. Inside he wrote, "To St. Elizabeths Hospital, where I worked, 1924–1933, to find some answers to the problem of mental disorder, and where more problems arose than were ever answered."

At the start of the 1953–1954 school year, Freeman informed the dean of the GWU Medical School that he intended to move to California at year's end. Although he was only fifty-seven years old and not yet ready to retire, he hoped that GWU would honor his quarter-century-long tenure and professional prominence by naming him professor emeritus of neurology. The university disappointed him. "Instead I was given a leave of absence for a year, and then I submitted my resignation, on request." Never one to hold a grudge for long, in 1957 Freeman changed his will to bequeath to GWU all of his psychosurgery records and $5,000 to fund an annual prize for the school's top undergraduate medical student.

Around Thanksgiving 1953 he and Marjorie visited several communities south of San Francisco to scout for a place to live. Limited to just a few days away from Washington, Freeman knew that their search would have to be efficient. He wanted to appraise the entire region at a glance. Like the human brain, his region of choice was more than just a collection of disparate parts. "We drove to the San Jose airport and had a pilot take us up in his four-seater for a birds-eye view of the area as a whole," Freeman wrote. "I had a map of it on my knees, and as we flew over the San Jose, Los Gatos, Saratoga, Los Altos and Palo Alto areas, I circled spots on the map that looked promising." They wanted a home far from industrial blight, large highways, and neighbors with ostentatious estates. Freeman also hoped for a house near hills that he could hike. They narrowed their choice to Los Altos and ended up renting a house that an airline pilot and his wife were vacating for five years. Later they bought another house nearby.

The day they selected their house, Freeman signaled his eagerness to open a new chapter of his life by making a dramatic change to his personal appearance. Upon arriving in California, he had noticed that most physicians were clean-shaven "while the boys with the beards were those who come down from the hills for the rodeos." After decades of wearing facial hair—he had grown his first timid goatee during his post–medical school year of research in Europe—he shaved off his beard and mustache. Through breakfast and some shopping that day, Marjorie failed to notice the change. At last a friend made a comment. "Marjorie looked at me and gasped," he wrote, a response that probably gave him some satisfaction. Eventually, though, they both decided that he looked incomplete without some facial ornamentation. First the mustache returned, then the beard.

Freeman used this same trip as an opportunity to appear before the state board of medical examiners in San Francisco, a necessary step in obtaining his license to practice medicine in California. A veteran of hundreds of oral examinations as a judge for the American Board of Psychiatry and Neurology, he felt no anxiety over submitting to such a test himself despite the notoriety that must have preceded him into the examination room. The board tested his knowledge on a few predictable aspects of his specialties before one examiner declared, "Dr. Freeman, we are here to test your fitness to practice not only neurology and psychiatry, but also general medicine and surgery." Freeman's blood pressure may have risen; it had been thirty-four years since his completion of medical school. "Suppose you tell us," the examiner continued, "the symptoms and signs of acute appendicitis." Freeman felt a wave of relief. Lorne and Paul had both undergone emergency removals of the appendix as children, and he vividly remembered their symptoms. "I knew I was in," he wrote.

Frequently away on transorbital lobotomy demonstration trips, Freeman left to Marjorie the chore of selling their Washington house. They temporarily moved into an apartment in Bethesda, Maryland, until the completion of the 1953–1954 academic year. Freeman found the apartment sterile and impersonal because of the loss of much inherited furniture, decor, and books that they had jettisoned. "We saved the really important pieces, but the sentiment had to go by the board," he wrote. Freeman was especially reluctant to dispose of a huge rubber plant that he had given to Keen years earlier, which had grown tremendously in the interim. The manager of his apartment building claimed it "almost with tears in her eyes, saying she would look after it tenderly and would wash its leaves with milk to keep them shiny." When the time at last came for Freeman to say goodbye to GWU, a younger faculty member replaced him as professor of neurology and Jim Watts took sole chairmanship of their department. Freeman noted with bitterness that his post as head of the laboratory of neurology, where he had studied lobotomy and dissected brains, was left vacant, as if all his work done there mattered for little.

His university colleagues did honor him with a farewell dinner at the Mayflower Hotel. Many friends, Watts included, testified to Freeman's accomplishments, and a representative of the Medical Society of the District of Columbia expressed gratitude for his contributions and leadership. This was the end of thirty years in Washington during which Freeman matured into an experienced clinician, academic, lobotomy pioneer, and father. The high emotion of the dinner seemed to have shaken his composure and disturbed his judgment. "I made a very stupid speech, reciting some limericks of my own composition, and omitted to open the going-away present of a Land Polaroid camera. I felt badly about this when the gathering broke up," he wrote. Although the limericks he recited are not known, David Shutts recorded this as one of his favorite verses that he composed during the 1930s:

> There once was a man named McKay,
> Who slept with a maid from Bombay,
> He got tabes and scabes,
> And sabershinned babies,
> And thinks he's the Queen of the May.

Freeman's departure saddened Watts, who continued to admire his old partner despite their conflict over the use of transorbital lobotomy. "When he left Washington, some of the sparkle left," Watts later declared.

In the last days of June 1954, after giving away whatever remaining possessions were not accompanying them to California, Freeman and Marjorie turned the nose of their Ford convertible west and began driving. A moving van loaded with their books, furniture, and clothes—as well as all of Freeman's records of transorbital lobotomy patients—followed a different route to California. Freeman anxiously tracked the moving van by phone. After several days of driving, they spent a last night on the road in the Sierras, near Donner Summit, before descending to Los Altos.

They moved into their rented house and Freeman pondered his future. He felt determined to steer clear of the politics and responsibilities of academe, not to mention the difficulty in obtaining a teaching position due to his notoriety. He wanted simply a private practice, opportunities to perform psychosurgery, and the chance to pursue his follow-up studies of his lobotomy patients. Earlier a friend who directed the staff of a medical clinic in Palo Alto had invited Freeman to join. But to Freeman's consternation, the controversy over his career had forced the friend to withdraw the invitation. Freeman never said so, but that rejection must have been painful, although he accepted it with his customary bravado. "It was the best thing that could have happened to me, threw me on my own," he wrote. "Palo Alto and San Jose were full of doctors, so I looked over the intervening towns. I postponed a decision for 6 months, being busy exploring the areas, often on foot, the hills, the parks, the streams, etc." He also undertook some head-hunting out of state.

The rejection from the Palo Alto clinic, however, represented the first gust of a chilly wind blowing against Freeman in his new home state. Practicing in Washington, D.C., Freeman had been a respected member of the neurological and psychiatric establishment: a physician who had helped found important organizations, had influential friends, worked from a solid academic base, and enjoyed a reputation as a gifted neurologist and neuropathologist, as well as an advocate of lobotomy. When Freeman abandoned all that to make his fresh start in California, at an age when many other doctors were considering retirement, he lost a dismaying amount of credibility.

To many California physicians, Freeman was simply a lobotomist, or even worse, the "ice pick lobotomist." They would not refer patients to him. And as psychoanalysis increasingly occupied the mainstream of psychiatry in California and elsewhere, Freeman felt the sting of being regarded as a heretic. He decided to end his membership in the East Bay

Psychiatric Society in part because of his philosophic opposition to the treatment that most other members advanced. "I just can't understand why there is so much preoccupation with psychoanalytic procedures when it has been so abundantly proven that they have such limited application to the problems of psychosis, alcoholism, addiction and violence. . . . I have said, mostly to myself, that one of these days the doctors may really have to go to work," he wrote.

Increasingly, Freeman found himself treating patients whom psychoanalysts would not or could not handle—those with major psychoses—and their desperation may have inspired him to tighten his grip on the one therapy he could offer. "Lobotomy is not the last desperate remedy when all else fails, but rather to be considered as the turning point in effective therapy," he continued to maintain in the late 1950s. "When milder measures are not sufficient, a beneficent change in the personality produced by lobotomy can enhance the effectiveness of those other measures to a gratifying degree." Freeman felt a responsibility not to deviate from his career path. "I think if he hadn't gotten hung up, in a sense obsessed with [lobotomy], he would have done more," his son Walter observed. "More of his energy would have gone into a payoff instead of a blind alley. . . . He felt that was his mission in life, that was what he was going to do, and that's what he did."

Another blow came late in 1955, when Egas Moniz's wracked body finally gave out at the age of eighty-one. Attacked and weakened by a raging patient that Moniz had previously resolved would be the last of his career—like the one who shot him in 1939, this patient was not the recipient of a leucotomy—the Nobel laureate died suddenly on December 13. Two years earlier Freeman had traveled to Portugal for the Fourth International Neurological Conference and for what he suspected would be his final opportunity to spend time with his mentor and idol. In Moniz's hometown of Avanca, "I spent a day with him and his wife," Freeman wrote. "His chauffeur drove us around the town, and in front of the church he pointed to an open space: 'This is where they will assassinate me.' He then went on to explain that they planned to erect a statue to him there." Although Moniz flared with anger when Freeman asked to photograph his crippled hands, the elder neurologist gave Freeman some autographed books as a parting gift. Several months later Moniz wrote to Freeman that repression often accompanies medical progress, a subtle appeal for Freeman not to become discouraged over the decline of psychosurgery.

Moniz's own achievements in the face of adversity were heroic, Freeman believed. "Neurologists, neurological surgeons and psychiatrists in the United States of America may well look back upon the period before the discoveries of Egas Moniz as equivalent to the Dark Ages," he wrote. Soon after moving to California, Freeman made his last visit to Portugal to attend a memorial meeting for Egas Moniz at the Academia das Ciencias in Lisbon. Marjorie accompanied him on the trip. Comfortable amid the palms and pines that reminded him of California, Freeman gave two speeches in French "and held my audience," he wrote. After spending some time in London, the Freemans returned from what would be their final international trip together.

Subduing his own impulse to work, Freeman meanwhile grabbed as many as seven or eight hours of sleep a day. He took additional interest in smoking, a habit he had begun as a medical student. "My start of smoking at 23 was done in earnest, 2 cigars, 2 pipes and a cigaret the same evening. Sweat poured off me, my stomach rumbled and my hands shook, while the room seemed to wander," he later recalled. The pipe became his favorite method of smoke delivery. "If some doctor should tell me to give up on my pipe, I would change doctors or try to educate him in greater tolerance for the evils of smoking." And as he had hoped, he was able to take advantage of the hills, beaches, and trails near his new home. He wrote to a friend in 1956, "Needless to say, the blue skies and cool summers in California have completely won me over." He later declared: "I like walking in the hills and bathing in the surf, opportunities for which are close at hand. I like fishing and digging for clams or prying abalones off the rocks. I like to roll out my air mattress and sleeping bag and sleep under the stars." Freeman usually took part in these activities all alone, a continuation of the solitary habits he had developed as a child.

In a certain way, his approach to the outdoors was similar to his approach to medicine. He was not afraid of solitude. "I like the sensation of hiking along the trails, the rhythmic swing of arms and legs, the gentle nudge of the pack on my back and the gradual change of scenery as I go along," he wrote. "If the woods are open I strike out away from the trail and climb over fallen treetrunks or work my way through brush." Despite Keen's death at Vernal Falls, or perhaps because of it, Freeman was drawn to the waterfalls of the High Sierras. "As the river runs over bare granite," he wrote of Waterwheel Falls on the Tuolumne River, "it goes faster and faster and when it meets with an obstruction it flies into the air, reminding me somewhat of the rooster tails kicked up by powerful speedboats. Fortunately, the ledges are not slippery so that it is possible to go fairly close without danger."

Moving to California, where he was more frequently able to take advantage of the mountains and trails, initially improved his health. He weighed about 185 pounds when he left Washington and soon dropped to 170 after several summers of hiking. No amount of exposure to a blissful climate, though, could stop the advance of age. During his first several years in California, Freeman would come down with a basal cell carcinoma on his shoulder, which was successfully treated by X-ray therapy; a cancerous spot of skin on his left eyebrow; a form of diabetes that threatened his eyesight; an enlarged prostate, which was surgically removed; and "a wart-like viral disease" that troubled him with anal itching. Freeman continued to believe, however, that the relocation to California had given him "a new lease on life."

Marjorie had initially favored the move west, as she hoped that relocation would calm her suspicions of Freeman's extramarital affairs. She received no such relief, because Freeman's infidelities were genuine and chronic. Eventually Majorie learned that one of her husband's lovers was planning on following him to California. Marjorie furiously called a meeting of her children and announced her intention to file for divorce. "I remember saying, 'We'll stand by you if you leave, and we won't hold it against you at all,'" said son Paul. In the end the other woman returned to Washington after a brief California appearance, and Marjorie lost her determination to follow through on her resolution. She still loved her husband, she told her children, and her life lay with him. "She couldn't make the break because she couldn't face living alone," Paul remembered.

The accumulation of disappointments weighed heavily on Marjorie. Although Freeman wrote to James Watts in 1955 that she "is on a pretty even keel," her course began to falter almost from the moment they first occupied their rented house in Los Altos. California's newly developed communities and their lack of such amenities as sidewalks and sewers repelled her. She felt gripped by isolation. "She wrote few letters in her later years, saw few people outside the family," Freeman observed, noting that her only friends appeared to be their gardener, the owner of a local grocery store, and their immediate neighbors. She sometimes expressed to Freeman her "hatred of me for the various infidelities of which she suspects me." Freeman believed that boredom was the cause of a mental and physical decline that afflicted Marjorie soon after their relocation. "She took progressively less interest in our home. There was a quite obvious depression, and she resented my efforts to get her to move about, to go places, take trips, etc.," he wrote. She refused to associate with the wives of other doctors and drank more frequently. "Alcohol numbed her feelings and was later her sole refuge. She would seldom reveal her thoughts, but when the topic of her life came up she said it had been a good one, but that it was coming to an end."

Freeman liked drinking wine, champagne, and martinis, but drunkenness and alcoholism disgusted him. "I dislike the immediate effects of intoxication beyond the stage of relaxation and maybe gaiety," he wrote. "I don't like to lose control either of my tongue or my limbs." He could recall getting intoxicated only three times, and one of those instances proved disastrous. "The last time I got drunk I neglected a patient upon whom I had operated," he remembered in the early 1960s, "and although the referring doctor [took over] and did all that could be done, the patient was dead when I put in an appearance the next morning."

His great prejudice against alcoholics was perhaps an inheritance from his devoutly Baptist grandfather, W. W. Keen. This antipathy, he realized, "has also disturbed my feeling for my wife. She may not be an alcoholic since she is alert and cheerful after sleeping it off, but the evenings are best when she is in bed by the time I have washed the dishes. I prefer the semistupor of the fifth glass of sherry to the caustic antagonism after the second." When Marjorie once broke her arm after their move to California and abstained from drinking for seven weeks during her recovery, Freeman saw a return of the woman he admired. She became "a charming, intelligent individual, a favorite of the doctors, nurses and others who recall her with pleasure a year or more later. The priest was making good headway at that time, but when she got home the pattern [of heavy drinking] was resumed."

Marjorie's problem soon worsened. She required hospitalization several times due to her alcoholism, and one time she ended up in a psychiatric ward. One period of sobriety ended at a family gathering, where "a glass of cheer" sparked a return of her habitual drinking. Meanwhile, her homemaking skills, meager as they were, vanished. Her frequent forgetfulness of food in the oven inspired Freeman to dub one of her dishes "Chicken à la Crematorium." After she once turned on the gas without lighting the oven, Freeman took responsibility for all the cooking and housekeeping and gave away her car. He cleared piles of newspapers out of their house and asked his daughter, Lorne, to buy her mother a replacement for her ratty bathrobe. Still, "Marjorie was spending about 20 hours a day either in bed or on the couch." In vain, Freeman called neighborhood liquor stores to stop them from serving her.

Freeman sought refuge in a return to his work. In 1955 he made his first tentative moves to reestablish himself as a practicing physician. He transferred his local medical association membership from the District of Columbia to the Santa Clara County Medical Society. Renting a modest office on Main Street in Los Altos, he felt far distant from his old position of academic prominence and activity at the center of Washington's sphere of medicine. He had to quickly dismiss two inept secretaries before finding a competent assistant. His practice rebuilt slowly, but most patients and referring physicians thought of him as a psychiatrist—one deeply associated with ice picks—so his neurological practice never regained strength.

Freeman resumed using a variety of office treatments for psychiatric problems, including electroshock therapy, carbon dioxide treatment, and intravenous doses of amytal. In addition, he operated a neurological clinic at Santa Clara County Hospital, where he served as the institution's chief of neurology for eight years. It took time, but he eventually joined the staff rosters of hospitals in Palo Alto and several other nearby communities. In addition, his old friend, A. E. Bennett, nominated him for honorary membership with full surgical privileges at Berkeley's Herrick Memorial Hospital. He took advantage of his status there to perform lobotomies for the remainder of his career, but in general he avoided facing critical appraisals of his career that might result if he applied for surgical privileges at the other community hospitals.

He also began touring state psychiatric institutions in California, several of which he had visited earlier to demonstrate transorbital lobotomy. In retrospect, he had to admit that the results of many of those operations "were not too promising." At Stockton, Napa, De Witt, Mendocino, Modesto, and Atascadero state hospitals, Freeman secured contracts to help train medical residents on staff in neurology and neuropathology. As in Washington, he enjoyed using patients as illustrative aids. He became well acquainted with these institutions and judged the hospital in Modesto to be the worst of the lot. Sometimes all he did during his visits to that hospital was conduct neurological examinations of patients without ever speaking with a doctor in charge. "The only reason I continued at Modesto was because of the ease with which I could earn $115 for a day's work," he wrote. Once in a while he was asked to perform transorbital lobotomies. At Atascadero, "I performed lobotomy on two of the most dangerous patients, with substantial improvement in ward adjustment, but because they were murderers there was no chance of their release." One of the killers was Denis "King" Huey, who "had been locked in a single room for many years and had to be shackled by 4 men when brought out," Freeman recalled. "After operation he became affable, though still King Huey."

For many years Freeman served as a touring ambassador of neurology and organic psychiatry in California's psychiatric hospitals, although he suspected the invitations to lecture were mainly "for the [medical] residents to see what an old-timer looks like. I often introduce myself as a surviving fossil." At some institutions, however, he took a more active role. During the late 1950s and early 1960s, he performed a series of controversial transorbital lobotomies on seven adolescents at the Langley Porter Clinic, a psychiatric hospital jointly operated by the state and the University of California at San Francisco. "A certain amount of anxiety is good for a boy or a girl, in the same way that a certain amount of fleas is good for a dog—keeps him from thinking about being a dog," Freeman wrote of these patients. "But immoderate anxiety can be harmfully disturbing to the adolescent and can lead not only to helplessness but to a variety of symptoms both physical (psychosomatic) and mental (psychotic). It is this type of anxiety that responds favorably to frontal lobotomy, the change in personality being usually in the right direction with better chances of rehabilitation."

Freeman arrived at Langley Porter in 1961 to deliver a talk on these patients, three of whom were present for observation, "and I thought I had made a favorable impression," he remembered. "Such was far from the case however. The staff and residents at the Institute [were] steeped in the Freudian tradition, and I was met with a barrage of hostile criticism." Freeman pointed out the satisfactory adjustment most of the youngsters had made at home, but "the specter of damaged brains prevailed." Frustrated and angry, partly because his son Paul (by then a practicing psychiatrist studying psychoanalysis as part of his professional training) was a witness to this attack on his credibility, Freeman spilled out a box containing more than five hundred Christmas cards he had received from lobotomy patients. "How many Christmas cards did you get from *your* patients?" he challenged his main critic in the audience.

Freeman's loss of temper and personal response to the Langley Porter criticism was uncharacteristic, perhaps a product of advancing age and the knowledge that lobotomy in any form had disappeared from his former home base in Washington, D.C. The incident led to another assault on his integrity. In 1961 Palo Alto Hospital ended his staff membership on the grounds of his advocacy of lobotomy in the treatment of psychotic children. His main antagonist, the hospital's chief of pediatrics, openly questioned Freeman's medical ethics, a slap that especially angered Freeman. "I had never applied for lobotomies there, and had no intention of doing so," he wrote. "I took the matter as far as the medical staff and the board of trustees, but the decision stood."

Many in California's medical community, though, did not regard Freeman as a pariah. In late 1954, just a few months after his arrival in the state, he attended a doctors' lunch at the Cook House in Los Altos Rancho. Without any consultation or discussion with Freeman in advance, a local pediatrician pointed a finger at him and declared: "Dr. Freeman, you're the most recent arrival and probably don't have too much to do." This presumption was true. "I'm going to appoint you chairman," he continued, "of a committee to see about a new hospital." When Freeman learned that this new facility, which would eventually be called El Camino Hospital, was going to be the first general hospital in the area to maintain a psychiatric service, he readily agreed to head up the planning.

For years he had believed that many patients with mental illness are best served in general hospitals like the one he had worked in and helped plan at George Washington University. General hospitals could better handle emergencies, had a mind-set of expecting speedy recovery, and were far better adapted than psychiatric institutions to the teaching of psychiatry. Although Freeman's son Walter considers this advocacy of community hospital psychiatry to be one of his father's most important contributions to medicine, Freeman was not alone in supporting this cause. President John F. Kennedy embraced the idea of community psychiatry several years later, when he proclaimed that recent "breakthroughs have rendered obsolete the traditional methods which imposed upon the mentally ill a social quarantine, a prolonged or permanent confinement in huge, unhappy mental hospitals where they were out of sight and forgotten. . . . We need a new type of health facility, one which will return mental health care to the mainstream of American medicine."

Freeman got right to work on the planning of El Camino Hospital, commissioning a study that demonstrated a need for such a facility, obtaining the support of the leaders of the surrounding cities, successfully shepherding a ballot measure through citizen voting, and raising funds. (He asked for and received fund-raising advice from his old lobotomy patient, Paul K. Hennessy.) While the hospital was under construction, Freeman served as chair of its medical advisory committee. "I boasted to my friends that I was chief of staff of a non-existent hospital," he wrote. He also directed the planning of the hospital's psychiatric department, which included a pastry kitchen in the ward for occupational therapy, and lobbied for a croquet green in the yard outside.

When the three-hundred-bed hospital opened in September 1961 near the common boundary of Los Altos, Sunnyvale, and Mountain View, Freeman's official role ended. But he ran into problems when he made his first patient referral to the facility, a woman found wandering the streets in a daze. "I escorted her to the psychiatric unit, examined her and decided to use electroshock as an emergency procedure," he wrote. The following day he discovered that his immediate treatment of the patient violated the hospital's policy on the use of electroshock therapy. The patient recovered, but some staff members viewed Freeman with suspicion. The director of the psychiatry service and an advocate of psychoanalysis, Allen "Kris" Kringel, lampooned Freeman with a limerick:

> A fellow named Freeman said: "I've
> A sharp little knife that I drive;
> If you want to be dead
> I'll bore holes in your head
> And then you won't know you're alive."

To be baited like this by a psychoanalyst sounded a trumpet call in Freeman's soul. The alarm spurred him to bait back. Freeman's low opinion of psychoanalysis had not changed for thirty-five years. He regarded his critic's favored form of therapy as "a wasteful procedure" that could be administered as successfully by "[p]sychologists, social workers, nurses, technicians, clergymen, even bartenders" as by psychiatrists. (Freeman was fond of relating the story that the influential psychiatrist Adolf Meyer once listened to the self-analysis and complaints of a patient and blandly responded, "What of it?")

Freud's teachings, however, had gained domination of the practice of psychiatry since Freeman's move to California, a development that aggravated him. Within a few years psychiatry would "come in the mind of the

American public to mean psychoanalysis. The seizure of power was virtually complete," noted Edward Shorter, a historian of the specialty, and it left biologically oriented psychiatrists like Freeman the task of treating severely psychotic patients who were of little interest to psychoanalysts. One such sufferer was Freeman's brother Norman, a brilliant cardiovascular specialist whose career ended because of manic depression and who once slugged Freeman during a violent episode and dislodged a few of his teeth. Freeman admired much about Freud and found the treatment and its founder seductively interesting—he often noted that like himself, Freud was an avid hiker and tireless walker—and Freeman sometimes tried to express the mechanics of lobotomy using the id, super ego, and other Freudian concepts. Whatever admiration he felt for Freud did not spare Kringel, though, and Freeman could not let the thrust of the psychoanalyst go by without a response. He parried in kind:

> Said Kris to the elegant maid:
> "Psychotherapy is a great aid
> At relieving your purse
> While making you worse,
> But for you I will take it in trade."

Predictably, Freeman's requests to perform transorbital lobotomies at El Camino Hospital were not approved. Yet his organic perspective still maintained a base of support, such as that of an admiring physician who wrote to Freeman during the planning of the hospital. "You know, of course, that you have made surgical history, and what is more important to me, you have added tremendously to our fundamental thinking about the relationship between neurological factors and physiological factors in emotional disturbances. . . . Thanks to you and to the endocrinologists, psychiatry will once more become a medical specialty," the fan declared.

Freeman was now receiving far fewer invitations than previously to perform or demonstrate transorbital lobotomies in state hospitals. Yet he believed his efforts as a traveling lobotomist had proven worthwhile. "When I visit the large state hospitals and see hundreds of idle patients," he wrote, "I am appalled at the waste of manpower and womanpower, and long to do something about it. I estimate that 10 percent of patients could be operated upon each year with substantial benefit. But nobody else sees it that way and it is very doubtful whether I shall ever be given the chance to prove my point. The state hospitals at Athens, Ohio, Spencer and Lakin in West Virginia and Eastern and Western State Hospitals in Virginia stand as monuments to the success of lobotomy."

The state hospital system of Delaware remained interested in Freeman's services longer than most. In 1964 it invited him to treat fourteen young institutionalized schizophrenics. Freeman realized how difficult it would be to bring improvement to such severe cases, so he experimentally followed their deep, frontal-cut transorbital lobotomies with the injection of hot water into their fresh incisions, "moving the needle tip from side to side in order to produce as much thermo-coagulation of the orbital white matter as possible." Freeman wrote that he was willing to accept the death of two of these patients—an astonishingly high fatality rate by his own or anybody else's standards—but all the patients survived the procedure. In these cases he set a low bar for success. "I don't see how any of these patients could improve, but at least one can now be cared for at home."

After moving to California, however, Freeman's work at out-of-state hospitals gradually receded into the past. He made just eleven head-hunting trips in 1955, and within a few years he was almost exclusively performing lobotomies in the California state hospital system. Freeman made his last entries in his traveling lobotomist notebooks in 1960, when he made a farewell visit to Athens, Ohio, to operate on a single patient. Nevertheless he continued adding mileage to his car's odometer in trips across the country—some forty-five thousand miles in 1956 and 1957—but with a different purpose in mind. His primary traveling accessories were no longer the transorbital leucotome and the portable electroshock machine. Now he devoted most of his attention to his former patients and their records. "Large individual cards with a summary of the patient's history and progress, with successive photographs fixed to the back, are filed in numerical order in loose-leaf volumes," he wrote in a description of his record-keeping methods. "In addition, I maintain 2 files on 3 × 5 cards, 1 alphabetical, with addresses, telephone numbers, addresses and phone numbers of relatives—many relatives—and physicians. The other is arranged geographically, with the patient's name and number, date and latest address." With these materials in hand, "I'll plan a head-and-shoulder hunt."

This type of hunt, however, featured lobotomized patients as his prey, not institutionalized psychotics needing treatment. Eventually he would amass the world's largest database of psychosurgical cases. He maintained a correspondence with many patients, especially through Christmas cards, but letters "are seldom very informative, particularly those written by the patients themselves," he observed. "The accounts of relatives are often adequate but apt to be tinged with too much optimism or pessimism to furnish a clear picture of the patient. Specific questions as to convulsions, obesity, incontinence and other disagreeable phenomena may be entirely neglected."

Photographs sent by patients also offered an incomplete picture. "The women tend to look glamourous in these pictures as compared with my snapshots of them. I believe that a great deal can be judged from the candid shots that will be revealed only with considerable expenditure of time in personal interviews." What Freeman really wanted was a face-to-face interview. He would prepare for a trip by sending out dozens of letters to patients. "I plan a trip from California to the East in a month or two and hope to stop by and see you, or at least talk with you by telephone," a typical letter went. "I would appreciate it if you would let me know on the enclosed card how I can get in touch with you." The card requested the patient's full contact information as well as the names and addresses of relatives and physicians. Many postcards came back filled out.

If there was no response—Freeman once observed that "psychiatric patients and their families are not proud to be enumerated"—or if the post office returned his letter as unforwardable, Freeman went into action as an indefatigable investigator. He took great pride in his ability to track down patients who had laid the faintest tracks. "She can read, write and speak English, French and German as well as ever, but otherwise her memory does not seem [too] good," noted the mother of a patient in Australia. "She helps me in the house, but I can't let her do it by herself, because she does everything only half. . . . I think Emmy is completely unchanged, only she is a much happier person and easy to get on with. I don't think there is any more I can tell you. I don't show her your letters, she does not think she should ever have been in hospital, she does not remember the operation."

Another time he set out after a patient whose most recent contact information was five years old and who had been operated on ten years previously. He began by going to the most recent address of which he had a record, then tracing his way back to the oldest. Finally, "at the earliest one I found a lady who remembered the girl as living with her family in the apartment above. The father had worked for a certain company but had died and the family moved away. The lady called the company, got the bookkeeper, and this woman [the bookkeeper] remembered receiving a card from the patient's mother the preceding Christmas. It was only a matter of minutes to get the present address in Indianapolis." Freeman arranged a rendezvous with the patient on a later trip, brought a camera, and snapped the trophy. One of the very rare patients who escaped Freeman's detection "had pawned a couple of our typewriters, and had done the same trick with his aunt's heirlooms, so I guess he won't turn up for a while."

Neighbors often proved indispensable allies. Freeman once lost track of a patient five years after her lobotomy. Both she and her mother had moved from their Virginia home without saying where. A former neighbor led him to an acquaintance of the mother, who passed along the name of the town in Florida where she had relocated. Freeman sent a certified letter with a request for any forwarding address. The post office returned it with another address written on the envelope. Later, "on my trip through Florida I found the [patient], deserted by her husband, at the home of her mother. I added another 10 year trophy to my picture album. The girl was regularly employed in a doctor's office and had matured very becomingly."

In another instance Freeman traced a former patient to a rooming house in Asheville, North Carolina. While the patient was out, the landlady showed Freeman his room, "which is a very pleasant one; he has a characteristic of buying all the magazines in town, particularly those dealing with nude art and female charms but he never seems to take any interest in flesh and blood and he probably doesn't even look at half the magazines and books that do pile up in his room," Freeman noted. The patient was neat and free of convulsions or incontinence, "but he seems rather childish and some of his ideas would do credit to a boy of 8 or 9 years. He relies on assertion rather than argument and, consequently, is not a brilliant conversationalist or really any sort of companion."

What Freeman especially sought, in addition to information on the subjects' medical history and periods of institutionalization, were clues as to how well patients were socially adapting, working at a job, keeping house, and helping out with chores. He craved evidence that he had helped restore patients to usefulness in society.

A few patients frustrated Freeman with their refusal to "meet the doctor part way." He recalled one who would not board a bus for a trip of twenty miles that would have made possible a meeting after Freeman had driven a thousand miles, and he felt discouraged by an Irish patient who would not "meet me at the Shannon Airport, which would have meant a trip of 50 miles for him. But he was too preoccupied with the menace of Russia (to judge from his yearly letters) to accommodate himself to my wishes." Others, though, invited him to dinner as if he were a family friend. "On achieving a visit, patients are almost invariably agreeable, pleased to receive a call, perhaps flattered that the doctor has taken such an interest in him. If not the patient, then the family is delighted to be able to show how well the patient is getting along." One transorbital lobotomy patient trailed to her home in Petersburg, Virginia, in 1960 "surprised me by recognizing my voice over the phone as soon as I spoke to her, and greeted me with great affection, saying she'd telephoned her husband to come home if he could but he was busy with a customer and couldn't make it."

Other times the patient's family complained bitterly of the patient's bad behavior "while the patient himself sits grinning, like an interested spectator, hearing tales of somebody else." In a 1964 letter to Freeman, the brother of a prefrontal lobotomy patient reported that his sister "has a marked tendency towards obesity, is prone to loud tantrums and is definitely unable to care for herself without supervision. Fortunately she has no children." An upset and overtaxed mother wrote of her daughter to Freeman in 1956: "Since the operation—lobotomy—she is a mental cripple. No desire to work and be useful. No ambition to improve her mind! She says, 'Nobody loves me! If I cannot get married, I shall commit suicide!' She does not have the intelligence and the will to make anything worthwhile with her life. I am under the impression that there must be a great many people in mental wards that would act or think or even behave more normally than she does, in many instances!"

Freeman complained that a few "relatives blame all the difficulties that the patient experiences upon the operation itself" without recalling the severity of the patient's disordered behavior and thinking before lobotomy. "They are not open to conviction—to the point where I sometimes think the wrong person was operated upon." The very worst reports Freeman filed among his follow-ups came from medical personnel at the psychiatric hospitals in which still-suffering patients—the failures of lobotomy—remained confined. "This 25-year-old white mental defective unfortunately shows no essential change," wrote a St. Elizabeths Hospital psychiatrist of one prefrontal lobotomy patient. "He spends most of the time pacing up and down the floor of the day room, being always mute and entirely out of contact with reality. He is also observed mumbling to himself, exhibiting a great many mannerisms and gesticulations. He is very untidy in his appearance and habits."

To some families, hospitalization was not necessarily a bad outcome. "All goes well with the boy. He has, of course, no emotional reaction whatsoever, so that he is spared any mental torment," wrote the father of one of Freeman's lobotomy patients institutionalized at Worcester State Hospital in Massachusetts. "We visit him once a month, taking him tidbits and entertaining him in the car where he has music and magazines. . . . When the afternoon is over, Fred goes back to his ward with no question as to why he must go back, why he is not taken home. . . . He behaves as if the entire environment was completely familiar and as if he had been at home all the time." In other instances, lobotomy patients remained in hospitals because their families simply could not tolerate their presence at home. "They would probably take him oftener and keep him longer, but Harry and his father do not get along very well," wrote an administrator at Wayne County General Hospital in Michigan of a Freeman prefrontal lobotomy patient in 1961. "Last year they had him home for one period of over two months, the visit terminating because of Harry and his father's getting into a fight."

Freeman sometimes sent his regrets to the families of such patients. "I am sorry to hear that she is not improving that extra 25 percent that would make it possible to keep her at home," he wrote to the mother of one. "When you look at it from her standpoint, however, it seems that she has her own little dream world to live in, and is not too unhappy. It will require a good deal of research in order to find the cause and cure of such a malady as dementia praecox [schizophrenia]. I am sorry that the operation, in her case, doesn't seem to have supplied the answer."

A well-adjusted or functioning patient was a far more satisfying find. He told the story of one veteran of prefrontal lobotomy, a victim of violent psychosis named Beauregard Travis who had been brought in shackles to Freeman and Watts for treatment at GWU Hospital. His threatening demeanor, violent behavior, and intimidating appearance—resembling the emperor Maximillian, he had mutton chop whiskers, long red hair, and an amputated leg, the result of an old roller-skating accident—so frightened the GWU staff that they denied him admission. He went instead to the psychiatric unit at Gallinger Hospital. After his lobotomy he became "rather vivacious, kindly disposed, and has had none of his rages." Travis was released to his sister's custody about one month after operation. His irresponsibility and foul language made him unable to return full-time to his former occupation as a fish grocer, and his wife divorced him. Over a period of years, however, the patient's mental state improved. For Freeman, who kept up with this patient until Travis's death in 1957, this case had a happy ending. "For the past 15 years . . . he has been living outside the hospital and [has been] at times employed," Freeman recorded. "He wrote me with considerable pride when his son was graduated from college and played professional football."

Freeman enjoyed telling the tale of another patient who had first appeared for treatment while a serviceman during World War II. Although obsessed with suicide, he could not receive a lobotomy because of his active status in the military. Freeman and Watts at last gave him a prefrontal lobotomy in 1946, but the patient proved frustrating in his ability to elude Freeman's efforts to obtain follow-up information. After two years of little or no contact, the patient suddenly appeared in Freeman's office; he was carrying "a long, narrow box," which he handed to the physician. "Take this, Doc," he told Freeman. "I've decided not to kill myself." After he left, Freeman opened the box and discovered a rifle and a supply of ammunition.

Over the next several years, the patient continued to avoid Freeman's follow-up communications, although he learned that the patient was employed. Finally reaching him on the phone, Freeman asked the patient if he would like his weapon returned. "No, you keep it, Doc" was the patient's reply. "It's mighty nice of you to offer, but I'd rather you kept it. I don't want it nohow." Freeman's quest for a face-to-face interview was foiled, but "I have a nice souvenir of a death weapon that might have been." (In 1968, when Freeman once again tracked down this patient, he discovered that the former servicemen lived alone, had held a job for twenty-five years, and was unfriendly with his neighbors.)

On another occasion in late 1956, Freeman visited a former prefrontal lobotomy patient living in Miami Beach. He found that her obsessive need to recall the names of people and places was confining her to home, where she ate too much angel food cake, but otherwise she appeared "animated, alert, witty, and makes a very good hostess. . . . She is really much more interested in her canary-colored parakeet."

Patients making contributions to their families or merely getting by outside the hospital regardless of their disabilities were Freeman's best discoveries. In 1957 he received a letter from the husband of a prefrontal lobotomy patient who had received her surgery in the early 1940s. "Lula has done remarkably well since the operation," he wrote. "There remain certain faults but these have more or less been overcome by tolerance and kindness. She performs her housework very well and has been an indispensable part of the home. The children, who are now almost 15 and 16, understand the case better, and better appreciate what [the patient] went through. . . . Lula has done a very acceptable job in the preparation of meals and the children's lunches for school which cannot be measured in dollars and cents. All in all I would say that you and Dr. Watts made a great contribution to society, well worthy of the highest in the medical profession." His wife signed her name at the bottom of the letter.

Likewise, Freeman must have felt cheered by the outcome of a prefrontal lobotomy patient living with her mother in 1956: "Lillian doesn't even wash out her undies, but her mother looks after her well nevertheless. Lillian takes in about five movies a week and the rest of the time is watching television," he noted. "She reads a little, seldom touches the piano although she played a somewhat complicated piece for me with some feeling but without brilliance. The most distressing social shortcoming, however, is a tendency to break out into rather uproarious laughter almost automatically and without good motivation. . . . There are no hallucinations or other signs of break with reality, nor have there been any convulsions or bed-wetting." Another patient, a ten-year veteran of a transorbital lobotomy, had contributed to her family's financial health by entering and winning several slogan-writing contests. One of her entries earned the prize of a $30,000 house. "She had never entered a contest before her lobotomy," Freeman observed.

Freeman found some former patients able to return to activities that made powerful demands on their mental functioning. These included a few musicians, including a player for the Detroit Symphony Orchestra during the late 1940s. Another, Violet Silver, was a violin prodigy who had showed symptoms of schizophrenia as a teenager and stopped playing her instrument for twelve years. After her lobotomy she returned to the violin, performed numerous recitals, and took several students. Nearly twenty years after her surgery, Silver still appeared to be making her living from music. Freeman clipped an advertisement promoting her stage appearances in 1960, which notes that the violinist "has been praised by the Prince of Wales" and "will give 49 percent royalties to someone who will help her publish the operetta she wrote." (Silver's case ended far better than that of the Polish violin prodigy Josef Hassid, praised by Fritz Kreisler as the greatest young talent of the previous two hundred years, whose chronic schizophrenia led English surgeons to leucotomize him in 1950. He died three weeks later from a cerebral infection.)

In 1963 Freeman heard from a lobotomized patient, a longtime cab driver, who wished to command an entirely different kind of vehicle. "If you have been driving a taxi all these years and have a good record for caution," Freeman wrote to him, "I don't see any particular objection, from the medical standpoint, why you should not be able to take lessons in piloting a helicopter. It sounds like an exciting venture, and there's no telling what it would lead to." But he ended with a caution: "If your wife has definite objections, I would be inclined to side with her." Freeman found other lobotomy patients working as lawyers, religious missionaries, government workers handling confidential data, and physicians—including "a psychiatrist [who] was promoted to chief of service in a large mental hospital." These high-achieving patients were a distinct minority, "but," Freeman was quick to observe, "the same can be said of other individuals whose frontal lobes have never been operated upon."

A few people required no detective work at all for discovery: Freeman received letters repeatedly from patients who occupied a gray zone of recovery between mental health and total disability. Some seemed as intent in tracking him as he was in locating less eager respondents. An early prefrontal lobotomy patient living in Massachusetts sent a typical letter in March 1957:

> Can do. Have time & will dabble in oils for a hobby. My Red Sox licked the invincible Yanks today in thirteen innings. You see my dear, you didn't destroy my brain exactly, have original ideas as yet. Aren't the new tranquilizer drugs the best so far? Wish I'd had same many years ago. Used to be in high gear all the time. Can't now, though. Get very exhausted & love it. . . . Please overlook my frequent letters. I need to confide in someone and am keeping my trap shut. People don't like talkative folks, do they?

Another such patient, an especially persistent correspondent recovering from lobotomy in Washington, D.C., machine-gunned her thoughts to Freeman in 1956:

> Did you read about the Grace Kelly and Prince Rainier wedding, at Monaco? Since then as you know the place where she married was a gambling place and there have been so many riots since the marriage did you know that??... Have you written to Bob Hope, if not you better had as he's going to Germany and real soon, so please do write to him for me, did you see his Tuesday night show, with that great singing star Pearl Bailey?... I've been reading quite a bit, since I've had about five or six books sent to me from the Doubleday One Dollar Book Club, and haven't paid for any of them.

Freeman usually dignified these letters with a serious reply. "Things seem to be going nicely," he wrote to the Washington patient. "I don't take much interest in the movies or television, and, consequently, can't tell you much about the private lives of Judy Garland, Grace Kelly, Eddie Fisher and so on. One of these days, however, maybe you will get to Hollywood and see the homes of the stars because they take you on a tour visiting the area and lots of people are accommodated." He noted in the file of another patient who traveled to visit him, "I think what she needs more than anything else is a bit of support."

In Christmas cards or letters, some confided mundane details of their lives that evidently seemed important at the time. "I had my teeth pulled in August," volunteered a former patient from West Virginia in 1965. "Won't get my dentures until after Christmas. Hard to eat without teeth." Another simply noted: "In the best of health—thanks to you." Other former patients expressed their gratitude mixed with puzzlement over their experience with mental illness and Freeman's treatment. "How many times we think of you—and how we were directed to you," wrote a Kentucky woman in 1957. "God gave you a wonderful brain and skills.... I feel good and am thankful I was saved. From what?"

Some searches ended with news of the patient's demise or the discovery of a death certificate. Usually the news came from relatives. "He enjoyed your cards so much & to think that you still remembered him," wrote the sister of a former prefrontal lobotomy patient in 1965, reporting the man's death from emphysema. "I miss him very much & he was a good man but suffered so much. May his soul rest in peace." Freeman regarded a patient's death, whatever the cause, as important. "There is a certain satisfaction in writing *finis* to a case when the patient dies," he noted. "That satisfaction is doubled when the cause of death is clearly stated, and redoubled when an autopsy is performed and the specimen [is] available for study." While this response may sound coldhearted, to Freeman a patient's death signified the conclusion of a medical narrative in which he had played a decisive role. Learning of a patient's passing was as satisfying as reading the final lines of an engrossing novel.

Although Freeman had been interested in follow-up studies of his lobotomy patients from the moment Adolf Meyer recommended them back in 1936, he acknowledged that his devotion to gathering data on his patients had swollen to a compulsion after his move to California. At a 1957 meeting of the Southern Medical Association where Freeman presented a far-reaching follow-up report on the first five hundred prefrontal lobotomy patients, Watts rose from the audience to hail what he termed Freeman's "magnificent obsession," a drive—perhaps unmatched by any other physician of the twentieth century—to accumulate every bit of information he could about those he had treated in the past. Freeman listened to this tribute and teased his audience and former colleague by alluding to the twenty-five hundred transorbital follow-ups he hoped to complete in the coming years.

Freeman compiled and published many follow-up reports on his patients, working on them through most of the 1950s and 1960s. In one of these papers, "Frontal Lobotomy 1936–1956: A Follow-Up Study of 3,000 Patients from One to Twenty Years," Freeman referred to the uniqueness of his tracking of patients over such a long period. The follow-ups were possible "because of the warm personal feelings that entered into the physician–patient relationship and made possible detailed knowledge of the long postoperative course in hundreds of cases." In addition, "the superintendents and staffs of state hospitals have exerted themselves to keep me informed about the progress of operated patients remaining in their care." Freeman wrote that he was most interested in the effect of lobotomy on the treated hospital population "rather than the effect of operation upon individual patients. It is as if attention has been focused upon the field of grain and its yield per acre, rather than selecting for particular observation the gaudy sunflowers—or the thistles."

Among the 2,454 prefrontal and transorbital cases he had handled through 1956, he knew of twenty-eight marriages and sixty-two children born. "The receipt of a wedding announcement or a pink or blue baby card from a former patient arouses a certain pleasurable reaction in the recipient that carries him back to Christmas Day in 1936," the day that a patient's unplanned departure from the hospital forced Freeman to miss the birth of his youngest child, Randy. "I recall one man who was operated upon after many years of a schizophrenic existence," Freeman wrote. "He later married and fathered a child. At the time of the evaluation, some 5 years after operation, the child was too young to be sure of, but of the family constellation, mama, papa, wife and parent, the patient appeared to us to have more common sense than the rest of them combined."

Freeman's research turned up one patient responsible for an accidental homicide, three who suffered violent deaths, and only a very small number who committed crimes. Many former patients became alcoholics, and four died in accidents caused by their driving while intoxicated. Most significant to Freeman, 80 percent of his schizophrenic patients who received lobotomies while institutionalized were out of the hospital six years after surgery, as were 90 percent of the patients with depression and psychoneuroses. Freeman concluded that "lobotomy is worthwhile, because it contributes notably to the social effectiveness as well as contentment of the patient and to the restoration of solidarity to many disrupted families." His results and conclusions conflicted with those of several other follow-up studies of lobotomy patients. In a five-year controlled study undertaken by K. G. McKenzie and G. Kaczanowski and published in 1961 in the *Canadian Medical Association Journal*, for instance, the authors concluded that "prefrontal leukotomy does not produce any rate of remission significantly beyond that to be expected without the operation." Freeman frequently criticized such studies as misleading because they often focused on patients who remained hospitalized—the failures of lobotomy. He offered the unverifiable explanation that "in the great majority of cases the lobotomy has failed not because of surgery but in spite of it."

Freeman reported some of his own results in 1961 at the Third World Congress of Psychiatry in Montreal and met a kindred spirit, Sadao Hirose, a Japanese surgeon who had performed hundreds of lobotomies in his homeland since the 1940s. Both men's names were synonymous with the practice of psychosurgery in their countries, and they formed a close bond. They maintained an active correspondence until Freeman's death. Soon after their meeting, Freeman invited Hirose and his wife—a scholar in her own right who published research on female murderers—on an outing to Yosemite, where the Japanese couple were impressed by, although somewhat unprepared for, a hike to Vernal Falls, where Keen had perished about fifteen years earlier. "Mrs. Hirose was wearing a tight skirt, and the high steps were a bit embarrassing to her," Freeman observed. Her husband inquired whether many visitors to the park committed suicide by throwing themselves from the top of El Capitan. "'No,' I said, 'because it's too hard to get to—8 miles by trail,'" Freeman recorded.

At home Freeman fought off boredom by remaining a prolific writer of articles for medical journals, for which he frequently authored papers on lobotomy and other subjects. In one article, titled "Bedside Neurology" and written for *Northwest Medicine*, he colorfully detailed his techniques for handling patients who suffered mortification from near-nude physical examinations. "Shorts for the men and bras and panties for the women are much more satisfactory than a Ku Klux Klan type of environment," he wrote. "Whatever embarrassment may occur at first can be quickly overcome by giving the patient something to do, particularly if the backside is examined first." One of his most unusual article topics was what he called "Hommigonna Syndrome," a term he derived from the utterances of "a fat slob of a patient" who repeatedly complained, "Hommigonna get up this morning?" and so on.

After two years as a solo practitioner in California, Freeman received an invitation to join Foothill Professional Corporation, a newly forming partnership of two dozen physicians and dentists. This time the offer was not rescinded, and Freeman became the group's only neuropsychiatrist. The partnership built a large clinic in Los Altos that was completed in 1959. Freeman signed a twenty-year lease and designed his own quarters, which covered more than 1,200 square feet and included three examination and treatment rooms, another room devoted to electroshock and amytal treatment, a lounge paneled with redwood, and offices for two physicians. "I was happy there, almost as much as I was in the grandiose office in [the] La Salle Building in Washington—even more so perhaps, since it was my own design." Freeman's monthly share for his space cost $576, or about $3,500 adjusted for inflation. This high overhead forced him to temporarily rent his lounge to other physicians.

Soon, however, he affiliated with a young neurosurgeon named Robert Lichtenstein. A former intern at Cook County Hospital in Chicago who had performed a few lobotomies during his residency, Lichtenstein had moved to California in 1960 and was looking for office space. "Somehow he got my name and heard that I was looking for a place, called me up and I met him," Lichtenstein remembered. "I vaguely knew Freeman as a lobotomist." Freeman offered him a space in his office. "When he offered it, it was such a wonderful gesture—a novice neurosurgeon coming into the community and being invited by the world-famous neurologist, with the proposal that I could be around or help him with psychosurgery. It was a little daunting."

As they became better acquainted, Lichtenstein found the elder physician to be an energetic, strong-spoken, and friendly partner happy to take Lichtenstein's children out on long country drives. Freeman rarely spoke of his wife. Lichtenstein frequently assisted him with transorbital lobotomies, always in a hospital and never in the office. "For me, it was interesting to see the patients before and afterwards—to see that a person could undergo a transorbital lobotomy and come out, for all intents and purposes, improved. It worked for certainly some patients," Lichtenstein said.

There in the Los Altos clinic, with a young neurosurgeon at his side, an aging Walter Freeman remained in practice. He still took his daily dosage of Nembutal. "I think that's what kept him from becoming truly manic and truly depressed again," Lichtenstein said. Diminished in professional stature and convinced that transorbital lobotomy still held a place in the treatment of mental illness, Freeman focused on his follow-up studies of the procedure on which he had staked his career, reputation, and happiness.

## Chapter 13

# Decline

IN A MOMENT of reflection in the early 1960s, Freeman divided his life into three phases: his childhood years in Philadelphia, which he labeled preparation; the following three decades in Washington, a period of accomplishment; and his final years in California, a time of decline. The terms he used in his appraisal of his personal life could just as easily characterize the stages of the flourishing of psychosurgery during the very same years. Such was the close association between the therapy and the man who championed it.

The years defining Freeman's childhood, the 1890s through the start of World War I, bracketed the earliest preparatory accomplishments in the history of psychosurgery: the efforts of Gottlieb Burckhardt and Ludwig Puusepp to manipulate the brain in a quest to treat mental disease. Freeman's tenure in Washington coincided with Moniz's initial accomplishments, the development of the Freeman–Watts techniques of prefrontal lobotomy, and the peak years of transorbital lobotomy. And almost from the moment Freeman headed west to California, important changes in the practice of psychiatric medicine—the introduction of new drugs, the gradual emptying of psychiatric hospital beds, and the ascent of psychoanalytically oriented psychiatry—sent lobotomy into a decline from which it never recovered.

Freeman's personal decline sprang from the flaws in his character—primarily ego and a preference for solitary work—that were evident from early adulthood. Once in California, he gave up the comforts and benefits of academic affiliation. Throughout the 1960s, while GWU students and colleagues heaped praise and honors on James Watts at the time of his sixtieth birthday, retirement, and other landmark life events, Freeman remained in the shadows. He failed to build a marriage that could offer him emotional succor in trying times. And although he could justify the risks and uncertainties of lobotomy in the era before chlorpromazine, Freeman persisted in advocating psychosurgery well past the time it offered the best choice to more than a tiny percentage of mentally ill patients.

No excess of hubris, however, compelled Freeman to recommend lobotomy to patients beyond the guidelines he had long followed. He nearly always performed lobotomy as a means of treating illness or pain, not as a tool for mind control. Nevertheless he acknowledged that lobotomizing patients often quelled their violent behavior, muted their emotional urges, and reduced property damage in psychiatric hospitals. In short, lobotomy changed their behavior in ways that benefited the medical establishment of caregivers—although not always in socially acceptable ways, as the families of patients discharged from hospitals could testify. A lobotomized patient traded one set of debilitating behaviors for another. At a time when chlorpromazine was about to unseat transorbital lobotomy as the best hope for untreatable patients, a few people outside of Freeman's circle lost sight of the needs of patients or the institutions that housed them. They began investigating the utility of psychosurgery as a means of controlling behavior in the interests of governmental authority, regardless of the previous mental condition of the patient. Lobotomy, speculated these investigators, might prove useful in controlling thought and behavior.

As early as 1947, Freeman's friend, Gösta Rylander, a Swedish neurosurgeon, noticed that some lobotomized patients experienced a reduction of the intensity of their political fervor. Speaking before a meeting of the Association for Research in Nervous and Mental Disease in New York City, Rylander—whose own household included a cook who had received a lobotomy—described one patient he had operated on in Stockholm. Before psychosurgery the patient had drifted left from socialism to communism. The patient had been concerned about the state of the world and viewed communism as a movement that could right many of society's wrongs. He voraciously read works by Karl Marx and Upton Sinclair. Psychosurgery amputated that drive to reform. After the operation the patient retained little of his interest in politics, and ceased even caring about the threat of nuclear war.

The tawdry *New York News*, which covered Rylander's presentation, speculated that psychosurgery represented "a sure way to rid the world of Communists. Make a hole in their heads, [Rylander] said—but by surgery, not bullets." Rylander, of course, made no such suggestion and merely reported the effect of psychosurgery on the political convictions of one patient. Freeman, who with Watts presented at the same meeting a paper on the use of psychosurgery to treat chronic pain, agreed with Rylander that lobotomy could prevent politically involved patients from following their convictions. "It seems to me that Dr. Rylander's conclusion that lobotomy takes something from an individual should be judged in the light of what lobotomy gives back to the individual in relation to his prepsychotic adjustability to his social surroundings," Freeman declared. Exhibiting his characteristic immunity to criticism, Freeman invited Rylander on a postconference hike along the Appalachian Trail.

In the following years rumors ran rampant that despite the Soviet Union's official ban on psychosurgery, it and other communist governments were using lobotomy and similar procedures to brainwash prisoners and render others politically apathetic. This fear spread into the figurative language of the era's Red-haters. In 1961 anticommunist activist Frank S. Meyer described a procedure he called "communist psychosurgery," a psychological technique of forcing party members to excise their devotion to family, romantic interests in people not dedicated to party causes, homosexuality, and other behaviors prohibited among followers—just as real psychosurgery cut out emotional response.

Others, however, advanced far more serious allegations of true psychosurgical abuse behind the Iron Curtain. In September 1956, for example, *Suppressed* magazine published a lengthy article covering allegations of the employment of psychosurgery in communist countries. The story maintained that the Soviet dictator Josef Stalin used lobotomy as a tool to smother dissent, and it suggested that twenty-one American prisoners of war who refused repatriation to the United States after the end of Korean War hostilities had been made docile with lobotomies administered by the Chinese. The article also conjectured that two prominent anticommunist figures of the 1950s, the West German intelligence chief Otto John, who claimed to have been abducted by East German agents in 1954, and Cardinal Joseph Mindszenty, primate of Hungary, received lobotomies from their communist keepers. "A comparison of pictures taken of the cardinal before and during his trial [on espionage and foreign currency charges] is like looking at two different and separate personalities," *Suppressed* observed. The surgical knife, it continued, may have diminished the political commitment of these and other prisoners. One proponent of this theory, the University of Texas Medical School professor Charles Pomerat, showed confusion on the techniques of lobotomy when he speculated that "the Communists may use a variation developed in Portugal which consists of removing an eyeball, severing two nerves in the back of the eyeball and replacing it."

These fantasies might have seemed even more far-fetched and paranoiac had the U.S. government not been conducting its own secret investigations of the utility of psychosurgery as a political tool. In 1952 the Central Intelligence Agency commissioned the psychiatrist Henry P. Laughlin to report on the potential of lobotomy and other psychiatric procedures to influence or disable communists. Laughlin finagled a two-day invitation to observe Freeman at work during his transorbital demonstrations in West Virginia hospitals.

In his classified report titled "Some Areas of Psychiatric Interest," Laughlin commented that the procedure would be adaptable to intelligence work and noted that he "watched Dr. Freeman perform 22 transorbital lobotomies with an average of about six minutes per operation. This included time for before and after photographs, as well as the keeping of notes and records. . . . From an empiric standpoint, the operative procedure is relatively simple and could be learned in a brief period of time by almost any intelligent person." In addition, he wrote, "there is not great outward evidence of injury or damage to the patient besides the behavior changes and the black eyes. . . . The average pathologist performing an autopsy would have to be a keen and careful observer to detect changes in the brain substance made by the operator." Because "I felt unable to disclose to Dr. Freeman the real basis of my interest," Laughlin noted, he could not solicit the lobotomy expert's opinions "as to how the procedure might be modified" for use by the CIA.

Laughlin, who also professed an interest in the possibilities of taking hypnotic control of patients during the period of unconsciousness following electroshock therapy, formed his own opinions on the potential lobotomy presented as an intelligence tool. "To date, there has been considerable discussion relative to the possible use of the lobotomy-type operation by this Agency as a neutralizing weapon," Laughlin wrote in prefacing his conclusions. He described the role of the frontal lobes as one that allowed a person to pursue a cause and feel devotion to it. "Certainly any crusading spirit is apt to be quenched," he reported. "Community enterprise, activities in the way of social uplift, leadership and executive abilities and activities, are apt to be lessened after operation. . . . On this basis, a zealous and fanatic communist, if lobotomized, might retain his interest in communism, but his drive, zeal, and ability to organize or direct would be substantially reduced."

Laughlin acknowledged the strangeness of his suggestions. "However, if it were possible to perform such a procedure on the members of the Politburo, the U.S.S.R. would no longer be a problem to us! . . . One must consider also that resorting to the use of such practices could contribute substantially to lowered morale in the general population because of its interpretation as an indication of desperation and fear. There would also be urgent and damning criticism, on the basis of the lowering of moral and ethical standards. Charges of germ warfare would be mild in comparison to charges that follow a proven case of the use of transorbital lobotomy." In the end, Laughlin declared that lobotomy was not a promising procedure for the CIA's use, given its risks, except in times of extraordinary danger to national security. Laughlin went on to attain prominence as the founder of the Eastern Psychoanalytical Association, which eventually grew into the American Society of Psychoanalytic Physicians.

Meanwhile, criminals had come into focus as possible subjects of psychosurgical experimentation. Freeman operated on murderers in the West Virginia and California state hospital systems and possibly elswhere, but only in an attempt to alter their psychiatric symptoms, not their criminal behavior. In the case of one prefrontal lobotomy patient who in 1955 went to trial for larceny, Freeman informed the judge that the lobotomy had left the patient still accountable for his acts. "A rather wide acquaintance with other patients of the same type, committing the same offenses . . . leads me to believe that the patient knows wrong from right and is responsible for his actions," he wrote. The judge agreed.

Other psychosurgeons took the view that lobotomy could relieve a patient's impulse to commit crimes. In one disastrous experiment of 1941, the renowned neurosurgeon Leo Davidoff lobotomized an inmate at Sing Sing Prison known as J.S., who had a history of pedophilia. His brief improvement prompted Davidoff to assert that psychosurgery could offer benefits to some sex offenders. After his release from prison the following year, J.S. remained in contact with his parole officer for two years, then vanished from sight. He resurfaced in 1948 in the psychiatric ward of the Veterans Administration Hospital in Hines, Illinois. J.S. had continued soliticing sex from young boys, and his lobotomy only succeeded in making him irresponsible, uncaring of his personal hygiene, and unable to hold a job.

In 1944 surgeons in Hawaii used lobotomy to secure a legal declaration of sanity and release from prison for a man convicted of violent crimes. The operation, the doctors argued, had cured the prisoner of his criminal impulses. Aage Nielssen of Wayne County General Hospital in Michigan reported in 1947 that he had given a lobotomy to a twenty-eight-year-old woman with a history of shoplifting, arson, drug peddling, jailbreaks, and assault dating from childhood. She was the one thousandth person in the United States known to undergo psychosurgery, and her case "is the first one, as far as can be learned, for the deliberate purpose of reforming a law violator," the *American Weekly* incorrectly speculated. One month after her surgery, noted her physicians, "She is more relaxed. In contrast to her previous defiant attitude, she shows genuine embarrassment about her past conduct."

The most notorious and ill-fated attempt to treat criminal behavior through psychosurgery began the same year. The patient was a thirty-seven-year-old habitual burglar named Millard F. Wright then facing a prison sentence in Pennsylvania of up to forty years. Wright, whose lengthy record included nonviolent crimes such as the theft of clothing, radios, and clocks, had not shown signs of psychiatric illness before an arrest in 1946. In a Pittsburgh jail, however, he refused to speak or eat, and he tried to take his own life. Authorities moved Wright for a year of observation to Fairview State Hospital. When the hospital pronounced him recovered and discharged him, he appeared in court to face his most recent burglary charges.

Wright's attorney devised a plan to keep his client out of prison. He requested a postponement of the trial, during which time Wright would receive a prefrontal lobotomy to treat his personality disorders. (Neither Freeman nor Watts would participate in the surgery.) The judge consented to this unprecedented proposal, and Wright returned to court two months after his operation. "Wright looked like a new man," *Time* magazine reported. "He was cheerful, sociable and relaxed. [His psychiatrist] thought there was a good chance that he had been cured of the urge to steal. But to complete the cure the prisoner would have to be set free and given a chance to live in a 'normal' environment."

Unluckily for Wright, the original judge had died during the postponement, and his replacement did not feel confidence in the lobotomy's power to cure a criminal. Fearful of a rush of convicts seeking lobotomies, the new judge convicted the defendant of burglary but gave him a light sentence of two to twelve years in prison. The sentence was not light enough for Wright, who used his eyeglasses to slash an arm, then hanged himself while in custody at the Butler State Police Barracks. "I am sentencing myself to death for my evil misdeeds," he wrote in a suicide note. Six months earlier the Ohio Pardon and Parole Commission had declined to release from prison an embezzler and forger named Frank Di Cicco who had voluntarily undergone psychosurgery in an attempt to "rid himself of criminal traits." Yet another case reinforced the belief that psychosurgery did not in itself prevent criminal behavior: the 1950 slaying of the Yale University psychiatrist Lewis Thorne by a former patient who had received a lobotomy several years earlier.

Freeman avoided operating on patients incarcerated in prison, but he seemed drawn to female patients. From the late 1940s through the introduction of antipsychotic medication in 1954, men slightly outnumbered women in the state hospital systems. Yet female patients constituted about 60 percent of those who underwent psychosurgery—and an even higher percentage of patients who were operated on with transorbital lobotomy. Two factors contributed to the gender disparity: a feeling within the psychiatric profession that it was easier to return women to a life at home than it was to rehabilitate men for a career as a wage-earner, and, most importantly, the abundance of women with diagnoses of the affective illnesses considered most responsive to lobotomy.

As lobotomy fell into disuse even among women patients during the late 1950s and 1960s, Freeman faced a swelling tide of change within his profession. From an all-time high of 559,000 patients in 1955, the population of state and county psychiatric hospitals fell an average of 15,000 per year through 1970. The use of electroshock therapy, largely replaced by pharmaceuticals, plummeted, and Utah became the first state to regulate its application in 1967. Twenty-five other states would follow suit over the next twenty years, a sign that the public refused to believe in the infallibility or even the competence of many in the psychiatric profession.

In such films as *Shock Corridor* (1964) and *A Fine Madness* (1965), psychiatrists were portrayed not as noble-minded physicians with a wealth of effective treatments at their disposal but as immoral and egomaniacal antagonists in command of few truly effective therapies. Meanwhile, women were enrolling in medical schools in increasing numbers. Freeman's old employer, GWU, viewed this achievement with mixed feelings: a 1965 medical alumni magazine cover showed a female diplomate with the headline "Medicine Never Looked Better." Freeman was not opposed to the increasing numbers of women physicians, but they sometimes caught him off guard. While in Hungary during a trip to Europe to visit his son Walter's family in 1965, he was astounded by the sight of one neurosurgeon he met there. "Here was a petite redhead," he observed, "in a surgical scrub-suit, with golden sandals and painted toenails!"

Even within the shrinking field of psychosurgery, in which the number of operations had declined from thousands a year to mere hundreds, Freeman was falling behind the times. New operations targeted not just the prefrontal lobes but also the cingulate gyrus, hippocampus, amygdala, and their connecting networks. Neuroscientists amassed evidence that an individual's emotional response was not simply a product of communication between the thalamus and the prefrontal lobes but involved the entire limbic system, the central regions of the brain that play a role in mood and attitude. They developed new operative techniques that made prefrontal and transorbital lobotomy seem crude and recklessly destructive by comparison.

There was psychosurgery by thermocoagulation of brain tissue, ultrasonic irradiation, implantation of radioactive particles, and gamma rays, and surgery preceded by electrostimulation. Freeman followed these developments, but he made no attempt to change his lobotomy technique or adopt new procedures. The great neurological innovator of decades earlier was unwilling to change—and was growing old. "I would say that my father was well aware in his last few years that the days of the hand-held lobotomy knife were over," Franklin Freeman remembered. Or maybe not. Freeman still maintained that lobotomy had suffered an eclipse, not extinction. "I believe it's due for adoption when the surgeons make up their mind for it," he said in 1968. "I think they're missing a good bet."

Although greatly reduced in number, Freeman's lobotomies continued. At Doctors' General Hospital in San Jose, for instance, he performed eight transorbital lobotomies in the first six months of 1965. Herrick Memorial Hospital in Berkeley, the site of some of Freeman's lobotomy demonstrations in the late 1940s and early 1950s, was the last institution to still grant him surgical privileges. In late 1966, however, staff members at Herrick began having second thoughts about hosting Freeman's psychosurgical activities. That year during a transorbital lobotomy at the hospital, an orderly who witnessed the surgery had fainted. "I had the impression that Dr. Freeman was acting somewhat like a lobotomized patient himself, withdrawn from his emotions," the orderly recalled.

A committee formed at Herrick to review the case histories of patients that Freeman had lobotomized there over the previous six years. "I think the committee is approaching this in a systematic way without a great deal of prejudice," wrote staff psychiatrist L. G. McKeever to Freeman. "We will appreciate anything you can do to make the task easier and although the committee has not requested it, if you could send us recent follow ups on any of the patients I am sure it would be helpful." Freeman responded with a promise to forward his follow-up data on the thirty-five patients he had operated on at Herrick between 1960 and 1966. He clearly still held faith in the power of the camera lens to produce evidence in support of transorbital lobotomy. "If you and the committee so desire," Freeman wrote back, "I would be glad to appear before the committee and bring with me photographs of the patients, before and after operation."

Freeman was seventy-two years old, but he felt fully capable of performing transorbital lobotomy. ("It's such an absurdly easy procedure to do," his son Paul has commented.) Given his age, his health was not markedly bad. He controlled his diabetes with medication and an adherence to a diet that avoided starches and sweets. Small skin cancers removed from his shoulder and eyebrow during the previous several years had not returned. He had undergone the removal of a troublesome prostate gland in 1964. Freeman remained an avid camper and hiker. During one trip to the Sierras with his son Walter and his grandson, the group had encountered an unexpected August snowstorm. They "trudged for miles through wet snow and pretty nearly froze our feet," Freeman wrote. "We had to sit up all night around a fire rather than go 'to bed' because our bedding was all wet, and then our food ran low." Yet Freeman came through fine, only losing 7 or 8 pounds from the experience. He regularly walked the beaches of the West Coast, where he loved catching abalone, and professed an interest in learning how to surf. He weighed about 155 pounds, giving his tall frame a decidedly lean look.

The end of Freeman as a psychosurgeon finally came in February 1967 when he scheduled two women for transorbital lobotomy at Herrick Memorial Hospital. Freeman had known one of the patients, Helen Mortensen, for more than twenty years—she had been among his first ten transorbital lobotomy subjects in 1946, and she had returned to him ten years later for a second psychosurgery, a transorbital lobotomy with the deep frontal cut. Now, following another eight-year interval, she believed she needed more work from Freeman. "This time, to my dismay, she had a hemorrhage and died in three days. . . . This was the last day I operated. The other patient done on the same day has done well." Soon afterward, Herrick stripped Freeman of his operating room privileges.

For Freeman a career without lobotomy was almost unthinkable. A great deal of his professional identity rested on his reputation as a lobotomist. His unhappy life at home kept him from embarking upon a retirement to leisure activities, as many of his colleagues had undertaken. So Freeman continued focusing his attention on various aspects of the practice of psychiatry. For years he had held a fascination with the lives of his colleagues. Using *Who's Who* and other reference books, he tracked and tried to detect patterns in psychiatrists' places of birth, gender, the number of their offspring, and their memberships in professional associations. ("Aloysius Church leads the list with memberships in 14 neurologic societies and 8 psychiatric, while Frank Fremont Smith is close behind with 10 of each," he discovered.)

Religious affiliation also engaged him. Using Cohen and Shapiro as representative Jewish names, for instance, he tried to measure their occurrence in the American Psychiatric Association directory against the number of Smiths. He found "44 Cohens, and 26 Shapiros, together with 75 Smiths. . . . Based on these somewhat tenuous data it would seem that psychiatry and particularly psychoanalysis attract Jewish physicians." Perhaps inspired by the medical propensities of his own family, he researched and wrote about American psychiatric dynasties in which father, son, and grandson practiced in the field. He also examined such magazines as the *New Yorker* and the *Saturday Evening Post* to determine the frequency with which psychiatrists appeared in cartoons.

The element that obsessed him most about psychiatrists was their high rate of suicide. Reading about the death by suicide of eight of Freud's associates first drew Freeman's attention to the topic. Attacking this subject with his usual alacrity, he combed back issues of the *Journal of the American Medical Association* for obituaries of psychiatrists in which suicide was the listed cause of death. Obtaining the death certificates of an additional two hundred psychiatrists, he "found that about a third of them had killed themselves. . . . It seems that the suicide rate for psychiatrists is about 8 times as high as for white males." Poisons, including overdoses of drugs, were the most common means of self-inflicted death. To explain this epidemic of self-destruction, Freeman laid the blame on psychoanalysis, whose practitioners had battled Freeman for four decades and who had so recently succeeded in helping to snuff out lobotomy. "What bears renewed emphasis is that personal analysis by no means always leads to a smoother adaptation to the life situations that arise," he wrote. "I suspect, though scientific proof is lacking, that insight is sometimes intolerable and that the emphasis on insight is sometimes misdirected in psychotherapy to the detriment or even disaster to the psychiatrist himself." Freeman submitted a paper on the topic to the *American Journal of Psychiatry*, whose article reviewer complained, "As a well-balanced scientific document, it leaves a lot to be desired. There really is no attempt at any sort of formal methodology. . . . I have no special comments for the author except that I admire his nerve."

Presenting what he termed a "toned-down" version of his findings on suicide at the annual meeting of the American Psychiatric Association in 1967, Freeman drew a laugh from the packed lecture hall by projecting on a screen a cartoon from *Playboy* showing a man perched on a high window ledge. The man says to a police officer: "Why should I wait and see a psychiatrist? I'm a psychiatrist." Freeman wrote, "When the laughter subsided, I paused for a moment and began: 'What I'm going to talk about is not funny. It's serious.' . . . The impact was definite if not frightening." A letter writer to the *American Journal of Psychiatry* offered the most obvious counterargument to Freeman's conclusions. "It seems to me that it was rather the author's bias than facts that inspired" them. Subsequent studies of suicide by psychiatrists have supported Freeman's hypothesis regarding the frequency of their self-inflicted death, although they have not endorsed his conclusions as to the cause.

Freeman also studied the frequency of suicide among his lobotomy patients. He found that only 22 of the 849 nonoperative deaths documented in his patient records were suicides. Most of the suicides had occurred among the first series of prefrontal lobotomies performed with Watts. Overall, lobotomy patients were about half as likely to take their lives as were patients in mental hospitals, although Freeman acknowledged that hospital inmates often lacked the means to commit suicide if they wished to do so.

Freeman's intense interest in suicide naturally raises the question of whether he was considering ending his own life in its waning phase. Certainly, Marjorie's deteriorating condition and his collapsing relationship with her greatly disheartened him. She drank daily. "We'd visit her, and nothing was prepared," her son Paul remembered, "and she was in sort of an alcoholic haze, usually from drinking sherry. She was kind of apologetic about it, but it was her refuge." Freeman, in response, plunged more deeply into his own work and projects. But the couple resisted ending what had evolved into a truly unhappy marriage. In addition, Marjorie felt she had done her duty as a parent and limited her role as a grandmother. "I don't know how much my mother wanted to be involved with her grandchildren—not very much," Paul remembered. "She was pleasant with them, but not ecstatic . . . I don't recall her making strenuous efforts to get to know the grandchildren or to make a fuss over them." Freeman, however, made more of an effort; he joined his grandchildren on a few of the same kinds of drives, camping trips, and hikes that he had shared with his own children. These shared experiences generated respect among the children for their grandfather, if not a strong connection.

One morning in late 1967 Marjorie stumbled in her bedroom after downing an entire fifth of sherry and broke her hip. Freeman, working in his study, heard the fall and took her to a hospital. Doctors nailed her hip, and Freeman soon checked her into a nursing home. There she fell once again and broke her other hip. When Freeman gave her a neurological examination, he found evidence of a deterioration of the nerves in her legs that could cause numbness or pain. By December 1967 he had to admit to a friend that Marjorie would probably never again be able to live at home. "She is quite confused and forgetful," he wrote. Her children noticed a gradually accelerating dementia. In her absence the Freeman house seemed empty.

His son Randy was also ill. Demoralized two decades earlier by the death of his brother and childhood partner in crime, Keen, Randy had grown into a likable though often lonely man. After he married he had been diagnosed with a malignant melanoma that began as a mole on his back. The melanoma was treated, although the effectiveness of the treatment was not certain. Soon after Marjorie entered the nursing home, Randy invited his father to accompany him on a guided boating trip on the Skagit River, not far from Mercer Island, near Seattle, where Randy lived and worked as a company president. "On this trip Randy seemed to be quite himself," observed Freeman, who remained concerned about his son's health.

On top of these worries, Freeman experienced a sudden deterioration of his own health. During the summer of 1967 his rectal discomfort had grown so severe that he was unable to take part in hikes with his children and grandchildren. That September his doctors told him he had perianal Paget's disease, a rare illness closely associated with colorectal cancer, which required him to undergo surgery to resect his colon and create a colostomy that passed his bodily waste into a bag attached to his abdomen for the next three months. Cancer, the scourge of his father, had returned to target him. "This was a very trying time," he wrote. "I couldn't get the colostomy functioning properly, and the pain of dilating the operated area twice a day was distressing. I stretched out on an air mattress in our patio at home and let the hot sunshine bake me. I became depressed at the thought that I had cancer and would die within a year, as my father had. I thought that the family would be better off if I didn't survive in complete helplessness." Soon, however, he was back to hiking up to ten miles in the hills near home and bathing in the Pacific Ocean.

Freeman acknowledged that his succession of personal and professional upheavals made him miserable. "The one thing that kept me from an act of desperation was the knowledge that my book *The Psychiatrist* was in course of publication." This volume, first conceived in 1961 and published seven years later, was Freeman's last book-length work. The idea of a book focusing on the lives of psychiatrists had struck him "when I had long evenings to myself since Marjorie was 'indisposed' and I had few professional responsibilities." He researched the life of Sigmund Freud and his circle of followers—many of whom committed suicide—and, frequently working in the library of the school of medicine at Stanford University, expanded his field of examination to other psychiatrists.

After several years of work, Freeman ended up with a book manuscript that eccentrically blended short biographies of such psychiatric pioneers as Freud, Moniz, Meyer, and Sakel with chapter-long investigations of such pet interests as psychiatric family dynasties and suicide. Clearly aware that *The Psychiatrist* represented his last chance to express his experiences and beliefs to what he hoped would be a wide readership, Freeman frequently inserted personal anecdotes and idiosyncratic ruminations that lent the book a strangely informal tone—something like listening to an aged uncle as he related tales of psychiatry around the campfire. Freeman completed the manuscript in 1966. He corrected the galleys and compiled the index while hospitalized for his cancerous colon. "I had high hopes that *The Psychiatrist* would be a best seller because of my personal knowledge of the biographees," he noted.

*The Psychiatrist* fell far short of best-seller status, and the meager attention accorded the book greatly dissatisfied Freeman. The *Journal of the American Psychiatric Association* "dismissed [it] as light bedtime reading, but neither good history nor good biography," Freeman observed. "The sales have been disappointingly small." One reviewer condemned Freeman for the book's shoddy construction: an error in binding resulted in the repetition of a set of thirty-two pages and the omission of thirty-two others. He received at least one review—a mixed one—from a psychosurgery patient. The patient, who had undergone a prefrontal lobotomy around 1940 and now lived in a nursing home in Minnesota, noted in a letter to Freeman that he found the book interesting "but punctuationally obnoxious as it causes me to wonder even more than previously about what became of all the proof readers!" Freeman blamed the book's lack of appeal on his controversial status as the promoter of lobotomy, as well as his dismissive stance toward psychiatry and many of its practitioners. "Psychiatrists, I think, take themselves and their specialty quite too seriously," he wrote. "At least I had an enjoyable time writing the book."

Alone, unable to operate, and facing a terminal illness, he grappled with his choices. In the end he decided to follow the example of those people who asserted their independence by embarking upon challenging journeys despite the threat of advancing age and weakening health. His aunt, Florrie Keen, was one such person. At age ninety-one she had set off alone on a trip around the world—a journey that ended unfinished in Hong Kong, where she died. Freeman shut down his medical office, which was losing money anyway, and sublet the space to a pediatrician and an internist. He sold his house, an act he found deeply depressing, but he consoled himself by buying a 1967 Clark Cortez camper bus—his first recreational vehicle since the 1940s—equipped with a file cabinet, compact galley, and bathroom. Freeman further stocked the bus with a rubber diving suit, a two-man life raft, a backpack, fishing rods, his camera, and a typewriter. His plan was to set off on a long journey to follow up on the fortunes of his lobotomy patients.

Despite his personal travails, Freeman remained intensely interested in keeping in touch with his lobotomy patients of the past. In 1964, after a camping trip in the Sierras, he adapted his son Walter's snapshot of him on a hiking trail—he is wearing a backpack—into a picture postcard that he mailed out to two thousand former patients. "This was an ambitious project but paid off well in the number of responses," Freeman wrote. The next year he bought two thousand Christmas cards in Europe and cast them into the postal system with a printed message reading, "Merry Christmas, Walter Freeman."

Half of these cards elicited responses. He faithfully wrote back to old patients who showered him with letters. "I do need money, Mr. Freeman," wrote one, a resident of Miami Beach, who received a prefrontal lobotomy during the 1940s. "I need to become married. Never did I realize that my need and desire for sexual intercourse could be so strong. It is carnal desire. My need and desire for sexual intercourse is rapidly driving me to point of insanity." To another former patient, Freeman offered the reassurance that he had recently seen a lobotomized woman who "had a difficult time of adjustment, going from job to job because of just the things you mention—lack of initiative and poor memory, as well as tactlessness, but she has been steady for a year now, steadily employed at $50 a week plus board and lodging, so I think you may find that things aren't too bad after all."

To Freeman a new head-hunting trip represented not only a chance to acquire more information about his patients, but it also gave him a sense of freedom during a time when his home life was dismal. While on the road, "you lose a sense of time or distance, you keep going without regard to bodily fatigue or hunger," he said. "It's such an absorbing preoccupation that these mere physical matters don't particularly halt a person in his pursuit." At the same time any expedition in the Cortez was a highly impractical venture. Freeman soon found out that the vehicle required constant babying and repair. But a long road trip in his cranky camper offered him periods of busy happiness.

His return to the road—a trip he called "the Great Man-Hunt of 1968"—carried him through the Southwest, South, Middle Atlantic states, and New England. Surviving on meals made up of coffee, hard meat, cheese, fruit, milk, and an occasional glass of wine, he displayed his usual enthusiasm for his pursuit of his quarry, once deviating two hundred miles from his course to interview and photograph a lobotomy patient from the 1940s.

Eventually, he reached Florida, where he received devastating news from Seattle. The sudden appearance of a massive brain tumor had sent Randy, only thirty-one years old, into a coma. "The neurosurgeons in Seattle wanted to operate on him, and dad said, 'No, don't,'" Paul said. They operated anyway, and Randy's tumor ruptured. His death rapidly followed. Three thousand miles away, Freeman decided to attend a medical conference in Florida instead of the funeral. "It was too painful, or [he thought] it wouldn't do any good," Paul speculated on this decision. "Deep down inside he suffered more than the rest of us," his son Franklin noted. "He preferred to endure his final grief alone, and I think that in a way his attendance at Randy's funeral would have been an ordeal of déjà vu. . . . There was a great love in that man. You just didn't mess with it." To add to his troubles, Freeman unexpectedly had to undergo surgery to treat the reappearance of his symptoms of cancer. He recuperated in the Florida Keys, then resumed his journey in the Cortez.

Freeman, now weighing only 140 pounds, pressed onward in the camper. He paused in Boston to deliver a talk about psychiatric family dynasties at the annual meeting of the American Psychiatric Association; his talk was scheduled during the final day of the meeting, and the audience was small. Then Freeman climbed back into the Cortez. After making a weekend stop at McLean Hospital in Massachusetts, where the staff refused to let him view patient records because of his unexpected arrival outside of normal office hours, he went to Washington to meet up with his son Walter's children, Rachael and Walter, who would accompany him during the return trip west. They played Monopoly while he drove.

This traveling party of an aged physician and his two young grandchildren must have appeared peculiar to the lobotomy patients and family members they encountered on their long drive through the Midwest and Great Plains. "On the whole I enjoyed having them with me," Freeman wrote, "though at times the welcome grew thin. They don't speak much about the trip but I suspect it was rather hard on them." One night they stayed in a cabin at Athens State Hospital that had once been occupied by patients. By the time he ended his trip in August 1968, Freeman had covered twenty-six thousand miles. Soon after returning to California, he reported to James Watts that his recent travels and correspondence had provided him with information on more than 600 former lobotomy patients, 230 of whom were living outside the hospital, 142 in hospital, and 235 of whom were deceased.

Despite his discovery of lumps in his groin and the subsequent surgical resection of twenty-nine intestinal nodes, by late 1968 Freeman felt ready for another highway excursion into the lives of his patients. He recovered from the surgery in Hawaii but wrote to Watts that "the foot still itches—for the throttle." This time he intended to cover ground that he had skipped over the previous year, especially the Carolinas, Virginia,

Maryland, and some regions of the Midwest. He relaunched the Cortez in January 1969. First, though, with his brother Jack as a passenger, he drove to Mexico City, where he spent a week with his former GWU resident, Manuel Velasco-Suarez, and tracked down four former patients. His careless driving habits resurfaced in Mexico when he slammed the Cortez into an abutment that nearly ripped off the camper's right door. Freeman was unable to make repairs until his arrival in Houston, Texas, a month later. From there he crossed the southern United States, though he had to make an unplanned stop in Mobile, Alabama, for an emergency resection of the apocrine glands near his rectum.

He continued through Georgia and South Carolina, and stayed with old colleagues and friends on those many occasions that the Cortez broke down. In Asheville, North Carolina, he visited a prefrontal lobotomy patient from the middle 1940s. "She is a pleasant faced, soft spoken lady in her middle fifties with grey hair nicely arranged, who served me some eggnog diluted with milk," Freeman wrote in his report. "She had been to service in the morning being an Adventist and she has a copy of the *Watch Tower* and is somewhat emphatic on the subject of her religious beliefs. . . . She weighs 142 lbs., had one seizure so far this year for which she was taken to the hospital overnight. She has no complaints."

In Oklahoma he caught up with the nineteenth patient in the Freeman–Watts prefrontal lobotomy series, a schizophrenic whose surgery had taken place thirty-three years earlier. Noting her history of postoperative relapses, hospitalizations, a second lobotomy at the hands of another surgeon, and a bout with spinal cord degeneration, Freeman commented, "Lura has had her ups and downs over the years." These reports on patients inspired James Watts to later comment on his old colleague's dedication to the follow-up. "In his records," Watts wrote, "his patients come to life and vividly portray their anxieties and depressions. . . . His fond concern for the patients and their families appears almost paternal."

In his six months of travel in the Cortez during 1969, Freeman had covered twenty-two thousand miles. He was ready to settle down "for a while." He prepared for the big job of compiling the new data he had collected on the road. "I removed the file case from the Cortez, 350 lbs., and installed it in my suite . . . and proceeded to go over all my records, preparing a code for IBM cards," he wrote. He would do this work in a downstairs apartment in his daughter Lorne's home in San Francisco. Over the years Lorne had remained loyal to her father and had supported him during his travels by typing his papers, transcribing his Dictaphone recordings, and handling some of his correspondence.

His days as a traveling proponent of lobotomy were not yet over, however. Toward the end of 1969 his former associate, Robert Lichtenstein, pointed out to Freeman a notice in a medical journal announcing a forthcoming event: the First International Conference on Psychosurgery planned for August 1970 in Copenhagen. "I was almost outraged," Freeman declared. He fired off a letter reminding the conference planners that the first such international congress had actually been held in Lisbon twenty-two years earlier and had been organized by none other than Walter Freeman. "I received an apology and an invitation to attend, and to present a paper." He spent the next spring and summer preparing his talk, which would offer follow-up data on 415 lobotomy cases in patients diagnosed with early schizophrenia.

Before making his final voyage overseas, Freeman took advantage of an opportunity to more satisfyingly close his career at GWU, which had ended with disappointment and bruised feelings in 1954. Although he had visited Washington nearly every year since moving to California, Freeman had never received or sought formal recognition from GWU for his three decades of service. In March 1970 he accepted an invitation from the university to attend the unveiling of four new portraits in the halls of the medical school. These canvases honored Freeman, Watts (who had retired as chair of the department of neurology and neurological surgery the previous year), an anatomy professor, and a dean. The artist Robert Fantuzzi painted the Freeman and Watts portraits, and in the arrangement of his subjects alone, succeeded in capturing essential elements of the personalities of the two men.

Watts, the team player and consummate partner, is shown sitting with seven of his most distinguished neurosurgical residents. "I suppose I'd like to be remembered as someone who worked with people," Watts remarked. Freeman is pictured sitting alone, one hand speculatively set on a skull, the other holding a pipe: a solitary figure in thought. He thought his likeness good, though not as good as the image Jonniaux had created years earlier, but was bothered by the depiction of his left hand shown as "long, straight, and smooth, quite incongruous in comparison with my own hand which is rather gnarled and hairy."

After the paintings were undraped, the honorees each spoke to the assembled audience. "You may be interested to note," Freeman said, "that this old building was the scene of the first lobotomy in the country. It was also the scene of the first insulin shock therapy, of the first Metrazol therapy, and the first electroshock therapy in the city. So remember, when this old building is torn down, some of the first activity along the lines of physical treatment for mental disorders was carried out right here."

If he intended these observations to inspire awe, Freeman was surely disappointed. To the medical students and instructors of 1970, who must have listened to him with some discomfort, psychiatry had already entered a new age that left dust on the lapels of Walter Freeman. Psychosurgery and the shock therapies of decades past were relics of a less enlightened era. In just two years a group of psychiatrists in St. Louis would publish the first guide establishing precise criteria for the diagnosis of psychiatric conditions, and thus ended much of the reliance on judgment and experience that characterized the work of Freeman and his colleagues.

But a last burst of glory awaited Freeman at the psychosurgery conference in Copenhagen. Traveling there with Lichtenstein, at the conference banquet Freeman delivered a lengthy chronicle of the career of Egas Moniz. He reunited with colleagues and friends from around the world, including the Hiroses from Japan. His presentation on schizophrenic patients, however, met an apathetic reception; it "provoked only one question and no discussion," Freeman noted. A sobering realization: Walter Freeman could no longer generate controversy. At the conclusion of the conference, however, the delegates elected him honorary president of the next international psychosurgery congress scheduled to be held in the United Kingdom in 1972. Freeman, as it turned out, would be unable to fulfill his obligations.

From Denmark, Freeman continued to Sweden for four days with his old friend Gösta Rylander, then proceeded to Paris to attend the International Congress of Neuropathology. Nostalgia—and not any desire to keep current in the field—must have been his main motivation for registering for this conference. "Most of it was over my head with enzyme and electromicroscopy, etc.," he wrote. The author of one of America's most important early textbooks on neuropathology now found himself woefully out of touch with the new techniques of his profession. But he did appreciate a "fine dinner in the hall where Mary Antoinette had been beheaded." After stops in London and Lisbon, Freeman returned home.

Meanwhile, Marjorie's health had continued its steady decline. Permanently living in a nursing home, she developed a circulatory blockage that resulted in gangrene of the left leg. "I refused amputation because it was obvious that she would never recover," Freeman wrote. His sons Walter and Paul disagreed with their father, however, and tried to make him change his mind. Freeman would not and even threatened to take the battle over Marjorie's limb into court. Eventually, however, the sons arranged for the amputation of their mother's leg. Relocated to a convalescent hospital, Marjorie developed dementia and lost a great deal of weight before succumbing to pneumonia on April 22, 1970. A proponent of autopsy since his days at St. Elizabeths Hospital, Freeman arranged for one before Marjorie's burial at Calvary Cemetery in Merced, California, near Keen's final resting place. "We gathered for a graveside service, and returned to Merced for luncheon, and then dispersed," Freeman wrote of the funeral.

By this time Freeman had formally cashed out of his medical partnership by accepting an offer of $20,400 for his share. It was about half of what he thought it was worth, but he "was pleased to be relieved of the responsibility for the lease." Freed of that burden, he took one last journey as a lobotomist. By now peppered with cancer and half blind—he had trouble reading street signs from his Cortez—he gathered his remaining strength in December 1971 to attend the Fifth World Congress of Psychiatry in Mexico City. There he delivered a paper that managed to combine two of his recurring interests: "Suicide After Prefrontal Lobotomy." Escorted by his former student, Manuel Velasco-Suarez, now the governor of the Mexican state of Chiapas, Freeman issued no complaints about his infirmities. He later visited the Mayan Temple of the Inscriptions at Palenque. Wielding his cane in a dapper manner, he gamely mounted the wall of steps at the temple, a feat one observer called a "triumph of will over physical infirmity."

Now at the close of his life he longed for the public recognition that had gone to so many of his medical idols and associates: W. W. Keen, Egas Moniz, James Watts, John Fulton. He did receive a sheepskin from the University of Pennsylvania honoring him as a distinguished alumnus, but this failed to give him satisfaction. "I have put it away in my bureau with a large number of other certificates, framed and unframed, that nearly fill two drawers." They were not enough. "I have waited in vain for an honorary degree, of which my grandfather had six or seven," he wrote.

Infirmity, however, triumphed over Walter Freeman in the end. Freeman suffered from multiple forms of cancer. He had lived far longer than he had predicted when his health took its first serious nosedive in 1967. "It is with distress that I write you[;] there has been a recurrence of the malignancy with a lot of pain in the lower back and abdomen," he informed Sadao Hirose in a deteriorated handwriting in February 1972. Freeman told his colleague that he was receiving opiates under the care of a nurse. The time for good-byes had come. "I look back on our friendship with great pleasure." He finished one last paper about the sexual lives of his lobotomy patients before he grew so weak that he was unable to leave his bed to present it to a meeting of the Society of Biological Psychiatry.

Six months after traveling to Mexico, in the late spring of 1972, Freeman fell under the assault of his colon cancer and slipped into a coma. Much of his family gathered around his hospital bed. Gradually, on May 31, 1972, his vital signs diminished to a whisper, and Walter Freeman died at age seventy-six.

The family sent telegrams to Freeman's associates. "REGRET TO INFORM YOU WALTER FREEMAN DIED TODAY," read the message that Hirose received in Japan. "MEMORIAL SERVICE SATURDAY 10AM GODEAU FUNERAL HOME 41 VAN NESS AVENUE, SANFRANCISCO."

"Working with Walter Freeman was a great experience but it was not always easy," James Watts later eulogized to colleagues. "His boundless energy and indefatigability imposed endless demands on his associates and staff. I do not know which was more stimulating, Walter Freeman, himself or our frontal lobe studies." At Freeman's funeral in Merced, near the graves of Marjorie and Keen, members of the Freeman family recited verses from Kipling's *Just So Stories*. The cat that walked by himself now lay silently, still alone. His tread, however, would continue to echo for many years in the halls of psychiatry.

# Chapter 14

# Ghost

After Walter Freeman's death in 1972, psychosurgery drifted further to the edge of medical acceptability. The procedure that Freeman had championed around the world and had succeeded in carrying into the mainstream of medicine became a rarity after his departure, a treatment advanced by a few zealots waning in number. The decline, of course, began long before Freeman was buried. An antipsychiatric movement had been ignited during the early 1960s that cast mental illness not as sickness but as a form of behavioral diversity. "The movement's basic argument was that psychiatric illness is not medical in nature but social, political and legal: Society defines what schizophrenia or depression is, and not nature," wrote the psychiatry historian Edward Shorter. "If psychiatric illness is thus socially constructed, it must be deconstructed in the interest of freeing deviants, free spirits and exceptional creative people from the stigma of being 'pathological.' In other words, there really was no such thing as psychiatric illness. It was a myth."

And if mental illness was a myth, the treatments designed to combat it were unnecessary at best, oppressive at worst. According to some members of the antipsychiatric movement, mental hospitals only succeeded in degrading and isolating patients. They viewed some hospital inmates as persecuted antiheroes, like Randle McMurphy in Ken Kesey's novel *One Flew Over the Cuckoo's Nest*, who had his spirit crushed by administrators antagonistic to nonconformity and willing to use lobotomy as a means of social control. Members of the antipsychiatric movement viewed any treatment that damaged the brain as highly unethical and possibly criminal.

Although not a member of the antipsychiatry movement because he is a psychiatrist, Peter Breggin emerged as a leader of the campaign against psychosurgery. Breggin was no stranger to Walter Freeman. They had once engaged in an acrimonious phone conversation that resulted in Breggin's commitment to assist a Freeman lobotomy patient who was preparing a medical malpractice suit at the time of Freeman's death. In the early 1970s Breggin began an effort to agitate against psychosurgery in the halls of the U.S. federal government. With the help of a U.S. representative, he entered into the *Congressional Record* the text of an article arguing that psychosurgery was on the increase. In addition, he cautioned that some researchers were proposing psychosurgery as a possible solution for the waves of inner city violence then rocking the nation. Organic brain disorders, these researchers hypothesized, might trigger violence in some ghetto dwellers and incite them to riot. Three prisoners at Vacaville State Prison in California, in fact, had been lobotomized with federal funding to control their violent behavior. The result was the formation of a U.S. federal commission to evaluate the safety and efficacy of psychosurgery.

In time the ideas of the antipsychiatry movement gained wide currency. The year after Freeman's death, Oregon enacted legislation limiting the practice of psychosurgery by setting up a governor-appointed review board to approve each proposed operation. Many other states followed suit by placing the power of approval in the hands of members of the medical community, lay people, or judges. Bans or restrictions on lobotomy were also mandated in such nations as Japan, Australia, and Germany. Meanwhile, a slew of organizations set up task forces to issue recommendations on acceptable (and unacceptable) uses of psychosurgery. The National Institute of Mental Health, the American Psychological Association, and the Society for Neurosciences, among others, jumped into the controversy.

A task force of the American Psychiatric Association condemned the use of psychosurgery in the treatment of children, while the National Commission for the Protection of Human Subjects of Biomedical and Behavioral Research approved of it with safeguards and court review. The latter group, which found only 321 confirmed cases of psychosurgery in the United States in 1973, cautioned that such operations should not be performed on patients involuntarily committed to hospitals or unable to give an informed consent to surgery. In 1978 the U.S. Department of Health, Education, and Welfare issued a statement agreeing with these procedural restrictions, but it refused to ban psychosurgery all together.

Meanwhile, the legal climate for psychosurgeons grew increasingly stormy as the perception of lobotomy outside medical circles—among judges, legislators, patients, and concerned citizens—deteriorated. In 1975 James Watts was called out of retirement to testify in the Pennsylvania case of *Chase v. Groff*, in which a psychiatric patient named Stanley Chase sought malpractice damages against the psychosurgeon Robert A. Groff. Chase underwent lobotomies in 1967 and 1969 from Groff—after previously receiving psychosurgery from another physician in 1956—and later said he suffered seizures, partial paralysis, the loss of his senses of taste and smell, and the disappearance of his emotions, intelligence, and judgment as a result of the operations. During the civil trial Watts gave four days of testimony in defense of Groff's practices. Groff, who died before the case went to the jury, was exonerated.

The entire concept of informed consent for psychosurgical procedures underwent a rapid upheaval immediately before and after Freeman's death. In 1972 a California court determined that physicians must give patients all the information that a reasonable person would need to arrive at a rational decision on whether to accept psychosurgical treatment. (This changed the previous standard, which called for doctors to simply provide the kind of information that their peers provided.) The following year, in the case of *Kaimowitz v. Department of Mental Health*, a Michigan court ruled that a sex criminal could not legally consent to psychosurgery as a condition of his early release from institutionalization. The criminal, identified in court documents only as John Doe, had raped and murdered a nursing student during the 1950s, and he spent nearly twenty years in Ionia State Hospital as a result of his civil commitment as a sexual psychopath. Scheduled for discharge in the fall of 1973 because of his exemplary behavior as a patient, Doe received a proposal from two Detroit researchers who hoped to accelerate his release by subjecting him to psychosurgery and destroying the region of his brain believed to be responsible for his past violence. Doe consented. When Gabe Kaimowitz, an attorney for the Michigan Medical Committee for Human Rights, heard about the arrangement, he tried to block it in court.

Doe's incarceration soon became a moot point when the statute under which he had been committed to the hospital was found to be unconstitutional. Doe was now a free man. But the Wayne County Circuit Court still addressed the issue of whether Doe, a diagnosed psychopath, had the judgment necessary to competently consent to psychosurgery. The court, consulting the Nuremburg Code for medical experimentation adopted a quarter-century earlier, determined that Doe was not competent to give consent because of the coercion unavoidable in a situation in which release from incarceration might be the result of the surgery. After reading the decision, some physicians abandoned psychosurgery rather than face malpractice suits from their patients. As research in psychosurgical techniques stalled, lobotomy now faced extinction in an environment suddenly much more hostile to the procedure than anything Freeman had experienced in his lifetime. Freeman's reputation dimmed along with the standing of his pet procedure.

While Freeman's predictions of the return of psychosurgery failed to come to pass in the years just after his death, another of his prophecies did come true. In a stunning development unimaginable just a decade earlier, Freudian psychoanalysis lost its grip on the practice of psychiatry all around the world. The force that vanquished psychoanalysis from its seat of power was the branch of psychiatry that Freeman had advocated all his life: the biological, organicist approach. Just as the arrival of tranquilizing drugs in the 1950s had tolled the end of psychosurgery, those same pharmaceuticals and the many that followed set into motion events and courses of thinking that boosted biological treatments for psychiatric illnesses and brought down psychoanalysis. "It could not be simultaneously true," wrote Edward Shorter, "that one's psychological problems were caused by an abnormal relationship to the maternal breast and by a deficiency of serotonin."

Although bartenders and members of the clergy never entered the practice of psychoanalysis, as Freeman had mischievously proposed, psychiatrists soon found that their monopoly on analysis was gone forever. Increasingly, psychologists, social workers, and other nonmedical therapists became trained in the techniques of analysis and other forms of psychiatric therapy, which multiplied in numbers. As these other varieties of psychotherapy blossomed—largely outside of the psychiatrist's office—psychoanalysis grew scientifically suspect.

As early as 1948, in fact, efforts to prove the medical effectiveness of psychoanalysis had failed or foundered because of a lack of cooperation from psychoanalysts. In one telling study at Camarillo State Hospital in California, researchers found in 1962 that schizophrenic patients receiving psychotherapy did worse than those treated with drugs or even the members of the control group who received no treatment. Twenty-five years later, just before it was about to reach trial, a malpractice suit that a patient had brought against the private Chestnut Lodge Hospital in Maryland, the former headquarters of Freeman's nemesis, Harry Stack Sullivan, was settled out of court. The patient had charged that Chestnut Lodge staff had treated his psychotic depression not with medications of known medical efficacy but with sessions of psychotherapy, a substandard course of treatment. This settlement, an apparent acknowledgment that psychoanalysis applied less to the realm of medicine than to psychological enlightenment, was a devastating blow against the remaining psychiatrists who supported psychoanalysis as a means of treating mental disease.

Into the leadership vacuum rushed psychiatrists who, like Freeman, promoted a biological basis to mental illness. To these physicians, the state of the mind was an expression of the functioning of the brain. Their cause had been strengthening since the introduction of the tranquilizing medications of the 1950s. Drugs became increasingly sophisticated in their interactions with the neurotransmitters of the brain involved in emotional response and psychiatric disorders. Starting in the 1970s, studies involving families, twins, and adopted children had amassed a pile of evidence that—as Freeman had hoped to find during his studies in pathology at St. Elizabeths Hospital in the 1920s—schizophrenia had a powerful biological component. Schizophrenia, it now seems certain, is an organic disease.

That inexorable, slowly moving pendulum of psychiatry—the cyclical, centuries-old transfer of power between promoters of the psychological origins of mental disorders and the supporters of biological origins—had swung again. For the first time since the 1930s, the strength was now on the side of the organicists, the believers in the fruitfulness of treating psychiatric disorders by studying and manipulating the chemistry and anatomy of the brain. Psychotherapy (although not usually psychoanalysis) still appears today in the treatment regimens of many people with common neuroses, but drugs now form the strongest weapons in the arsenal of psychiatrists, and by the mid-1990s, Prozac, an antidepressant pharmaceutical, had become the second-best-selling prescription drug in the world.

Walter Freeman has been dead for decades, and all but a handful of his lobotomy patients have followed him to the grave, but in the current environment of biological psychiatry it is not surprising that psychosurgery is undergoing a renaissance. The use of surgery to treat tough psychiatric problems never truly vanished; it remained alive at institutions such as Massachusetts General Hospital as well as others in London, Sydney, Stockholm, and Madrid, where neurosurgeons have used a variety of tools—stereotactic techniques, gamma rays, sonic blasts—to perform cingulotomies, limbic leucotomies, and anterior capsulotomies on small numbers of obsessive-compulsive and depressive patients resistant to more conventional treatments through the 1980s and 1990s. "Surgery is then offered as a therapy of last resort when all other treatments options have been exhausted," wrote G. Rees Cosgrove of Massachusetts General Hospital in 2001, echoing the declarations of Walter Freeman and James Watts of sixty years earlier.

Although fewer than three hundred brain operations are now conducted annually worldwide to treat psychiatric disorders, the number is certain to rise, perhaps dramatically. These new procedures are not lobotomies; they most often use lasers or radiation to produce tiny lesions in narrowly targeted regions of the brain, especially the regions most closely implicated in the development of obsessive-compulsive disorder. The most recently developed forms of psychosurgery employ implanted devices that deliver minute amounts of electrical current to small regions of the brain suspected to be overactive or underactive. Doctors can easily halt the flow of electricity if undesirable side effects appear or the treatment fails. New techniques of brain imaging give neurosurgeons a view of circuits in the brain that appear to be malfunctioning in some patients. "In OCD, for example, a circuit linking portions of the orbital frontal cortex, which is behind the eyes, to deeper structures, such as the thalamus, appears to be more active than normal," the *Los Angeles Times* has reported. Freeman, were he alive, would nod knowingly.

In many ways Freeman haunts the work of this new generation of physicians. In general, they continue to hold to his conception of the roles of the thalamus and frontal lobes in collecting, distributing, and responding to emotional stimulation. But they also stand apart from his promotion of lobotomy to treat a variety of unrelated disorders, his patients' permanently altered personalities and emotional lives, and his pride and craving for public attention. While the dominant psychoanalytic wing of psychiatry used to hinder the spread of psychosurgery, a half century later it is Freeman himself who inhibits the work of contemporary surgeons. The words *psychosurgery* and *lobotomy*, stigmatized by their close association with ice picks and assembly-line operations, have vanished from the vocabulary of today's practitioners. Instead, they speak of *functional neurosurgery, psychiatric surgery,* or *neurosurgery for mental disorders.* To be sure, they have conflicting feelings about Freeman, the old lobotomist. (And so do Freeman's few patients who are still alive. "He seemed like such an honest man, such a truthful man," Ellen Ionesco, Freeman's first transorbital lobotomy patient, recently said. "Well, I'm sitting here just as confused as I was then.")

"In some ways he was a pioneer," said Ali Rezai, head of the section of stereotactic and functional neurosurgery at the Cleveland Clinic, "but in others he did a disservice and slowed the pace of development by being too much of a cowboy and acting too exuberantly without scientific foundation." Others involved in functional neurosurgery programs will not even allow Freeman that much credit, and his specter dissuades them from allowing their names to be mentioned in news accounts of the new forms of psychosurgery.

So the specter of Walter Freeman remains. "Despite advances in our understanding of psychiatric disease, we remain largely uninformed as to the neurobiological mechanisms that underlie intractable depression or obsessive compulsive disorder," Cosgrove wrote. "Until we understand these processes or have more convincing empirical outcome observations, it will be difficult to overcome the social, moral, and political resistance to surgery for intractable psychiatric illness." In at least one nation, Norway, the stigma against these procedures burns so strongly that former psychosurgery patients of decades past have received official apologies and government compensation.

Psychiatric surgery holds the promise to wipe out many of the symptoms of the most recalcitrant forms of obsessive-compulsive disorder and some varieties of depression. This was not the case in Freeman's lifetime. Now that it might happen in ours, Freeman's presence is unwelcome. He flits about the consciousness of physicians and patients, a pesky spirit looking for the recognition he believes he is due, an unwanted ghost causing sighs and regret.

Walter Freeman lies buried in Calvary Cemetery in Merced, California, next to the graves of Marjorie and their son Keen. Freeman's career may seem bizarre to many people today, but it is playing out again—with better technology, a better understanding of the functioning of the brain, and better ethical guidelines—in the burgeoning world of the new psychosurgery. The lifelong proponent of a close patient–physician relationship, a man who followed some of his patients for thirty-five years, Freeman continues to offer guidance to those who look for it. At least one person has sought ideas and answers to questions from the spirit of Walter Freeman at the physician's grave site. "I used to go back fairly often," said Freeman's son Paul, now in his seventies and retired from the practice of psychiatry. After Lorne's death from breast cancer in 1975, Paul and his brothers Walter and Franklin were left as Freeman's surviving children. "I learned that I could talk with my dad. I put myself in a state where I could sit and visualize that person. . . . Before long I would get some kind of response, something that [he] might say, that would help me to have a better perspective."

The life of Freeman is a lesson in perspective. "Put simply," wrote the psychosurgery historian Jack Pressman, "psychiatry is the management of despair." Despair is a constant in the human drama, but the settings in which it appears and our responses to it are infinitely varied. To Freeman the despair of psychiatric illness demanded a decisive, drastic remedy. He clung to that belief for thirty-six years, from the age of shock therapies to the era of psychiatric pharmaceuticals. Today our remedies are different, even though some of them echo Freeman's. The master teacher of the medical school classroom—the man who enthralled students with his tricks and demonstrations—can still put over a lesson or two. We should not allow Walter Freeman's ghost to flicker unnoticed in the shadows.

# Acknowledgments

Many people and organizations helped me as I worked on this book. Although I take full responsibility for any errors in the text, I am grateful for the assistance that others gave me in providing suggestions, information, and inspiration. My apologies to anyone I have inadvertently left out.

The three living children of Walter Freeman—brothers Franklin, Paul, and Walter Freeman—generously spent hours with me and lent their memories, scrapbooks, and speculations. Their help made this book much stronger than it otherwise would have been.

I am indebted to three writers who preceded me in investigating the life of Walter Freeman. The books of David Shutts, Jack Pressman, and Elliot Valenstein offered a solid foundation for my research into Freeman's career and the origins of psychosurgery. Although my interpretation of Freeman's life differs in many respects from theirs, I have benefited from the skills of each of these authors as historians and chroniclers of the phenomenon of Walter Freeman.

Several institutions gave me invaluable assistance during the course of my research. In particular, I want to thank the archive of George Washington University and its director, G. David Anderson, as well as staff members Lyle Slovick and Amy Stempler. They tolerated my long spells of research in their compact offices and fulfilled my requests for access to the large volume of papers and other materials in their Freeman–Watts collection. I also appreciate the research assistance I received from the staffs of the Bio-Medical Library of the University of Minnesota; the College of Physicians of Philadelphia; the Mütter Museum; the medical library of the University of California at San Francisco; the Bakken Library and Museum; the Minnesota Historical Society; the National Archives; the Carnegie Hero Foundation; the Federal Bureau of Investigation; and the U.S. Army.

A special thanks to Sound Portraits Productions, especially to Dave Isay and Piya Kochhar, for the research materials, interview transcripts, and enthusiasm they shared with me during the course of our work together on a radio documentary about Freeman's transorbital lobotomy patients. I am also indebted to the people at River Road Productions, particularly Bill Pohlad, Linda Flynn, and Lynn Anderson, for their help and support in the early stages of my work.

Two magazines published my articles about Walter Freeman and psychosurgery before this book was even a faint spark. At *Minnesota Medicine*, editor Meredith McNab listened to my unorthodox pitch and corrected some of my early errors. Articles editor Margaret "Pooh" Shapiro at the *Washington Post Magazine* asked tough questions that demanded good answers. I am grateful to both.

This manuscript also benefited from discerning readings by Peter Skutches and Arnold E. Aronson, for which I am appreciative.

My thanks also to a pair of coffee houses in my Minneapolis neighborhood, Caribou Coffee and Sebastian Joe's, where—seated amid the caffeinated smells and generally uninquisitive customers—I wrote nearly the entire manuscript of this book.

I wish to add my thanks to the following people, for reasons they may or may not understand: David Ahrendts, Cheryl Alementi, Amy Anderson, Sylvia Aronson, Paul Arshawsky, Vern Bailey, Harvey Beigel, Ellen Benavides, Carol Bly, Tim Brady, Laurie Brickley, Lillian Bridwell-Bowles, Bill Contardi, Tom Cook, Guy Cooper, C. Michael Curtis, Craig Davidson, Scott Edelstein, Scott Egerer, Jenney Egertson, Karla Ekdahl, Estelle El-Hai, Robert Elhai, Mary Ann Feldman, Colleen Frankhart, Phil Freshman, Amy Funk, Harry Funk, Susan Gaines, Nancy Gardner, Burl Gilyard, Conrad Goeringer, Eleanor Grohosky, John Habich, Prince Haleya, Lynn Hargreaves, Anne Hodgson, Joel Hoekstra, Bonnie Hoffman, Eugene Hoffman, Laura Hoyt, Peter Hutchinson, Steve Kaplan, Jeff Kauffman, Suzanne Krupp, David Lebedoff, Jeanne Lee, Dina Levine, Marvin Levine, Sally Levine, Cathy Madison, David Mahoney, M. McManus, Mary Meehan, Terry Monahan, Dan Olson, Susan Perry, Kyle Radcliffe, Mike Ravnitzky, Cheryl Reed, Gerry Richman, Rebekah Rising, John Rosengren, Kate Sandweiss, Beth Scanlon, Brad Schultz, Patty Sherburne, Stephen Smith, Jeff Spiegel, Marx Swanholm, Ron Tarrel, Ruth Taswell, Ana Viray, Steve Weinberg, Mark Weisberg, and Rebecca Welch.

My family has tolerated several years of Walter Freeman's unusual presence at the dinner table. My wife, Ann, read early drafts of the book and made many astute suggestions. To her and my daughters, Natalie and Sasha, I can only observe that their love has kept me living well. I'd be a lost man without them.

I deeply appreciate the enthusiasm, patience, and professionalism of the people at John Wiley & Sons, especially publisher Kitt Allen, who made me feel at home within the company; my editor, Eric Nelson, who took wonderful care of my manuscript as he guided it toward publication; and Chip Rossetti, whose early interest in the book helped me start work.

Finally I wish to thank Laura Langlie, a remarkable literary agent I feel lucky to have on my side. From the very beginning of this project, she has improved it through her suggestions, editorial talents, and attention.

# Notes

## Prologue

5 *"What manner of man"* Freeman, *The Psychiatrist,* ix.

## 1: September 1936

7 *Her expression placid* Freeman and Watts, 1950, p. xix.
*That evening, Hammatt was able to name* Freeman and Watts, 1936, p. 327.
*Curious about her emotional state* Freeman and Watts, 1950, p. xix.
8 *As she spoke* Freeman and Watts, 1936, p. 327.
*Her condition deteriorated* Freeman and Watts, 1950, p. xviii.
*She did not care about the shaved areas* Ibid., 1950, p. xix.
*Freeman and his partner* Menninger, 1988, p. 224.
*A native of Emporia* "Mrs. T. D. Hammatt, Former Topekan, Dies in Washington."
9 *Her parents spoiled her* Freeman and Watts, 1950, p. xviii.
*In time, the Hammatts* "Mrs. T. D. Hammatt, Former Topekan, Dies in Washington."
*The murder-suicide* "Mystery in Hammatts Dual Death."
*She developed a crush* Freeman and Watts, 1936, p. 327.
*Sometimes she would grimace* Freeman and Watts, 1950, p. xviii.
*But her insomnia had worsened* Ibid.
*Hammatt continued to entertain* Freeman and Watts, 1936, p. 327.
10 *Freeman, who had read Moniz's descriptions* Menninger, 1988, p. 224.
*Freeman had known Menninger* Freeman, *The Psychiatrist*, p. 220.
*"Of course," Menninger later wrote* Menninger, 1988, p. 223.
*At the last minute* Freeman, *Autobiography,* p. 14-3.
11 *"Who is that man?"* Freeman and Watts, 1950, pp. xviii–xix.
*The anesthetist bolstered* Shutts, 1982, p. 62.
*"I realized when I did the first"* Peery, 1979, p. 68.
*Behind their surgical masks* Peery, 1979, p. 68.
*They first cleaned* Shutts, 1982 p. 62; Freeman and Watts, "Prefrontal Lobotomy in the Treatment of Mental Disorders" (manuscript), GWU Archives.
12 *Word got back to Menninger* Menninger, 1988, p. 223.
*A few days later she was able* Freeman and Watts, 1936, p. 328.
*Freeman was concerned that Hammatt* Menninger, 1988, p. 224.
*and she resumed the odd* Freeman and Watts, 1936, p. 328.
*He believed it was too soon to determine* Ibid.

13 *"The result was spectacular"* Freeman, *Autobiography*, p. 14-3.
*Freeman found that Hammatt could direct* Freeman and Watts, 1950, p. xix.
*"I can go to the theatre now"* Ibid., p. xx.
*The changes were also noticeable to her husband* Freeman and Watts, 1936, p. 328.
*Others, roaring their agreement* Shutts, 1982, pp. 64–65.
14 *The anxiety was still there* Freeman and Watts, "Notes on Prefrontal Lobotomy" (manuscript), GWU Archives.
*He acknowledged certain side effects* Ibid.
*Five years after her lobotomy* Freeman and Watts, 1950, p. xx.
15 *"I felt somehow that we were in the presence"* Fulton, *Diary*, November 16, 1936.

## 2: RITTENHOUSE SQUARE

16 *and his topic was his technique* Valenstein, 1986, p. 122.
*Burckhardt directed the Prefargier Asylum* Joanette, 1993, pp. 573–576.
17 *It might be possible, Burckhardt thought* Berrios, 1997, p. 69.
*Burckhardt sliced out an inch-wide strip* Joanette, 1993, p. 581.
*Burckhardt later operated three more times* Freeman and Watts, 1950, p. xv.
*His report at the Berlin conference* Joanette, 1993, p. 585.
*and generated no strong interest* Feldman, "Psychosurgery: A Historical Overview," p. 6.
*After the conference he returned to Switzerland* Berrios, 1997, p. 71.
*He retired five years after* Feldman, "Psychosurgery: A Historical Overview," p. 6.
18 *Rush gave his patients alcoholic drinks* Freeman, *The Psychiatrist*, pp. 151–152.
*(Rush also gave Meriwether Lewis)* Frayer, 2001, pp. 185–189.
*and in 1965, the American Psychiatric Association* Shorter, 1997, p. 15.
19 *During his travels in Europe* Freeman, *The Psychiatrist,* p. 157.
*"We think your hospitals"* Ibid., p. 3.
*and soon afterward another physician, Emory Lamphear* Valenstein, 1986, p. 118.
20 *"It is remarkable," Walter Freeman would write* Freeman and Watts, 1950, p. xv.
*Two prominent physicians* Freeman, *Autobiography,* p. 1-9.
*At one point, stress* Ibid., p. 1-10.
*Another time he listened to a description* James, 2002, p. 151.
21 *He later remembered Rittenhouse Square* Freeman, *Autobiography*, p. 3-3.
22 *The red brick* Freeman, "Collection of Themes: Student Essays by Undergraduates," GWU Archives.
*(He knew that somewhere in his lineage)* Freeman, *Autobiography*, p. 1-2.
*Once in school he penned* Ibid., p. 3-9.
23 *But after they grew up* Ibid., p. 3-4.
*He found girls "bothersome"* Ibid., p. 3-8.
24 *Years later, paging through an edition of* Who's Who Freeman, *The Psychiatrist*, p. 168.
*"The chair was in a far corner"* Freeman, "Collection of Themes: Student Essays by Undergraduates," GWU Archives.
25 *Once when a truant officer* Shutts, 1982, p. 151.
*When it came time for Walter to hear from his father* Freeman, *Autobiography*, p. 1–12.
*(Later, in his own medical practice)* Freeman, "Democracy Versus Autocracy in the Home" (manuscript), GWU Archives.
*Emotional distance even tarnished* Freeman, *Autobiography*, p. 1-12.
26 *He fumbled with the cello* Ibid., p. 3-5.

26 *"Actually, I suppose my worst handicap"* Ibid., p. 2-7.
*"I was a watcher rather than a performer"* Ibid., p. 3-10.

27 *"Here was a noble feast"* Ibid., p. 3-6.
*"Seven professors, one demonstrator"* Rovit, 2002, p. 3.
*Service in the Civil War* Ibid.
*After Keen's return from Europe* Ibid., pp. 3–5.

28 *Among his many early successful cases* Kandela, 1999, p. 937.
*and of Queen Victoria, from whom Lister* Gropper, 1997, p. 1331.
*who during the Civil War had seen* James, 2002, p. 222.
*(One of his favorite expressions)* Rovit, 2002, p. 6.

29 *Keen felt proud* James, 2002, p. 176.
*He was part of the medical team* Rovit, 2002, p. 16.
*Keen's fame made him wealthy* James, 2002, p. 58.

30 *Keen's accomplishment as the first American surgeon* Rovit, 2002, pp. 2–3.
*He operated "with a missionary zeal"* James, 2002, p. xiii.
*Keen suffered a grief so intense* Ibid., p. 145.
*"Every year, on the anniversary of her death"* Freeman, *Autobiography*, p. 1-5.

31 *he penned such a large number of letters* James, 2002, p. xv.
*Once, after Keen had unsuccessfully used* Ibid., p. 64.
*Similarly, in 1912* Ibid., pp. 65–66.

32 *Once, when Keen mentioned that he was shrinking* Ibid., p. 192.

## 3: The Education of a Lobotomist

33 *Although Freeman lamented* Freeman, *Autobiography*, p. 1-13.
*"In prep school a daily stint"* Ibid., p. 4-1.

34 *When he joined a swim team* Ibid., p. 4-2.
*His classmates sensed Freeman's social discomfort* Ibid., p. 4-1.
*Dissatisfied with himself* Ibid., p. 4-3.
*Freeman's sophomore year was nearly as bad* Ibid.

35 *His failure of descriptive geometry* Ibid., pp. 4-7–4-8.
*After his discharge* Ibid., p. 4-8.
*Freeman's weakened condition* Ibid.

36 *Before returning to Yale* Ibid., p. 4-10.
*He viewed his father as a failure* Ibid., p. 4-9.
*After attending a chorus rehearsal* Letter from Freeman to Mr. Goodwin, November 28, 1918, GWU Archives.

37 *Meanwhile, the intensifying of World War I* Freeman, *Autobiography*, p. 4-11.
*That spring, on a solitary* Ibid., p. 4-10.
*"I learned less at college"* Ibid., p. 4-13.
*There, to his own astonishment* Ibid., p. 5-1.
*He rented a room* Ibid.

38 *Once again, however, Freeman's health* Ibid.
*There, in one of the first lectures* Ibid.
*An even closer call came* Freeman, *The Psychiatrist*, pp. 23–24.

39 *He studied psychiatry with Professor Charles W. Burr* Freeman, *Autobiography*, pp. 13-2–13-3.
*(Freeman later visited Burr's home)* Ibid., p. 6-14.

40 *At 200 yards, Freeman led* Ibid., p. 5-5.

41 *During the next five months* Shutts, 1982, p. 8.

41 *"It was a frightening experience"* Freeman, *Autobiography*, p. 5-11.
*He speculated that a head cold* Ibid., p. 2-2.
*"None of the frills"* Ibid., p. 5-12.
*"Finally one night"* Ibid., p. 5-14.
42 *When the official announcement of an armistice* Ibid.
*"I remember clinging to the Statue of Liberty"* Letter from Freeman to Dani, May 9, 1945, GWU Archives.
*"I made occasional weekend trips"* Freeman, *Autobiography*, p. 5-17.
*Hunched and gaunt, Spiller* Ibid., p. 5-16.
*While other students grew restless* Ibid., p. 5-15.
43 *He spent hours abstracting cases* Ibid., p. 5-20.
*Freeman also spent two weeks in the home delivery service* Ibid., p. 5-15.
*When he graduated from medical school* Shutts, 1982, p. 9.
*Medicine "held my interest"* Freeman, *Autobiography*, p. 5-21.
*In Freeman's mind this was regrettable* Ibid., p. 5-22.
44 *Freeman's one regular activity* Ibid., p. 6-2.
*To Freeman, the most vivid moment* Ibid., p. 6-5.
*a trip in which they broke a hole* Ibid., p. 6-1.
45 *Though he gained valuable experience* Ibid., p. 6-2.
*The hospital's antique ledgers* Ibid., p. 6-4.
*He had no patience for what he called* Freeman, *Autobiography*, p. 6-6.
*Yet the teacher's abrasive manner* Freeman, "Charles H. Frazier" (manuscript), GWU Archives.
46 *He marveled at the teacher's endurance* Ibid.
*At the same time, Freeman discovered* Freeman, *Autobiography*, p. 6-6.
*Soon another patient commanded* Ibid., pp. 6-9–6-10.
*One of the first patients* Ibid., p. 6-10.
47 *Freeman also visited her home* Ibid., p. 6-13.
48 *In May 1923 Madeleine accompanied* Ibid., p. 6-14.
*La Salpêtrière initially struck Freeman* Letter from Freeman to Phil Brown, June 22, 1923, GWU Archives.
49 *In one such assessment* Freeman, "A Clinic of Pierre Marie" (manuscript), GWU Archives.
*"The seal ring from Madeleine"* Freeman, *Autobiography*, p. 7-3.
*Remembering an incident eleven years earlier* Ibid., p. 2-8.
*His inexperience with alcohol* Ibid., p. 2-5.
*After his return to America* Ibid., p. 2-6.
*"Beards have always interested me"* Freeman, "On Beards, with Special Reference to My Own" (manuscript), GWU Archives.
*"Those who have never grown beards"* Ibid.
50 *Near the end of his life* Freeman, *The Psychiatrist*, p. 217.
*In fact, he dated his break* Freeman, *Autobiography*, p. 7-3.
*and where Freeman had his pocket picked* Ibid., p. 7-8.
*"My grandfather had been president"* Ibid., p. 12-3.
*During the fall of 1923 at the meeting* Letter from Freeman to Tommy Fitz-Hugh, November 16, 1923, GWU Archives.
*His six months in Paris ended dismally* Freeman, *Autobiography*, p. 7-5.
51 *Mingazzini also served* Valenstein, 1986, p. 128.
*whom the American described as* Letter from Freeman to Dr. Weisenburg, December 8, 1923, GWU Archives.

51 *Once, while attempting to sever* Letter from Freeman to Dr. Weisenburg, March 1924, GWU Archives.
*One odd highlight of his time* Freeman, *Autobiography*, pp. 7-9–7-10.
52 *including a political rally* Ibid., p. 7-8.
*In the spring of 1924, Freeman finished an article* Ibid., p. 7-7.
*Needing a vacation* Ibid., p. 7-8A.
53 *Freeman intended to stay for a while* Letter from Freeman to Winkelman, GWU Archives.
*"I had a pair of dancing slippers"* Freeman, *Autobiography*, p. 7-8A.
*Freeman felt embarrassed when he belatedly discovered* Letter from Freeman to W. W. Keen, May 17, 1924, GWU Archives.
*William Alanson White, director of St. Elizabeths* Letter from E. R. Stitt to W. W. Keen, April 28, 1924, GWU Archives.
*"I have never known a golden apple"* Letter from W. W. Keen to Freeman, April 29, 1924, GWU Archives.
54 *("Only once have I been anywhere near")* Letter from Freeman to Madeleine James, 1923, GWU Archives.
*Although he cautioned Madeleine* Letter from Freeman to Norman Freeman, September 26, 1923, GWU Archives.
*Freeman decided that the $5,200* Letter from Freeman to Laila, August 30, 1924, GWU Archives.
*"You will be surprised to hear"* Letter from Freeman to W. W. Keen, May 17, 1924, GWU Archives.
*To his mother, Freeman confessed* Letter from Freeman to Corinne Freeman, May 18, 1924, GWU Archives.
*and he found that the elephant had* Freeman, *Autobiography*, p. 7-10.
*Freeman later learned* Freeman, *The Psychiatrist*, p. 76.
*(Customs officials had raised concerns)* Ibid., p. 7-4.
55 *The following day, Freeman made the nerve-wracking trip* Ibid., p. 7-10.
*Time healed their wounds* Letter from Freeman to Corinne Freeman, August 25, 1924, GWU Archives.
*He paid a call to Spiller* Freeman, *Autobiography*, p. 5-16.

## 4: In the Hospital Wards

56 *The first time, in 1908* Freeman, *Autobiography*, p. 8-1.
57 *At the Navy Laboratory* Ibid., p. 8-4.
*Arriving in Washington, "I fell"* Freeman, "The Development of Neurology at George Washington University" (manuscript), GWU Archives.
*White had been running St. Elizabeths* Freeman, *The Psychiatrist*, p. 7.
58 *White believed that physically restraining* Shutts, 1982, p. xv.
*In 1917 von Jauregg* Shorter, 1997, p. 192.
59 *whom Freeman described as "not a friendly"* Freeman, *The Psychiatrist*, p. 20.
*Debilitated victims accounted for as many* Pressman, 1998, p. 33.
*From the U.S. Public Health Service in Puerto Rico* Freeman, *Autobiography*, p. 9-6.
*This fever therapy, introduced two years* Ibid.
*In 1963 Freeman heard from the daughter* Freeman, *The Psychiatrist*, p. 19.
60 *"Here I was, not yet 29"* Freeman, *Autobiography*, p. 8-2.
*they had, for instance, broken* Valenstein, 1986, p. 7.

60 *like Central State Hospital* Shorter, 1997, p. 190.
*"One could cure nothing"* Ibid., p. 192.
61 *Freeman admired Karpman's sense of humor* Freeman, *Autobiography*, p. 12-14.
*They made him experience "a weird mixture"* Ibid., p. 14-1.
*He was initially thrilled to have at his disposal* Ibid., p. 9-2.
*He was nearly drowning in paperwork* Ibid., p. 9-3.
62 *"I am also learning a lot about the mind"* Letter from Freeman to Dot Freeman, September 2, 1924, GWU Archives.
*A few weeks after starting work* Letter from Freeman to Laila, August 30, 1924, GWU Archives.
*Marjorie's initial impression of this newcomer* Freeman, *Autobiography*, p. 8-4.
*she invited him to accompany her* Ibid., p. 8-5.
63 *and she was beginning to doubt* Paul Freeman, Interview, 2002.
*Soon Marjorie was bobbing her hair* Letter from Freeman to Dot Freeman, September 25, 1924, GWU Archives.
*Just a few weeks earlier* Letter from Freeman to Laila, August 30, 1924, GWU Archives.
*"No longer would a pretty face"* Letter from Freeman to Florrie Keen, September 29, 1924, GWU Archives.
*"Am I not lucky?"* Letter from Freeman to Dot Freeman, September 25, 1924, GWU Archives.
*"I really monopolized"* Freeman, *Autobiography*, p. 8-5.
64 *Their meeting with this clergyman* Ibid., p. 8-6.
*Later, at the wedding* Ibid.
*Twelve days after their wedding* Letter from Freeman to Dot Freeman, November 16, 1924, GWU Archives.
*His wife was, as Freeman termed it* Freeman, *Autobiography*, p. 8-8.
*"We ate our first meal"* Ibid., p. 8-7.
*Although they could have lived in staff housing* Walter Freeman Jr., Interview, 2002.
*While they were at work, their maid* Letter from Freeman to Virginia Freeman, December 3, 1924, GWU Archives.
65 *Freeman celebrated by bringing cigars* Letter from Freeman to Dot Freeman, August 8, 1925, GWU Archives.
*They celebrated their anniversary* Letter from Freeman to Corinne Freeman, November 8, 1925, GWU Archives.
*he instantly declared Walter an extrovert* Freeman, *Autobiography*, p. 1-9.
*(Virgie later proved to be a carrier)* Paul Freeman, Interview, 2002.
66 *On weekends the Freemans led their tribe* Freeman, *Autobiography*, p. 8-10.
*When lecturing, Freeman took the position* Freeman, "Democracy Versus Autocracy in the Home" (manuscript), GWU Archives.
*Although she once confessed* Franklin Freeman, Interview, 2002.
*"She really taught me how to love"* Walter Freeman Jr., Interview, 2002.
*"We all knew that he loved us"* Franklin Freeman, Interview, 2000.
*Once, when the kids were at home* Walter Freeman Jr., Interview, 2002.
67 *"I remember being cautioned"* Ibid.
*Freeman recalled with horror the time* Freeman, *Autobiography*, p. 9-5.
*One neurological question that puzzled him* Ibid.
*"I went through the autopsy material"* Ibid., p. 9-8.

68 *He admitted that the procedure* Ibid.
*Ultimately, Freeman reported* Ibid., p. 9-9.
*His paper summing up the results* Ibid., p. 9-10
*"I came to St. Elizabeths imbued"* Ibid., p. 9-1.
*For a while he grew excited by his discovery* Ibid., p. 9-2.
*The apparent normality of these patients' brains* Freeman, "History of Psychosurgery" (manuscript), p. 6-5, GWU Archives.
*He came to harbor the perverse thought* Freeman, *Autobiography*, p. 14-1.
*"I approached the study from the standpoint"* Ibid.

70 *A 1931 issue of* Time *Time*, 1931.
*He startled many of his colleagues* Valenstein, 1986, p. 201.
*In 1928, for instance, the editor-in-chief* Freeman, *The Psychiatrist*, p. 118.

71 *"I looked around me at the hundreds of patients"* Freeman, *Autobiography*, p. 14-1.
*"Fame and fortune really beckon me"* Letter from Freeman to Corinne Freeman, June 14, 1926, GWU Archives.

72 *(Washington had seen the birth)* Shorter, 1997, p. 164.

73 *The psychiatric historian Edward Shorter wrote* Ibid., p. 160.
*Freeman paid a call on Stitt* Freeman, *Autobiography*, p. 8-2.

74 *"As soon as the knife cut through"* Ibid., p. 11-2.
*"I was often in for a surprise"* Ibid., p. 11-1.
*"it's a wonder nobody lost a finger"* Ibid., p. 11-2.
*"I heard the unmistakable click"* Ibid.
*the Council on Medical Education and Hospitals* Ibid., p. 10-6.

75 *"The effect was electric"* Ibid., p. 10-2.
*"This proved an excellent challenge"* Ibid., p. 10-2.
*Freeman knew ahead of time* Ibid., p. 11-3.

76 *(A dozen years later)* Freeman, Untitled paper for Medical Society of the District of Columbia, October 14, 1942, GWU Archives.
*continuing to work even while recovering* Freeman, *Autobiography*, p. 9-11.

77 *Most strikingly, Freeman declared that pathological studies* Freeman, *Neuropathology*, p. 256.
*"Friends tell me that the chapter"* Freeman, *Autobiography*, p. 9-12.
*That pride revisited him when he traveled* Freeman, *The Psychiatrist*, p. 200.
*Part of the thirteen-stanza poem read* Freeman, "Psychological Plagues."
*Aboard the liner, Freeman breathed* Freeman, *Autobiography*, p. 2-3.

78 *He began swallowing Nembutal* Ibid.
*In 1933, 43 percent of all patients* Grob, 1994, p. 187.
*The American Psychiatric Association did not organize* Shorter, 1997, p. 298.
*Patients living in psychiatric institutions* Ibid., p. 190.
*Otto Loewi, a faculty member* Ibid., p. 246.

80 *"While papers are more enduring"* Freeman, *Autobiography*, p. 12-8.
*Many of the physicians drawn by the 1931 exhibit* Ibid., p. 12-7.
*"I was a few minutes early at the house"* Ibid., pp. 8-17–8-18.

81 *he was left "disoriented and querulous"* Ibid., p. 1-4.
*Freeman's son Franklin remembered* Franklin Freeman, Interview, 2002.
*Pushed in a wheelchair* Ibid., p. 12-9.
*"A full, rich life"* Freeman, *Autobiography*, p. 1-4.
*During Freeman's years at St. Elizabeths* Ibid., p. 1-8.
*When they traveled together in 1929* Ibid., p. 12-5.

82 *"My eyes were moist"* Ibid., p. 1-8.

## 5: A PERFECT PARTNER

83 *Freeman quickly set to work at building* Freeman, *Autobiography*, p. 14-17.
84 *Dangling from the vest pocket* Shutts, 1982, pp. 13–14.
*"a rather crabby bunch"* Freeman, *Autobiography*, p. 13-2.
*who sat at the long meeting table* Freeman, *The Psychiatrist*, p. 181.
85 *his successor would declare* Freeman, *Autobiography*, p. 13-3.
*The board agreed to start examining* Freeman, *The Psychiatrist*, p. 179.
*"Sad to relate, he died within a short time"* Ibid., p. 178.
*"The fact that we were nominated"* Ibid.
*The goal was "to certify"* Freeman, *Autobiography*, p. 13-11.
86 *"I expected of the candidates"* Freeman, *The Psychiatrist*, p. 184.
*one, after receiving his failure notice* Ibid., p. 187.
*Among the earliest to be examined* Ibid., p. 185.
*Freeman viewed Bullard's position* Ibid., p. 182.
*"My main recollection of him is the sweating"* Ibid., p. 188.
*"When appropriate patients appeared"* Freeman, *Autobiography*, p. 10-12.
87 *As Wright's condition deteriorated* Ibid., p. 10-16.
*The second patient was Lillian Murphree* Ibid.
88 *By his own admission, Freeman had little to contribute* Ibid., p. 10-17.
89 *Watts first sighted Freeman* Watts, James W.,"Psychosurgery: A 20 Year Follow-Up of the Freeman-Watts Series" (manuscript), GWU Archives.
*One morning, "I had left for the hospital"* Shutts, 1982, p. 25.
*Watts had medicine and money* Peery, 1979, p. 1.
*(where he took advantage)* Ibid., p. 136.
90 *Married to a wealthy heiress* Horwitz, 1998.
*he published more than one hundred thirty articles* Ibid.
*Over the next several years, Watts and Fulton* Fulton, *Diary*, entries for January 19, January 20, and June 13, 1934.
*Once, while attending a medical meeting* Shutts, 1982, p. 37.
*Watts's patients often displayed* Fox, John L.,"Professor James W. Watts: Teacher, Scientist, Physician," GWU Archives.
*Watts was a deliberate, nonconfrontational* Valenstein, 1986, p. 140.
*he took "nearly everything seriously"* Peery, 1979, p. 25.
*As a medical student* Ibid., p. 5.
91 *In later years* Ibid., p. 26.
*When he grew convinced* Ibid., p. 3.
*Watts wanted to return to the South* Peery, 1979, pp. 43–44.
*Freeman's first move was to invite Watts* Shutts, 1982, p. 28.
*The X-ray facilities particularly alarmed* Peery, 1979, pp. 45–46.
*"I don't use the word[s]"* Ibid., p. 54.
*"The worst thing about Walter Freeman"* Shutts, 1982, p. 59.
92 *Watts considered Freeman "a great"* Ibid.
*Medical students "often brought"* "Salute to Teaching," p. 2.
*Watts recalled watching his colleague* Watts, 1965, p. 226.
*Freeman knew the power of unpredictable* Freeman, "The Development of Neurology at George Washington University 1924–1954," p. 10, GWU Archives.
*Another time Freeman nearly set* Freeman, *Autobiography*, p. 10-3.
*"I wrote that what the teacher had to say"* Freeman, "The Development of Neurology at George Washington University 1924–1954," p. 8.

92 *In 1941, for instance* Letter from Freeman to Marjorie Freeman, September 11, 1941, Paul Freeman personal collection.

93 *(Marjorie also miscarried)* Paul Freeman, Interview, 2002.

*"sewed up the pockets of the boys"* Freeman, *Autobiography*, p. 2-11.

*After spending a summer with friends* Ibid., p. 2-12.

94 *"I think this was expecting too much"* Ibid., p. 8-11.

*For the remainder of his life* Ibid., p. 20-16.

*"a rather short, erect old man"* Freeman, *The Psychiatrist*, p. 65.

95 *Freeman had to content himself* Freeman, *Autobiography*, p. 16-5.

*"Moniz seemed to me a kindly old"* Freeman, *The Psychiatrist*, p. 52.

96 *Richard Brickner of the New York Neurological Institute* Macmillan, 2000, pp. 237–239.

*He may have concluded from Penfield's* Ibid., p. 247.

*(Previously these animals had been used)* Shutts, 1982, p. 37.

98 *(Fulton later added, a bit regretfully)* Fulton, "Progress Report: Lobotomy Project," October 6, 1952, Yale Archives.

*Freeman wrote that only Moniz "envisaged"* Freeman, *The Psychiatrist*, p. 52.

*Moniz peevishly responded* Letter from Egas Moniz to Freeman, January 29, 1952, GWU Archives.

*"My paper and exhibit were soon forgotten"* Freeman, *Autobiography*, p. 16-9.

99 *Moniz's monograph was for a time banned* Shutts, 1982, p. 34.

*(It would later go through nineteen)* Freeman, *The Psychiatrist*, p. 50.

100 *"Here was a new method for exploration"* Ibid., p. 51.

*This invention of cerebral angiography* Valenstein, 1986, p. 73.

*Before attending the London conference* Macmillan, 2000, p. 247.

*"For awhile, he stayed at his rococo"* Shutts, 1982, p. 49.

101 *(Leucotomy derives from the Greek)* Ibid.

*Moniz and Lima made one practice attempt* Feldman, "Psychosurgery: A Historical Overview."

*the Santa Marta Hospital in Lisbon* Shorter, 1997, p. 226.

*He began using a device called a leucotome* Macmillan, 2000, p. 243.

102 *Moniz's ninth leucotomy patient* Feldman, "Psychosurgery: A Historical Overview."

*but Sobral Cid, the director* Shutts, 1982, pp. 54–55.

*Within weeks he published an article* Feldman, "Psychosurgery: A Historical Overview," p. 10.

*"which, if not organized"* Partridge, 1950, p. 4.

103 *Sobral Cid of the Bombarda Asylum* Macmillan, 2000, p. 244.

*"Since no written document corroborating"* Berrios, 1997, p. 61.

*One trephined skull discovered in France* Feldman, "Psychosurgery: A Historical Overview."

*Malcolm Macmillan convincingly shows* Macmillan, 2000, p. 5.

104 *In 1889 William Macewan* Ibid., pp. 230–231.

*That same year and in 1890* Berrios, 1997, pp. 63–64.

*with one of the participating surgeons* Ibid., p. 66.

*In 1895 Emory Lamphear* Macmillan, 2000, p. 233.

*"the question of the propriety of excising"* Ibid., p. 231.

*"a climate of meddlesomeness"* Shorter, 1997, p. 226.

105 *The surgery had done no good* Feldman, "Psychosurgery: A Historical Overview."

*François Ody, a Swiss disciple* Valenstein, 1986, p. 119.

*Walter Dandy, noted of a patient in 1922* Macmillan, 2000, p. 235.

105 *"This was too much even for me"* Sargant, 1967, p. 65.

106 *Soon afterward, Ramirez Corria* Shutts, 1982, p. 78.
*His first communication to the Portuguese* Letter from Freeman to Egas Moniz, May 25, 1936, GWU Archives.
*Moniz replied in unidiomatic* Letter from Egas Moniz to Freeman, June 24, 1936, GWU Archives.
*Moniz had inscribed the book* Freeman, *The Psychiatrist*, p. 53.

107 *"I could never get interested in the speculations"* Freeman, *Autobiography*, p. 14-2.
*"Yet careful study of Moniz' case reports"* Freeman and Watts, "Frontal Lobotomy in the Treatment of Certain Mental Disorders" (manuscript), GWU Archives.
*He found himself struck by* Freeman, *Autobiography*, p. 14-2.
*He regarded Moniz's creative leap* Ibid., p. 16-8.
*With Freeman's recommendation* Letter from Freeman to Egas Moniz, August 6, 1936, GWU Archives.
*"It is truly a fascinating study"* Ibid.

108 *Moniz responded to the review with warmth* Ibid.
*Watts "also was ready"* Freeman, *Autobiography*, p. 16-7.
*"I was interested in the frontal lobes"* Shutts, 1982, p. 60.
*Admissions to psychiatric hospitals were growing* Ibid., p. ix.
*Freeman "was concerned about the lack"* Ibid., pp. 59–60.

109 *"Now sometimes it'd be five members"* Ibid., p. 61.
*but before the advent of widespread air conditioning* Ibid., p. 60.
*"I used to say that the children cared little"* Freeman, *Autobiography*, p. 15-2.
*"We were beyond civilization"* Ibid., p. 15-3.
*"they seemed just as much impressed by a squirrel"* Ibid., p. 15-4.

110 *"We arrived home lean, tanned"* Ibid., p. 15-6.

## 6: REFINING LOBOTOMY

111 *suffered from agitated depression* Laurence, 1937, p. 10.
*For the first time, Freeman photographed* Shutts, 1982, p. 69.
*When Freeman examined the seventy-two-year-old patient* "Proceedings of the First Postgraduate Course in Psychosurgery."

112 *"He had spent eighteen months in bed"* Freeman and Watts, 1950, p. 331.
*"You know, I have been talking a blue streak"* Freeman and Watts, "Prefrontal Lobotomy in the Treatment of Mental Disorders" (manuscript), 1936, pp. 6–8, GWU Archives.
*"He makes his home with his mother"* Freeman and Watts, 1950, pp. 332–333.
*Freeman and Watts's next psychosurgery patient* Ibid., p. 84.

113 *A month after surgery* Letter from Freeman to Egas Moniz, November 26, 1936, GWU Archives.
*In this case, Freeman decided to evaluate* Shutts, 1982, pp. 66–67.
*She suffered a temporary* Freeman and Watts, 1950, p. 392.
*"grossly obese, untidy and mannerless"* Ibid., p. 396.
*but by early November, Egas Moniz had heard nothing* Letter from Egas Moniz to Freeman, November 1, 1936, GWU Archives.

114 *"New cases, the choice of the patients"* Letter from Egas Moniz to Freeman, November 6, 1936, GWU Archives.
*For Freeman considered Moniz a genius* Letter from Freeman to Almeida Lima, March 22, 1937, GWU Archives.

114 *Freeman replied that he and Watts would promptly* Letter from Freeman to Egas Moniz, October 20, 1936, GWU Archives.
*Freeman offered to send one* Letter from Freeman to Egas Moniz, January 5, 1937, GWU Archives.
*When the journal rejected the paper* Letter from Editor, *Journal of the American Medical Association*, to Freeman, February 1, 1937, GWU Archives.
*where the editor overcame his reservations* Letter from C. B. Farrar to Freeman, February 22, 1937, GWU Archives.
*"I trust that by that time the procedure"* Letter from Freeman to Egas Moniz, January 5, 1937, GWU Archives.
*"All hail to its originator!"* Letter from Freeman to Almedia Lima, March 22, 1937, GWU Archives.

115 *One of the other conference presenters* Letter from Freeman to Egas Moniz, November 26, 1936, GWU Archives.
*He had long ago learned that* Freeman, "History of Psychosurgery" (manuscript), p. 4-1, GWU Archives.
*Henry made quick use* Henry, "Brain Operation by D.C. Doctors Aids Mental Ills."
*"As was to be expected"* Freeman, *Autobiography*, p. 14-4.
*When four other reporters ambushed him* Macmillan, 2000, p. 245.

116 *He showed his "before and after" photographic* Shutts, 1982, p. 69.
*"We are able to say, with Moniz"* Freeman and Watts, "Prefrontal Lobotomy in the Treatment of Mental Disorders" (manuscript), 1936, pp. 2–3, GWU Archives.

117 *(Less than a year later)* Shutts, 1982, p. 76.
*"I am not antagonistic to this work"* Freeman, *The Psychiatrist*, p. 122.

118 *"one that might join the separate fields"* Pressman, 1998, p. 8.
*"Meyer studied the* person" Freeman, *The Psychiatrist*, p. 126.
*Meyer's directive to "follow up"* Ibid., p. 122.
*"Had it not been for his sympathetic"* Freeman, *Autobiography*, p. 14-5.
*"The results that you and Dr. Watts"* Letter from Egas Moniz to Freeman, December 24, 1936, GWU Archives.

119 *Freeman's letters to Moniz voiced his expectation* Letter from Freeman to Egas Moniz, November 26, 1936, GWU Archives.
*"As a fund-raiser I was a flop"* Freeman, *Autobiography*, p. 10-9.
*He blamed part of his lack of success* Ibid., p. 14-7.
*"He has already obtained a leucotome"* Letter from Freeman to Egas Moniz, November 26, 1936, GWU Archives.
*They operated on a new patient* Shutts, 1982, p. 71.

120 *A disastrous outcome was the case* Freeman and Watts, "Psychosurgery: Effect on Certain Mental Symptoms of Surgical Interruption of Pathways in the Frontal Lobe" (manuscript), 1937, GWU Archives.
*After his recovery from the lobotomy* Freeman, *Autobiography*, pp. 14-5–14-6.
*In 1957 Hennessy reported to Freeman* Letter from Paul K. Hennessy to Freeman, November 27, 1957, GWU Archives.

121 *"This, of course, distressed us"* Letter from Freeman to Egas Moniz, January 5, 1937, GWU Archives.
*Although the patient's family refused to give* Shutts, 1982, p. 72.
*Of the first twenty patients* Letter from Freeman to Egas Moniz, January 5, 1937, GWU Archives.
*(Freeman was delighted after he allowed)* Freeman, *Autobiography*, p. 15-11.
*"I had a feeling that he wanted"* Shutts, 1982, p. 75.

122 *On the day of one early lobotomy* Ibid.
*(Ten years later, a review)* Freeman and Watts, 1950, p. 487.
*and he and Watts operated on only twelve* Freeman, *Autobiography*, p. 14-6.
*"We may recall the response of Faraday"* Freeman, "Frontal Lobotomy 1936–1956: A Follow-Up Study of 3,000 Patients from One to Twenty Years" (manuscript), GWU Archives.

123 *On June 6, 1937, the* New York Times Laurence, 1937, p. 1.
*"Watts and I had made the headlines"* Shutts, 1982, p. 84.
*"Our work was roundly criticized"* Letter from Freeman to Egas Moniz, August 11, 1937, GWU Archives.
*Freeman noted with satisfaction that his display* Freeman, "History of Psychosurgery" (manuscript), pp. 4–10, GWU Archives.
*Freeman exhibited at every AMA convention* Valenstein, 1986, p. 160.
*sometimes by using a clacker* Watts, James W., "Psychosurgery: A 20 Year Follow-Up of the Freeman–Watts Series" (manuscript), p. 7, GWU Archives.
*William A. White, the hospital superintendent* Letter from Freeman to Egas Moniz, October 20, 1936, GWU Archives.

124 *When he at last broached the subject* Letter from Freeman to Egas Moniz, November 26, 1936, GWU Archives.
*White questioned the ability of a mentally* Shutts, 1982, pp. 81–82.
*"Freeman, it will be a hell of a long"* Freeman, *The Psychiatrist*, p. 57.
*"Such a radical procedure is not to be"* "The Surgical Treatment of Certain Psychoses" (manuscript), GWU Archives.

125 *In 1758 William Battie* Shorter, 1997, p. 27.
*Johann Riel further refined these notions* Ibid., p. 28.
*in Charcot's case because his target* Ibid., p. 85.
*By the end of the nineteenth century* Ibid., p. 103.

127 *Freud, in fact, favored removing* Freeman, *The Psychiatrist*, p. 85.
*As early as 1894* Shorter, 1997, p. 162.
*A psychoanalytic society was launched* Ibid., p. 163.
*Propelled by the financial opportunities* Shorter, 1997, p. 181.
*"If in the end analysis won"* Ibid., p. 154.
*Only in a few institutions clustered* Ibid., p. 176.
*"Insight is a terrible weapon"* Freeman, *Autobiography*, p. 14-10.

128 *"I am forced to the conclusion"* Freeman, *The Psychiatrist*, p. 18.
*He called the neural network in the brain* Fulton, John F., "Summary of Thomas W. Salmon Lectures, Lecture III" (manuscript), January 11, 1951, Yale Archives.
*such as bowssening* Partridge, 1950, p. 1.

129 *As Edward Shorter noted* Shorter, 1997, p. 208.
*The end of the coma or seizures* Freeman, *The Psychiatrist*, p. 29.
*who claimed to be a direct descendant* Ibid., p. 37.
*"He had a withdrawn, rather defensive"* Ibid., p. 31.
*and a hundred American hospitals* Shorter, 1997, p. 212.

130 *Freeman met Meduna several times* Freeman, *The Psychiatrist*, p. 42.
*Two American psychiatrists who experimented* Shutts, 1982, p. 44.
*His first patient, a thirty-nine-year-old* Shorter, 1997, p. 219.
*"He started to sing abruptly"* Freeman, *The Psychiatrist*, p. 47.
*[In another account]* Shorter, 1997, p. 220.

131 *Electroshock, which arrived in the United States* Ibid., p. 222.
*In 1934 he began treating an explorer* Shutts, 1982, p. 121; Shorter, 1997, p. 223.

131 *As a graduate student in Paris* Letter from Freeman to Dr. Winkelman, August 3, 1923, GWU Archives.
*While doctors set up shock units* Shutts, 1982, p. xx.
*Freeman initially mocked the theory* Ibid., pp. 77–80.

132 *becoming the first physician in the Washington area* Ibid., p. 83.
*whose drive to break new ground* Freeman, *The Psychiatrist*, p. 28.
*Florence had lived with her father* Shutts, 1982, p. 83.
*Within ten seconds "she began twitching"* Freeman, *The Psychiatrist*, pp. 41–42.
*Electroshock, which he began using* Freeman, *Autobiography*, p. 8-19.
*("Maybe it will be shown that a mentally ill")* Shutts, 1982, p. 112.

133 *"Electroshock treatment seems to be the answer"* Letter from Freeman to "Folks" (family letter), March 6, 1945, Franklin Freeman collection.
*One patient in particular* Freeman, *Autobiography*, pp. 8-19–8-20.

## 7: The Lines of Battle

134 *"The psychiatric doctors professing the classic"* Letter from Egas Moniz to Freeman, December 24, 1936, GWU Archives.
*"In the depth of the problem"* Letter from Egas Moniz to Freeman, May 30, 1949, GWU Archives.
*One of the most aggravating* Freeman and Watts, 1944, pp. 1–2.

135 *The English neurosurgeon William Sargant* Sargant, 1967, pp. 66–70.
*started with an attempt to shout down* Pressman, 1998, p. 80.
*In the summer of 1941 Bullard and his wife* Letter from Marjorie Freeman to Freeman, July 21, 1941, Paul Freeman collection.

136 *(Freeman once accepted Bullard's invitation)* Freeman, *The Psychiatrist*, p. 247.
*"Which is better," Watts once replied* Shutts, 1982, p. 107.
*Catholic doctrine of the time* Ibid., p. 105.
*Eventually, the Catholic Church decided* Kelly, 1957, p. 270.
*Watts recalled that in an informal survey* Shutts, 1982, p. 106.

137 *In general, he wrote* Letter from Freeman to Egas Moniz, March 15, 1938, GWU Archives.
*including the psychiatrist Loyal Davis* Valenstein, 1986, p. 146.
*"One of my friends observed"* Macmillan, 2000, p. 245.
*Giving a public lecture in 1944* "Dr. Freeman Lectures on Neurology."
*The previous year, during a trip to Toronto* Letter from Freeman to "Folks" (family letter), January 18, 1943, Franklin Freeman collection.
*"I try to bring in a little vulgarity"* Ibid.
*Partly on the basis of his reception* Letter from Freeman to Norman Freeman, June 30, 1944, Franklin Freeman collection.
*"My grandfather, W. W. Keen"* Letter from Freeman to Egas Moniz, July 24, 1946, GWU Archives.

138 *He calmly explained to one antagonist* Letter from Freeman to Smith Ely Jelliffe, January 6, 1940, GWU Archives.
*Writing sixteen years later* Letter from Freeman to Magnus C. Petersen, September 25, 1956, GWU Archives.
*The first, an alcoholic* Sargant, 1967, pp. 65–66.
*Wilder Penfield, the Canadian physician* Freeman, *The Psychiatrist*, p. 60.

139 *At last, five years after the first lobotomies* Shutts, 1982, p. 113.
(This was the only instance) Valenstein, 1986, p. 187.

139 *"Psychosurgery was a form of human salvage"* Pressman, 1998, p. 10.
*"We noticed that considerable dullness"* Letter from Freeman to Egas Moniz, March 15, 1938, GWU Archives.
*Two of the early lobotomy patients* Letter from Clara Hoye to Freeman, March 7, 1962, GWU Archives.

140 *(Freeman did not consider Hanshew's)* Freeman, "History of Psychosurgery" (manuscript), p. 6-58, GWU Archives.
*By August 1937, nearly a year* Letter from Freeman to Moniz, August 11, 1937, GWU Archives.
*(Eventually eight of the first twenty patients)* Valenstein, 1986, p. 148.
*"When a patient has improved following lobotomy"* Freeman and Watts, 1950, pp. 83–84.
*One such patient was the woman* Ibid., pp. 84–85.
*Among those receiving repeat lobotomies* Shutts, 1982, p. 82.
*The family of the woman known as Case 6* Ibid., pp. 82–83.
*When those treatments failed to produce* Ibid., p. 128.

141 *"But such things"* Letter from Freeman to Egas Moniz, July 24, 1946, GWU Archives.
*Freeman, in his follow-ups on this patient* Shutts, 1982, p. 85.
*Surprisingly this patient returned in 1941* Ibid., p. 114.
*By 1956 Freeman had tallied* Freeman, "Frontal Lobotomy 1936–1956" (manuscript), GWU Archives.
*"A recent letter from a patient"* Ibid.
*By the spring of 1938 he had performed* Letter from Egas Moniz to Freeman, April 5, 1938, GWU Archives.
*Freeman updated Moniz on the first forty* Letter from Freeman to Egas Moniz, March 15, 1938, GWU Archives.

142 *By this time, the signals of impending war* Shutts, 1982, p. 112.
*a young man suffering from hormonal problems* Ibid., p. 109.
*"I knew at the fifth shot"* Ibid., pp. 109–110.
*His blood sprayed the patient record* Freeman, 1956, p. 771.
*Moniz did not die* Valenstein, 1986, p. 221.
*Writing years later, he noted* Freeman, "Head-and-Shoulder Hunting in the Americas," p. 336.
*Pressure in Portugal* Feldman, "Psychosurgery: A Historical Overview," p. 12.
*He also believed that his Portuguese colleagues* Letter from Egas Moniz to Freeman, July 9, 1946, GWU Archives.

143 *"We studied brains of patients who died"* Freeman, *Autobiography*, p. 14-7.
Case 5, for example Shutts, 1982, p. 137.
*"are the most embarrassing sequel"* Freeman and Watts, 1950, p. 103.
*"During the operation, I stood at a distance"* Freeman, "History of Psychosurgery" (manuscript), p. 7-5, GWU Archives.
*"That's pretty damn dramatic"* Shutts, 1982, p. 90.

144 *"Whenever I felt resistance"* Ibid., p. 88.
*In these cases, they "found that inertia"* Freeman and Watts, 1950, p. 32.

145 *Any patient risked "a considerable sacrifice"* Ibid., p. 33.
*(Years earlier Moniz had observed)* Shutts, 1982, p. 134.
*"The bizarre 'china doll'"* Ibid., p. 133.

146 *(Freeman, who usually took charge)* Freeman and Watts, 1950, p. 114.
*"An operation under local anesthesia"* Ibid., p. 113.

146 *"a grinding sound that is as distressing"* Ibid., pp. 113–114.
*When Freeman once asked* Ibid., p. 115.
*One woman identified as Mrs. A.* Ibid., pp. 116–124.

148 *Freeman: Who am I?* Kaempffert, 1941, p. 72.
*Freeman [after cuts made in the left side]* Freeman and Watts, 1950, p. 129.
*a "disorientation yard stick"* Freeman, "History of Psychosurgery" (manuscript), p. 7-4, GWU Archives.
*He cited the case of one psychosurgery patient* Freeman and Watts, 1950, p. 18.
*Instead of damaging intelligence* Ibid., p. xxii.

149 *one physician noting that it "would appear"* Partridge, 1950, p. 78.
*one of the lasting effects of the psychosurgery era* Pressman, 1998, p. 12.
*"Prefrontal lobotomy has the effect of a surgically induced"* Ewald, 1947, p. 210.
*Some male patients exhibited a kind of* Shutts, 1982, p. 103.
*In instances that demanded discipline* Ewald, 1947, p. 212.
*"On one occasion the nurse supplied her"* Freeman and Watts, 1950, p. 533.
*(These side effects prompted)* Franklin Freeman, Interview, 2000.
*Freeman had formulated the conviction* Shutts, 1982, p. 127.

150 *"some of the most difficult"* Letter from Freeman to Egas Moniz, April 17, 1944, GWU Archives.
*Freeman confessed to his family* Letter from Freeman to "Folks" (family letter), January 18, 1943, Franklin Freeman collection.
*Over the next seven years* Freeman, *Autobiography*, p. 9-14.
*One of the most unusual of these patients* Patient Record, A.W., March 19, 1945, GWU Archives.
*"'There is not an ounce of fat on her body'"* Operative Record, Oretha Henley, February 2, 1944, GWU Archives.

151 *and in 1949 she wrote to Watts* Letter from Oretha Henley to James Watts, February 21, 1949, GWU Archives.
*"Recently we have become even more radical"* Letter from Freeman to Egas Moniz, April 17, 1944, GWU Archives.
*He painted a brighter picture* "Surgery Advocated for Schizophrenia."
*Watts, in fact, confessed that despite* James Watts, "Psychosurgery: A 20 Year Follow-Up of the Freeman–Watts Series" (manuscript), p. 4, GWU Archives.
*Franklin Freeman believed that his father* Franklin Freeman, Interview, 2002.
*(During a 1942 investigation)* FBI file on Walter Jackson Freeman.

152 *"Marjorie did most of her Christmas shopping"* Freeman, *Autobiography*, p. 16-10.
*Over the next several days* Ibid., pp. 16-10–16-12.

153 *One member of Freeman's audience* Valenstein, 1986, p. 281.
*The war would prevent Freeman from receiving* Freeman, "History of Psychosurgery" (manuscript), p. 6-29, GWU Archives.

154 *After their return, Freeman remembered* Freeman, *Autobiography*, p. 8-12.
*"Therefore we have to fall back upon cripples"* Letter from Freeman to "Folks" (family letter), July 5, 1942, Franklin Freeman collection.
*(In a letter, to his brother, Norman)* Letter from Freeman to Norman Freeman, June 27, 1942, Franklin Freeman collection.
*"I found myself giving less and less care"* Letter from Freeman to "Folks" (family letter), March 6, 1945, Franklin Freeman collection.
*a palatial penthouse in which patients* Peery, 1979, p. 53.
*There they netted $12,300* "Income from Lobotomy," 1942, GWU Archives.

154 *"I don't believe in letting on to her"* Letter from Freeman to Norman Freeman, September 26, 1942, Franklin Freeman collection.

155 *he occasionally strolled a dozen miles* Letter from Freeman to "Folks" (family letter), January 18, 1943, Franklin Freeman collection.

*While attending a medical conference* Letter from Freeman to Norman Freeman, May 18, 1943, Franklin Freeman collection.

*"When their muscles are working"* Untitled article, El Kharj, Saudi Arabia, Franklin Freeman collection.

*"We must develop a sure cure"* "Activity a Safeguard."

*"It isn't the ease and the pleasures of life"* Letter from Freeman to Lorne Freeman, February 4, 1945, Franklin Freeman collection.

## 8: ADVANCE AND RETREAT

157 *with Freeman noting that one boyfriend* Letter from Freeman to Norman Freeman, June 30, 1944, Franklin Freeman collection.

*"She really looked up to my father"* Paul Freeman, Interview, 2002.

*"Almost every hour there is a yell"* Letter from Freeman to Lorne Freeman, July 5, 1942, Franklin Freeman collection.

158 *Soon they stove the boat* Freeman, *Autobiography*, p. 15-19.

*"I recalled an old song"* Letter from Freeman to Norman Freeman, June 30, 1944, Franklin Freeman collection.

*The trip hit its nadir* Freeman, *Autobiography*, pp. 15-6–15-7.

*After visiting San Francisco* Ibid., pp. 15-8–15-10.

*("even though the chow mein resembled")* Letter from Freeman to Marjorie Freeman, August 28, 1941, Paul Freeman collection.

*On the way home, the car broke down* Freeman, *Autobiography*, pp. 15-12–15-14.

*The Freemans launched their trailer on July 31* Ibid., p. 15-18.

159 *While still at Western State* Ibid., pp. 15-18–15-19.

*Freeman's detailed ledger* "Travel Report," Paul Freeman collection.

*"There is something essentially healing"* Freeman, *Autobiography*, p. 15-11.

160 *Freeman and his sons visited Shenandoah* Letter from Freeman to "Folks," January 18, 1943, Franklin Freeman collection.

*"Everything is damp with the dampness"* Letter from Freeman to "Folks," September 13, 1944, Franklin Freeman collection.

*"This is a very masculine household"* Letter from Freeman to "Folks," May 24, 1944, Franklin Freeman collection.

*all six-footers who wore size 13* Letter from Freeman to Bill Freeman, September 18, 1943, Franklin Freeman collection.

*Months later he checked up on the work* Letter from Freeman to "Folks," May 24, 1944, Franklin Freeman collection.

*Freeman suggested that he get in touch* Letter from Freeman to Franklin Freeman, April 1, 1945, Franklin Freeman collection.

161 *Keen and Randy, despite their two-year age difference* Letter from Freeman to "Folks," May 31-July 14, 1942, Franklin Freeman collection.

*Together the family celebrated Christmas* Letter from Freeman to Norman Freeman, December 1943, Franklin Freeman collection.

*Once a week this large and energetic family* Letter from Freeman to Bill Freeman, January 1, 1945, Franklin Freeman collection.

161 *"Marjorie is tired and worn out"* Letter from Freeman to Bill Freeman, September18, 1943, Franklin Freeman collection.
*She took dancing lessons and developed* Paul Freeman, Interview, 2002.
*Overall, she found her absence* Letter from Freeman to "Folks," December 1943, Franklin Freeman collection.
*In a year, she decided never to go back* Letter from Freeman to Bill Freeman, January 1, 1945, Franklin Freeman collection.
*the available assistants "were leftovers"* Freeman, *Autobiography*, p. 8-12.

162 *"If I can get Mother accustomed"* Letter from Freeman to Jack Freeman, July 8, 1945, Franklin Freeman collection.
*Freeman had a limited appreciation of the demands* Freeman, "Democracy Versus Autocracy in the Home" (manuscript), GWU Archives.
*Freeman enjoyed mixing a signature cocktail* Franklin Freeman, Interview, 2002.
*"I think she was not exactly on his cheering bench"* Ibid.
*"Always when you go away"* Letter from Marjorie Freeman to Freeman, February 22, 1945, Franklin Freeman collection.
*The family's finances were now secure* Letter from Freeman to "Folks," December 1943, Franklin Freeman collection.
*"Occasionally when I wake up"* Letter from Freeman to Norman Freeman, May 18, 1943, Franklin Freeman collection; Letter from Freeman to "Folks," December 1943, Franklin Freeman collection.
*During a five-day stretch in 1944* Letter from Freeman to "Folks," May 24, 1944, Franklin Freeman collection.
*That spring they began work on a motion picture* Freeman, "History of Psychosurgery" (manuscript), p. 6-14, GWU Archives.

163 *So far, the medical literature in English* Ibid., p. 6-31.
*A Macmillan representative wrote* Letter from Macmillan Co. to James Watts, July 31, 1940, GWU Archives.
*Watts later learned the real reason* Letter from James Watts to Freeman, August 6, 1940, GWU Archives.
*"There is considerable opposition to our publishing"* Letter from Charles C. Thomas to James Watts and Freeman, July 25, 1941, GWU Archives.
*"Library [research] time had to be stolen"* Freeman, *Autobiography*, p. 14-8.

164 In the book, for example, he described one patient Freeman and Watts, 1942, p. 145.

165 *"a mark of your kindness"* Letter from Egas Moniz to Freeman, October 30, 1940, GWU Archives.
*Sick in bed with a cold* Letter from Freeman to "Folks," July 5, 1942, Franklin Freeman collection.

166 *was painted by Mary Lawrence* Letter from Freeman to Charles Thomas, December 14, 1941, GWU Archives.
*(Later questioned whether a finger painting)* Shutts, 1982, p. 124.
*although five hundred copies bound for Europe* Ibid.
In England, a supportive writer called it Partridge, 1950, p. 4.
*"The book by Freeman and Watts hit us like a bomb"* Valenstein, 1986, p. 172.
*John Fulton asked if he could have* Letter from Freeman to Charles Thomas, January 6, 1942, GWU Archives.
*A writer for the* New York Times *called it* Kaempffert, 1942, p. D7.

167 *they fronted a substantial sum* Publication contract, July 14, 1941, GWU Archives.

167 *When in 1943 he received an invitation* Letter from Freeman to Norman Freeman, December 1943, Franklin Freeman collection.
*"a sort of scrapbook"* Letter from Freeman to "Folks," July 5, 1942, Franklin Freeman collection.
*Of this large group of patients* Feldman, "Psychosurgery: A Historical Overview," p. 14.
*"I have been reviewing all the cases"* Letter from Freeman to Mr. Spivak, March 31, 1942, GWU Archives.
168 *An Associated Press report* McDonough, 1941.
*Another Associated Press story* McDonough, 1942.
*Only John Fulton's intervention* Freeman, "History of Psychosurgery," p. 4-14, GWU Archives.
169 *Kaempffert's article was later condensed* Ibid.
*One of Freeman's early lobotomy patients* Dannecker, 1942.
170 *Ten years after his lobotomy* Freeman and Watts, 1950, p. 470.
171 *"Gentle, clever your surgeon's hands"* Pressman, 1998, unpaginated front page.
*"Some patients come to operation"* Freeman and Watts, 1950, p. 113.
*Seventeen of their first twenty* Shutts, 1982, p. 104.
*"Women with involutional depression"* Freeman and Watts, 1950, p. 66.
*(Freeman and Watts once expressed)* Shutts, 1982, p. 123.
172 *In one case, a patient died* Freeman and Watts, 1950, p. 100.
*Another early patient died* Ibid., p. 101.
*Their 193rd patient died from dehydration* Ibid., p. 102.
*"one from [a] convulsive state"* Letter from Freeman to Norman Freeman, May 5, 1945, Franklin Freeman collection.
*During one surgery conducted under local* Freeman and Watts, 1950, pp. 95–96.
*In 1943 Freeman told Norman about* Letter from Freeman to Norman Freeman, December 1943, Franklin Freeman collection.
173 *When May confirmed that he had murder* Shutts, 1982, p. 111.
*the police brought him to a nearby* Freeman, "Head-and-Shoulder Hunting in the Americas," p. 336.
*Freeman also wrote that twice* Freeman, *The Psychiatrist*, pp. 286–287.
*As a precaution against attack* Shutts, 1982, p. 110.
*although not enough to keep her* Gibson, 1995, p. 61.
*participating in sailboat races* Ibid., p. 102.
*or learning to read* Ibid., p. 56.
*There, according to Kennedy family* Leamer, 2001, p. 169.
*He first took Rosemary* Gibson, 1995, p. 60.
174 *(Watts, in fact, diagnosed Rosemary)* Leamer, 2001, p. 169.
*several times she wandered off* Gibson, 1995, p. 72.
*Rose Kennedy did not see her daughter* Ibid., p. 71.
*when John F. Kennedy secretly* Leamer, 2001, p. 280.
*Leamer encountered a retired nurse* Ibid., pp. 170, 766.
*The first child patient* Shutts, 1982, pp. 99–100.
175 *Four years later* Freeman and Watts, 1950, pp. 441–444.
*Another child, a twelve-year-old boy* Ibid., p. 102.
176 *"most of the patients have shown"* Ibid., p. 437.
*"the army is interested in disposition"* Letter from Freeman to Bill Freeman, September 18, 1943, Franklin Freeman collection.

176 *"Within this new framework"* Pressman, 1998, p. 28.
177 *"I was in the lounge car"* Freeman, *Autobiography*, pp. 8-14–8-15.

## 9: WATERFALL

178 *psychiatric cases filled more than half* Shutts, 1982, p. 139.
179 *To further worsen the problem* Valenstein, 1986, p. 229.
*American hospitals and physicians* Grob, 1994, p. 183.
*"It took a number of dramatic"* Freeman, *Autobiography*, p. 14-12.
180 *"Lobotomy, instead of being the last resort"* Freeman, "Transorbital Lobotomy in State Mental Hospitals."
*"There is an enormous inertia"* Letter from Freeman to Dot Freeman, December 29, 1946, Franklin Freeman collection.
*The neurosurgeon Harold Buchstein* Pressman, 1998, p. 127.
181 *Freeman praised Schrader's work* Shutts, 1982, p. 131.
*"Personally," Petersen wrote Freeman* Letter from Magnus C. Petersen to Freeman, September 21, 1956, GWU Archives.
*"limits the value of this type"* Freeman, "Lobotomy: A Comparison of Prefrontal Lobotomy (Freeman and Watts). with Transorbital Lobotomy" (manuscript), GWU Archives
182 *"In the present state of affairs"* Freeman, *The Psychiatrist*, p. 48.
*As a result, few such leucotomies* Valenstein, 1986, p. 201.
*Freeman judged Fiamberti's results as poor* Ibid., p. 204.
183 *Unlike the neurosurgeons performing* Freeman, "Transorbital Leucotomy," pp. 372–373.
*"a light tap with a hammer"* Freeman and Watts, 1950, p. 53.
184 *"The transorbital method brings the possibilities"* Ibid.
*"I pull the handle of the instrument"* Ibid., p. 100.
*"consists of knocking them out"* Valenstein, 1986, p. 203.
185 *"an audible crack"* Ibid., p. 214.
*"Just as in the case of Moniz"* Freeman and Watts, 1950, p. 53.
*"Electro-shock appears to have"* Freeman, "Transorbital Leucotomy," p. 373.
*He also believed ECT aided in the clotting* Letter from Freeman to Egas Moniz, September 9, 1952, GWU Archives.
186 *But lobotomy patients with tuberculosis* Cheng, 1956, p. 32.
*"I was there to hold"* El-Hai, 2001, p. 16.
*"98 percent of the time"* Freeman and Watts, 1950, p. 56.
*His later partner* Valenstein, 1986, p. 215.
*During a surgery they were working* Shutts, 1982, p. 205.
187 *For a long time* Isay, 2004.
*"I explained to her and her husband"* Freeman, *Autobiography*, p. 14-13.
*Decades later, her daughter* Isay, 2004.
188 *Four years afer Ionesco's lobotomy* Freeman and Watts, 1950, p. 53.
*Freeman and Ionesco corresponded* Isay, 2004.
*"This manoeuver not only makes"* Letter from Freeman to Egas Moniz, September 9, 1952, GWU Archives.
189 *"It remains to be seen how these cases"* Valenstein, 1986, pp. 203–204.
*"In fact he didn't like it so much"* Letter from Freeman to "Folks," October 5, 1947, Franklin Freeman collection.
190 *Later Watts informed his partner* Valenstein, 1986, p. 205.

190 *Watts once explained* Peery, 1979, p. 79.
*In 1979 he told* Ibid., p. 76.
*"dog-in-manger attitude"* Freeman, *Autobiography*, p. 14-14.
*He still tried to engage Watts* Valenstein, 1986, p. 205.
191 *"His lively temperament"* Freeman, *Autobiography*, p. 14-16.
*"Well, he knew more anatomy of the brain"* Shutts, 1982, p. 160.
*"because all you're going to do is pull up"* Ibid., p. 174.
*In 1950, after a patient had three times* Ibid., p. 205.
*Jonathan Williams recalled* Valenstein, 1986, p. 213.
192 *"The surgeon is offended by the spectacle"* Freeman, "Lobotomy: A Comparison of Prefrontal Lobotomy (Freeman and Watts) with Transorbital Lobotomy" (manuscript), GWU Archives.
*"no question about it"* Shutts, 1982, p. 150.
*In discussions with Watts* Letter from Freeman to "Folks," October 5, 1947, Franklin Freeman collection.
*Examining his first ten patients* Freeman, "Transorbital Leucotomy," p. 373.
193 *(Returning to their cases)* Freeman, "History of Psychosurgery" (manuscript), p. 7-15, GWU Archives.
*"Recovery from transorbital lobotomy is spectacular"* Freeman and Watts, 1950, p. 511.
*Infections were rare* Letter from Freeman to Egas Moniz, September 9, 1952, GWU Archives.
*"He believes that each case should be studied"* Letter from Freeman to Egas Moniz, September 9, 1952, GWU Archives.
*"Freeman liked to do it in what you might call"* Shutts, 1982, p. 173.
*Nevertheless, Watts performed twenty-eight transorbital* James Watts, Chart tracking transorbital lobotomies, undated, GWU Archives.
*"As you probably know, it is my opinion"* Letter from James Watts to David C. Wilson, March 14, 1952, GWU Archives.
194 *In another instance, Watts* Letter from Freeman to Egas Moniz, September 9, 1952, GWU Archives.
*"I watched two lobotomized women during delivery"* Freeman, "Air Castles from Diseased Imaginations" (manuscript), 1958, p. 6, GWU Archives.
*Watts did not object* Peery, 1979, p. 74.
*he accompanied Freeman* Shutts, 1982, p. 193.
*"This procedure can be carried out"* Letter from Freeman to Egas Moniz, June 21, 1949, GWU Archives.
*One patient, whose terminal illness had left* Letter from Freeman to Egas Moniz, June 21, 1949, GWU Archives.
195 *"Jim made a fancy operation out of it"* Letter from Freeman to "Folks," October 5, 1947, Franklin Freeman collection.
*One of the earliest, a woman with throat* Freeman and Watts, 1950, pp. 363–364.
*But he conceded that the new procedure* Ibid., p. 373.
*Freeman, in contrast, gave a reporter* Lerch, 1948.
196 *Of the first four hundred transorbital operations* Freeman and Watts, 1950, p. 511.
*In 1948 Freeman reported that three patients* Freeman, "Transorbital Leucotomy," p. 375.
*"cheerful to the point of elation"* Freeman and Watts, 1950, p. 511.
*"a certain indolence and tactlessness"* Freeman, "Transorbital Leucotomy," p. 375.
*"a certain lack of subtlety"* Freeman and Watts, 1950, p. 511.

196 *"the Boy Scout virtues in reverse"* Freeman, "Psychosurgery: Present Indications and Future Prospects," p. 430.

197 *"Relapses are fairly common"* Freeman and Watts, 1950, p. 511.

*some of his patients had emerged* Freeman, "Transorbital Leucotomy," p. 376.

*"My parting words are usually"* Freeman and Watts, 1950, p. 57.

*"Anxiety and emotional tension"* Freeman, "Transorbital Leucotomy," p. 376.

*This more radical procedure* Shutts, 1982, p. 180.

*Meanwhile, Freeman's enthusiasm for the transorbital* Beam, 2001, p. 91.

198 *The new instrument* Freeman, 1952, p. 826.

*"I felt safe in using maximum power"* Freeman, "History of Psychosurgery" (manuscript), p. 7-24, GWU Archives.

*With pride, Freeman told Moniz* Letter from Freeman to Egas Moniz, September 9, 1952, GWU Archives.

199 *Elliot Valenstein reports* Valenstein, 1986, p. 258.

*While Fulton acknowledged* Letter from John Fulton to A. Earl Walker, January 19, 1951, Yale Archives.

*Freeman denied that he considered* Pressman, 1998, p. 342.

200 *Fulton later tried to dissuade* Valenstein, 1986, p. 219.

*He belonged to sixty-three medical and scientific* Pressman, 1998, p. 511.

*None of these reservations* Ibid., p. 336.

202 *"Keen . . . is doing well in school"* Letter from Freeman to "Folks," March 6, 1945, Franklin Freeman collection.

*sometimes chew his cod liver oil* Walter Freeman Jr., Interview, 2002.

*Keen had already told everyone* Letter from Freeman to "Folks," March 16, 1947, Franklin Freeman collection.

*"On the rest of the trip"* Letter from Freeman to "Folks," March 16, 1947, Franklin Freeman collection.

*When a storm on the Great Plains* Letter from Freeman to Franklin Freeman, June 27, 1946, Franklin Freeman collection.

*He was especially taken by the way* Letter from Freeman to "Folks," May 24, 1944, Franklin Freeman collection.

*This time, two years later* Freeman, *Autobiography*, p. 15-22.

203 *It was a hot day, but Freeman* Ibid., pp. 15-22–15-23.

*discharged form the navy five days earlier* "Dayton Navy Veteran Dies in Futile Attempt to Rescue Boy."

*The sailor could not reach* Carnegie Fund Commission minute.

*"There is so much water going over"* Wire from Frank A. Kittridge to Mr. and Mrs. O. C. Loos, July 8, 1946, Franklin Freeman collection.

*While the recovery efforts proceeded* Freeman, "Collection of Themes: Student Essays by Undergraduates" (manuscript), GWU Archives.

204 *Instead, two days after the accident* Letter from Freeman to Franklin Freeman, July 22, 1946, Franklin Freeman collection.

*Marjorie and Freeman decided* "Freeman Boy to Rest Near Scene of Death."

*"Somehow," Freeman told son Franklin* Letter from Freeman to Franklin Freeman, July 22, 1946, Franklin Freeman collection.

*"Mr. Kittridge telephoned to say"* Handwritten note, July 15, 1946, Franklin Freeman collection.

*Loos's body surfaced* Letter from Freeman to Franklin Freeman, July 22, 1946, Franklin Freeman collection.

205 *"Of course being in the water"* Ibid.
*The things his mind could manage to consider* Ibid.
*"Her home is far from beautiful"* Letter from Freeman to Jack Freeman, August 7, 1946, Franklin Freeman collection.
206 *In Yankton, South Dakota* Letter from Freeman to Franklin Freeman, August 16, 1946, Franklin Freeman collection.
*"The rest of that trailer trip"* Freeman, *Autobiography*, pp. 15-23–15-24.

## 10: FAME

207 *"Marjorie is still deeply affected"* Letter from Freeman to Aunt Dodie, September 29, 1946, Franklin Freeman collection.
*"Marjorie personifies Keen"* Letter from Freeman to Dot Freeman, December 29, 1946, Franklin Freeman collection.
*"I would say that in some respects her life ended"* Walter Freeman Jr., Interview, 2002.
*"The children grow from birth"* Freeman, Handwritten document, July 1946, Franklin Freeman collection.
208 *"There seems to be nothing but work"* Letter from Freeman to Paul Freeman, January 13, 1947, Franklin Freeman collection.
*"I knocked myself out with Nembutal"* Letter from Freeman to "Folks," October 5, 1947, Franklin Freeman collection.
*"I suspected it was just one of those"* Franklin Freeman, Interview, 2002.
*"Whatever work I have to do"* Letter from Freeman to Dot Freeman, December 29, 1946, Franklin Freeman collection.
209 *"Keen was marked"* Freeman, *Autobiography*, p. 15-24.
210 *"There was a certain amount of horror"* Ibid., p. 14-15.
*Freeman even demonstrated transorbital* Shutts, 1982, p. 178.
*During one dramatic demonstration* Ibid., p. 212.
*"The psychiatrist is in a position"* Freeman, "Transorbital Lobotomy in State Mental Hospitals."
*"is being used in psychiatrists' offices"* Valenstein, 1986, p. 257.
*"He cited as examples eight of his"* "Personality Shift is Laid to Surgery," undated and unsourced newspaper article, Franklin Freeman collection.
211 *"Is the quieting of the patient"* Feldman, "Psychosurgery: A Historical Overview," p. 18.
*Freeman wanted a less stilted* Freeman, *Autobiography*, p. 14-18.
212 *Cartier tried to join the Carmelite order* Letter from Freeman to Gladys Cartier, September 9, 1955, GWU Archives.
*He found inspiration for a possible* Freeman, *Autobiography*, p. 14-25.
213 *The psychiatric historian Elliot Valenstein* Valenstein, 1986, p. 199.
*he operated on 2,400 more* Pressman, 1998, p. 340.
*Eight years earlier* Letter from Freeman to Mr. Spivak, March 31, 1942, GWU Archives.
*"I believe that transorbital lobotomy"* Freeman, "Lobotomy: A Comparison of Prefrontal Lobotomy (Freeman and Watts) with Transorbital Lobotomy" (manuscript), p. 7, GWU Archives.
*Fifteen percent of the prefrontal* Valenstein, 1986, p. 256.
214 *He concludes that "the fact that psychosurgeons"* Pressman, 1998, p. 16.

215 *By the spring of 1945* Letter from Freeman to "Folks," March 6, 1945, Franklin Freeman collection.
*"we can report results"* Letter from Freeman to Dot Freeman, December 29, 1946, Franklin Freeman collection.
*"fifty beds of our own"* Ibid.
*After the hospital administration ignored* Peery, 1979, p. 77.
216 *and he regarded this breaking of Jim Crow* Shutts, 1982, p. 175.
*"If this is fame"* Letter from Freeman to Dot Freeman, December 29, 1946, Franklin Freeman collection.
*although he refused to let them* Shutts, 1982, p. 152.
*He told one patient* Ibid., p. 151.
*In one case, a middle-aged bachelor* Freeman and Watts, 1950, p. 178.
*Another patient had sex with his wife* Ibid., p. 176.
217 *Freeman also treated a homosexual* Ibid., p. 178.
*VA Hospital staff* Shutts, 1982, p. 148.
*By 1949 the VA was performing* Ibid., p. 199.
*"The surgeon sees what he cuts"* Ibid., p. 94.
218 *"an effort to determine which part"* Freeman, *Autobiography*, p. 14-16.
*Freeman maintained that the benefits* Shutts, 1982, p. 165.
*Pool, who noted* Ibid., pp. 216–217.
*Pool eventually acknowledged* Letter from J. Lawrence Pool to Freeman, October 1, 1956, GWU Archives.
*and he even admitted that the main advantage* Pressman, 1998, p. 348.
219 *Freeman called Poppen's technique* Shutts, 1982, p. 192.
*"vacuum cleaner over a tub"* Valenstein, 1986, p. 257.
*"These brains, at least the one or two"* Shutts, 1982, p. 192.
*"I don't think the various techniques"* Ibid., p. 192.
*earlier than Freeman and Watts reached* Freeman and Watts, 1950, p. xxi.
*and by 1950 had performed more* Partridge, 1950, p. 16.
*McKissock, who used a plain cannula* Freeman and Watts, 1950, p. 64.
220 *Traveling more than sixty thousand miles* Partridge, 1950, p. 7.
*including one man who after seven years* Ibid., p. 473.
*"Bizarre illnesses may require"* Ibid., p. 1.
*"On one occasion in a district"* Ibid., p. 10.
221 *"One mother, after a private"* Ibid.
*"One woman, in general somewhat"* Ibid., p. 33.
*In June 1947 Hirose* Hirose, 1963.
*"I brought some nylons with me"* Freeman, *Autobiography*, p. 16-17.
222 *"five of them fainted"* Ibid., p. 14-17.
*"a particular high point in my evangelistic"* Freeman, "History of Psychosurgery" (manuscript), p. 6-26, GWU Archives.
*"there was general agreement"* Letter from Freeman to Egas Moniz, March 31, 1946, GWU Archives.
223 *"We flew the ocean"* Letter from Freeman to "Folks," September 19, 1948, Franklin Freeman collection.
*"This is my revenge"* Ibid.
*"I found out later"* Freeman, *Autobiography*, p. 16-18.
*and Freeman's demonstration of two* Freeman, "Comment: International Conference on Psychosurgery" (manuscript), GWU Archives.

223 *"the dark continent of Africa"* Letter from Freeman to Egas Moniz, April 27, 1946, GWU Archives.

224 *At the congress, Kleist* Freeman, *Autobiography*, pp. 16-19–16-20.
*"He shook his head in doubt"* Letter from Freeman to "Folks," September 19, 1948, Franklin Freeman collection.
*"one of the young assistants who reported"* Freeman, *Autobiography*, p. 16-20.
*A Soviet physician cited these ethical* Shutts, 1982, pp. 213–214.

225 *"While it is essential to have an assessment"* Freeman, "Transorbital Lobotomy in State Mental Hospitals."
*where he performed that country's first* Letter from Freeman to "Folks," September 19, 1948, Franklin Freeman collection.
*Dosed with Nembutal* Freeman, *Autobiography*, p. 16-27.
*"one could overload the circuits"* Ibid., p. 16-21.

226 *"When I think of your magnificent"* Letter from Freeman to Egas Moniz, September 9, 1952, GWU Archives.
*Moniz had raised the issue* Letter from Egas Moniz to Freeman, February 4, 1946, GWU Archives.

227 *"My father said something to the effect"* Franklin Freeman, Interview, 2002.
*Freeman estimated that five thousand* Letter from Freeman to Egas Moniz, December 12, 1946, GWU Archives.
*Twenty thousand people in the United States* Valenstein, 1986, p. 229.
*Psychosurgery became a form of treatment* Shorter, 1997, p. 228.
*"This has, to my mind"* Freeman, *Autobiography*, pp. 8-16–8-17.

228 *One weekend in August* Ibid., pp. 14-11–14-12.

229 *One of their lobotomy patients completed* Shutts, 1982, p. 176.
*Freeman clashed with his publisher* Ibid., p. 232.
*Freeman explained to Thomas his obsession* Letter from Freeman to Charles C. Thomas, July 17, 1950, GWU Archives.

230 *(They also presented evidence)* Freeman and Watts, 1950, p. 515.
*whom Freeman and Watts hired for a year* Shutts, 1982, p. 163.
*"She is enthusiastic, says she wonders"* Letter from Freeman to "Folks," October 5, 1947, Franklin Freeman collection.

231 *which remained in print until 1956* Letter from Freeman to James Watts, September 28, 1956; Royalty statement, 1956; GWU Archives.
*"As I approach my 52nd birthday"* Letter from Freeman to "Folks," October 5, 1947, Franklin Freeman collection.
*Freeman "was internationally known"* Shutts, 1982, p. 178.

232 *"A slender rather intense man"* Freeman, *Autobiography*, p. 8-18.
*"I don't know what better investment"* Letter from Freeman to Dick Freeman, December 12, 1945, Franklin Freeman collection.
*he boasted to his son Paul* Letter from Freeman to Paul Freeman, May 19, 1946, Franklin Freeman collection.

233 *"was a man who complained of asthma"* Freeman, "History of Psychosurgery" (manuscript), p. 6-6, GWU Archives.
*One batch of correspondence that year* Letter from Freeman to Lorne Freeman, August 25, 1946, Franklin Freeman collection.
*"In recent years much has been written"* Valenstein, 1986, p. 257.
*In the spring of 1948* Freeman, *The Psychiatrist*, p. 137.

234 *(The* Post *had retitled the story)* Wallace, 1965, p. 157.

234 *"I am not overly sensitive to censure"* Letter from Freeman to Irving Wallace, May 16, 1951, GWU Archives.
*Writing about the experience years later* Wallace, 1965, pp. 191–192.

## 11: ROAD WARRIOR

236 *even today, most new strategies* Pressman, 1998, p. xiv.
*especially in the United States, where the disorder* Shorter, 1997, p. 296.
237 *"To the hardpressed hospital superintendents"* Valenstein, 1986, p. 207.
*Otto Poole, a patient at the state hospital* Freeman, "History of Psychosurgery" (manuscript), p. 6-1, GWU Archives.
*"Suppose it should happen"* Ibid., p. 6-2, GWU Archives.
*"The extinction curve of schizophrenic"* Freeman, "Head-and-Shoulder Hunting in the Americas," p. 343.
238 *"To have abandoned the enterprise"* Pressman, 1998, p. 14.
*"Transorbital lobotomy is a simple"* Freeman, "Transorbital Lobotomy in State Mental Hospitals."
*"The definition of genius"* Freeman, *The Psychiatrist*, p. 73.
*"I may have some opportunity to carry it"* Letter from Freeman to Franklin Freeman, May 19, 1946, Franklin Freeman collection.
*"What I wanted was a large group"* Letter from Freeman to Franklin Freeman, August 16, 1946, Franklin Freeman collection.
239 *He "did five transorbitals"* Letter from Freeman to James Watts, June 28, 1946, Franklin Freeman collection.
*"The patients I worked on the way out"* Letter from Freeman to Franklin Freeman, August 16, 1946, Franklin Freeman collection.
*"Frustrated again"* Letter from Freeman to "Folks," October 5, 1947, Franklin Freeman collection.
*"I felt many a time as though I were out"* Freeman, "History of Psychosurgery," p. 7-18.
*"Whoever undertakes to perform psychosurgery"* Freeman and Watts, 1950, p. 513.
240 *"One of the men had such shaky hands"* Letter from Freeman to "Folks," October 5, 1947, GWU Archives.
*Calling the surgeries "super-orbital" lobotomies* "Brain Operations on 13 Mental Patients."
*The following year, Freeman reviewed* Freeman and Watts, 1950, p. 513.
*"Dr. Freeman . . . explained to a staff meeting"* Cohen, "New Surgery Used on 5 Patients."
241 *Western State doctors reported* Jones, 1948.
*"It's her all right"* Shutts, 1982, p. 184.
*In her family memoirs* Elliot, 1978, p. 153.
*In their account of their lobotomy program* Jones, 1948.
242 *He came to Berkeley at the invitation* Shutts, 1982, p. 181.
*"An extremely revolutionary, spectacular"* "Rare Operation Used on 2 Berkeley Women by Noted Brain Surgeon."
*Returning to Herrick* Shutts, 1982, pp. 212–213.
243 *"I tell them by all means yes"* Freeman, "History of Psychosurgery," p. 5-17, GWU Archives.
*"Freeman thrived on the 'horror'"* Pressman, 1998, p. 340.
244 *During the next eight months* Shutts, 1982, p. 187.

244 *"Although transorbital lobotomies comprised"* Valenstein, 1986, p. 229.
*In 1948 Patricia Derian* Scheflin, 1978, pp. 247–249.

245 *A minor stroke that summer* Freeman, *Autobiography*, p. 2-4.
*(One of the Milledgeville staff)* Shorter, 1997, p. 228.
*At Cherokee State Hospital* Valenstein, 1986, p. 231.

246 *The English neurosurgeon William Sargant* Sargant, 1967, p. 130.
*at the request of one of its neurosurgical consultants* Valenstein, 1986, p. 258.
*In these institutions, Freeman declared* Freeman, "Transorbital Lobotomy in State Mental Hospitals."
*One ugly incident marred* Valenstein, 1986, pp. 235–236.
*one such mishap had left a Maryland* Ibid., p. 231.
*Sixty percent remained hospitalized* Virginia State Hospitals, "Report on Transorbital Lobotomy Project," 1960, GWU Archives.

247 *"apparently thought"* Freeman, "History of Psychosurgery," p. 5-4, GWU Archives.
*Freeman maintained there were few outright refusals* Letter from Freeman to Egas Moniz, September 9, 1952, GWU Archives.
*He contradicted this assertion* Freeman, 1954, p. 939.

248 *"Actually only twelve days"* Letter from Freeman to Egas Moniz, September 9, 1952, GWU Archives.
*(Freeman repeated with pride)* Ibid.
*Watts later acknowledged* Shutts, 1982, p. 214.
*(He earned as much as)* Ibid.
*But summing up the results* Letter from Freeman to Egas Moniz, September 9, 1952, GWU Archives.

249 *A month after the end* Ibid.
*There were four deaths* Ibid.

250 *In 1958 he reported on his attempts* Freeman, "Head-and-Shoulder Hunting in the Americas," p. 340.
*He concluded that the "personality"* Freeman, 1962, p. 1134.
*The sole exception was* Letter from Freeman to Egas Moniz, September 9, 1952; "History of Psychosurgery," p. 6-17, GWU Archives.

251 *"My only regret"* Letter from Freeman to Hubert Fockler, August 3, 1961, GWU Archives.
*On that same expedition* Freeman, *Autobiography*, p. 18-8.
*He acquired IBM card punching* Ibid., p. 17-2.

252 *Freeman recorded all of the travel details* Ibid., p. 18-5.
*"thousands of chronically ill and violent"* Freeman, "Frontal Lobotomy, 1936–1956: A Follow-Up Study of 3,000 Patients from One to Twenty Years," p. 11, GWU Archives.
*"At the week's end"* Ensz, 1948.
*but he told a reporter that even if transorbital* Patrick, 1951.
*"Thank God for men such as you"* Letter from John T. Ferguson to Freeman, September 22, 1958, GWU Archives.

253 *"like a chemical lobotomy"* Shorter, 1997, p. 252.
*The state hospital system in New York* Ibid., p. 280.
*By the end of its first year* Feldman, "Psychosurgery: A Historical Overview," p. 19.
*Starting in 1955* Shorter, 1997, p. 280.
*"a drug that seemed to have a selective"* Freeman, *The Psychiatrist*, p. 74.

253 *The number of papers published* Josephine Ball, C. James Klett, Clement J. Gresock, "The Veterans Administration Study of Prefrontal Lobotomy" (undated manuscript), p. 3, GWU Archives.

254 *"Virtually every patient who might qualify"* Beam, 2001, p. 90.

*"The procedure faded away"* Shorter, 1997, p. 228.

*"the same phenomenon occurred in 1937"* Letter from Freeman to Sadao Hirose, July 7, 1955, Paul Freeman collection.

255 *"a backwater far from the great mainstream"* Slaughter, 1958, p. 94.

*"Once the connection between"* Ibid., p. 16.

*"I still maintain lobotomy is wrong"* Ibid., p. 31.

## 12: Leaving Home

257 *"I think I fell in love with California"* Freeman, *Autobiography*, p. 20-1.

258 *In 1952 the* Washington Evening Star Ibid., pp. 9-14–9-15; Letter from Freeman to Egas Moniz, September 9, 1952, GWU Archives.

*"To St. Elizabeths Hospital"* Shutts, 1982, p. 223.

*"Instead I was given a leave of absence"* Freeman, *Autobiography*, pp. 17-2–17-3.

*in 1957 Freeman changed his will* Letter from Freeman to Walter Bloedorn, January 22, 1957, GWU Archives.

*Around Thanksgiving 1953* Freeman, *Autobiography*, p. 17-4.

259 *and ended up renting a house* Ibid., p. 21-1.

*Upon arriving in California* Ibid., p. 2-8.

*The board tested his knowledge* Ibid., p. 17-3.

260 *"We saved the really important pieces"* Ibid., p. 17-5.

*Freeman noted with bitterness* Ibid., p. 17-6.

*"I made a very stupid speech"* Ibid.

*"There once was a man named McKay"* Shutts, 1982, p. 223.

*Freeman's departure saddened* Peery, 1979, p. 59.

261 *In the last days of June 1954* Freeman, *Autobiography*, pp. 17-6–17-7.

*"It was the best thing"* Ibid., p. 21-1.

262 *"I just can't understand why"* Ibid., pp. 21-12–21-13.

*"When milder measures are not sufficient"* Freeman, 1958, p. 433.

*"I think if he hadn't gotten hung up"* Walter Freeman Jr., Interview, 2002.

*Attacked and weakened* Shutts, 1982, p. 228; Ligon, 1998.

*"I spent a day with him"* Freeman, *Autobiography*, pp. 16-23–16-24.

*Although Moniz flared with anger* Shutts, 1982, p. 220.

263 *"Neurologists, neurological surgeons and psychiatrists"* Freeman, 1957, p. 147.

*"My start of smoking at 23"* Freeman, *Autobiography*, pp. 2-4–2-5.

*"Needless to say, the blue skies"* Letter from Freeman to Rex Blandinship, October 15, 1956, GWU Archives.

*"I like walking in the hills"* Freeman, *Autobiography*, p. 2-6.

*"I like the sensation of hiking"* Ibid., p. 20-9.

264 *"a new lease on life"* Ibid., p. 19-14.

*"I remember saying, 'We'll stand by you'"* Paul Freeman, Interview, 2002.

*"is on a pretty even keel"* Letter from Freeman to James Watts, August 3, 1955, GWU Archives.

*"She wrote few letters"* Freeman, *Autobiography*, p. 2-6.

265 *"She took progressively less interest"* Ibid., p. 19-13.

*"I dislike the immediate effects"* Ibid., pp. 2-5–2-6.

265 *Her frequent forgetfulness of food* Ibid., pp. 19-10–19-11.

266 *He had to quickly dismiss two inept secretaries* Ibid., p. 21-1.

*"The only reason I continued at Modesto"* Ibid., p. 21-8.

267 *At Atascadero, "I performed lobotomy"* Ibid., p. 21-10.

*"for the [medical] residents to see"* Ibid., p. 21-11.

*"A certain amount of anxiety"* Freeman, 1961, p. 555.

*Freeman arrived at Langley Porter* Freeman, *Autobiography*, p. 21-11.

*partly because his son Paul* Shutts, 1982, p. 233.

268 *lobotomy in any form had disappeared from* Letters to Freeman from O. Hugh Fulcher, August 27, 1956; James Peter Murphy, August 31, 1956; GWU Archives.

*"I had never applied for lobotomies"* Freeman, *Autobiography*, p. 21-2.

*Although Freeman's son Walter* Freeman, 2000.

*President John F. Kennedy* Freeman, *The Psychiatrist*, p. 15.

*Freeman got right to work* Freeman, *Autobiography*, pp. 19-1–19-6.

269 *(He asked for and received)* Letter from Freeman to Paul K. Hennessy, November 27, 1957, GWU Archives.

*"A fellow named Freeman said"* Freeman, *Autobiography*, p. 21-3.

*"a wasteful procedure"* Ibid., p. 21-13.

*(Freeman was fond of relating)* Freeman, *The Psychiatrist*, p. 129.

*"come in the mind of the American public"* Shorter, 1997, p. 174.

270 *he often noted that like himself* Freeman, *The Psychiatrist*, p. 82.

*Freeman sometimes tried to express the mechanics* Freeman, "With Camera and Ice-Pick in Search of the Super Ego" (manuscript), 1960, GWU Archives.

*"Said Kris to the elegant maid"* Freeman, *Autobiography*, p. 21-3.

*"You know, of course"* Letter from David Harold Fink to Freeman, September 29, 1958, GWU Archives.

*"When I visit the large state hospitals"* Freeman, *Autobiography*, p. 14-26.

271 *In 1964 it invited him to treat* Letter from Freeman to Sadao Hirose, December 27, 1964, Paul Freeman collection.

*"Large individual cards with a summary"* Freeman, "Head-and-Shoulder Hunting in the Americas," p. 337.

272 *"are seldom very informative"* Ibid., p. 338.

*"She can read, write and speak"* Letter from patient's mother to Freeman, undated, GWU Archives.

*Another time he set out after a patient* Freeman, "Head-and-Shoulder Hunting in the Americas," p. 337.

273 *One of the very rare patients* Letter from Freeman to "Folks," October 5, 1947, Franklin Freeman collection.

*Freeman once lost track of a patient* Freeman, "Head-and-Shoulder Hunting in the Americas," p. 338.

*In another instance Freeman traced a former* Freeman, Follow-up notes, November 25, 1956, GWU Archives.

*He recalled one who would not board* Freeman, "Head-and-Shoulder Hunting in the Americas," p. 344.

274 *One transorbital lobotomy patient trailed* Freeman, Follow-up report, May 4, 1960, GWU Archives.

*Other times the patient's family* Freeman, "Head-and-Shoulder Hunting in the Americas," p. 344.

*In a 1964 letter to Freeman* Letter from patient's brother to Freeman, May 18, 1964, GWU Archives.

274 *An upset and overtaxed mother* Letter from patient's mother to Freeman, April 29, 1956, GWU Archives.

*Freeman complained that a few* Freeman, "Head-and-Shoulder Hunting in the Americas," p. 341.

*"This 25-year-old white mental"* Letter from Jay L. Hoffman to Freeman, March 7, 1956, GWU Archives.

275 *"All goes well with the boy"* Letter from patient's father to Freeman, December 21, 1955, GWU Archives.

*"They would probably take him oftener"* Letter from John Belisle to Freeman, April 14, 1961, GWU Archives.

*"I am sorry to hear that she is not improving"* Letter from Freeman to patient's mother, February 8, 1956, GWU Archives.

*He told the story of one veteran* Patient records, Beauregard Travis, GWU Archives.

276 *"For the past 15 years"* Freeman, "Head-and-Shoulder Hunting in the Americas," p. 340.

*Freeman enjoyed telling the tale* Ibid., p. 342; "Results of Frontal Lobotomy Psychosurgery" (manuscript), 1968, p. 1, GWU Archives.

*On another occasion in late 1956* Letter from Freeman to A. M. Ornsteen, November 23, 1956, GWU Archives.

277 *In 1957 he received a letter from the husband* Letter from patient's husband to Freeman, February 4, 1957, GWU Archives.

*"Lillian doesn't even wash out"* Letter from Freeman to Hans Wassing, November 23, 1956, GWU Archives.

*Another patient, a ten-year veteran* Freeman, "Results of Frontal Lobotomy Surgery," pp. 5-6, GWU Archives.

*After her lobotomy she returned to the violin* Freeman and Watts, 1950, pp. 250–252.

278 *(Silver's case ended far better)* Feinstein, 1997, pp. 55–58.

*In 1963 Freeman heard from a lobotomized* Letter from Freeman to patient, December 30, 1963, GWU Archives.

*Freeman found other lobotomy patients* Freeman, "Frontal Lobotomy 1936–1956: A Follow-Up Study of 3,000 Patients from One to Twenty Years," p. 3, GWU Archives.

*An early prefrontal lobotomy patient living in* Letter from patient to Freeman, March 13, 1957, GWU Archives.

*Another such patient* Letter to Freeman from patient, May 6, 1956, GWU Archives.

279 *"Things seem to be going nicely"* Letter from Freeman to patient, May 29, 1956, GWU Archives.

*He noted in the file of another patient* Freeman, Follow-up report, April 2, 1956, GWU Archives.

*"I had my teeth pulled"* Christmas card from patient to Freeman, December 10, 1965, GWU Archives.

*Another simply noted* Christmas card from patient to Freeman, 1965, GWU Archives.

*"He enjoyed your cards so much"* Christmas card from patient's sister, 1965, GWU Archives.

*"There is a certain satisfaction"* Freeman, "Head-and-Shoulder Hunting in the Americas," p. 341.

280 *At a 1957 meeting of the Southern* Valenstein, 1986, p. 271.

*"The receipt of a wedding announcement"* Freeman, "Head-and-Shoulder Hunting in the Americas," p. 344.

281 *Many former patients became alcoholics* Freeman, "Frontal Lobotomy 1936–1956: A Follow-Up Study of 3,000 Patients from One to Twenty Years," pp. 9–10, GWU Archives.
*"lobotomy is worthwhile, because it contributes"* Freeman, "Head-and-Shoulder Hunting in the Americas," p. 345.
*In a five-year controlled study* McKenzie, 1961.
*He offered the unverifiable explanation* Freeman, "Results of Frontal Lobotomy Psychosurgery," p. 5, GWU Archives.
*a scholar in her own right* Letter from Freeman to Sadao Hirose, February 4, 1959, Paul Freeman collection.
*"Mrs. Hirose was wearing a tight skirt"* Freeman, *Autobiography*, p. 20-14.

282 *In one article, titled "Bedside Neurology"* Freeman, 1960, p. 491.
*One of his most unusual article topics* Freeman, *Autobiography*, p. 21-5.
*A former intern at Cook County* Lichtenstein, Interview, 2001.

## 13: DECLINE

284 *Freeman divided his life into three* Freeman, *Autobiography*, p. 1-1.

285 *whose own household included a cook* Pressman, 1998, p. 329.
*The tawdry* New York News Desmond, 1947.

286 *"It seems to me that Dr. Rylander's"* Pressman, 1998, p. 329.
*Freeman invited Rylander on a postconference* Freeman, "History of Psychosurgery," p. 6-36, GWU Archives.
*In 1961 anticommunist activist* Meyer, 1961, pp. 80–84.
Suppressed *magazine published* Lamb, 1956, pp. 33–35, 60–61.
*One proponent of this theory* "'Confessions' to Reds Are Laid to Surgery."

287 *These fantasies might have seemed even* Laughlin, 1952.

288 *"A rather wide acquaintance"* Letter from Freeman to Judge J. Philip Perry, June 8, 1954, GWU Archives.
*In one disastrous experiment* Valenstein, 1986, pp. 247–250.

289 *In 1944 surgeons in Hawaii* Shutts, 1982, p. 154.
*Aage Nielssen of Wayne County* Lal, 1947.
*The patient was a thirty-seven-year-old* Shutts, 1982, p. 154.
*Wright's attorney devised a plan* Ibid., pp. 155–156.

290 *"I am sentencing myself to death"* "Burglar Uncured by Operation Kills Himself" (undated and unsourced newspaper clipping), GWU Archives.
*Six months earlier* "Parole Is Denied."
*Yet another case reinforced the belief* "Scholar Is Seized as Yale Murderer"; "Trent-Lyon Committed."
*the abundance of women with diagnoses* Shutts, 1982, p. 257.
*From an all-time high of 559,000* Shorter, 1997, p. 280.
*The use of electroshock therapy* Ibid., p. 284.
*Utah became the first state* Ibid., p. 283.
*In such films as* Grob, 1994, p. 275.

291 *"Medicine Never Looked Better"* Cover, *GW Magazine*, summer 1965, GWU Archives.
*While in Hungary during a trip* Freeman, *The Psychiatrist*, p. 200.
*Neuroscientists amassed evidence* American Psychiatric Association 1984, pp. 257–258.
*"I would say that my father"* Shutts, 1982, p. 259.
*"I believe it's due for adoption"* "D.C. Surgeon Pioneered Lobotomy."

291 *At Doctors' General Hospital* Freeman, "Transorbital Lobotomy at Doctors' General Hospital" (undated manuscript), p. 6, GWU Archives.
*That year during a transorbital* Isay, 2003.

292 *"I think the committee is approaching"* Letter from L. G. McKeever to Freeman, December 13, 1966, GWU Archives.
*Freeman responded with a promise* Letter from Freeman to L. G. McKeever, December 16, 1966, GWU Archives.
*("It's such an absurdly easy")* Paul Freeman, Interview, 2002.
*During one trip to the Sierras* Letter from Freeman to Ruth Donahue, September 9, 1964, GWU Archives.
*Freeman had known one of the patients* Freeman, *Autobiography*, pp. 21-3–21-4.

293 *("Aloysius Church leads the list")* Freeman, *The Psychiatrist*, p. 173.
*"44 Cohens, and 26 Shapiros"* Ibid., p. 175.
*He also examined such magazines* Ibid., p. 160.
*"found that about a third of them"* Ibid., p. 274.

294 *"What bears renewed emphasis"* Ibid., p. 164.
*"As a well-balanced scientific document"* Memorandum of article reviewer, *American Journal of Psychiatry,* undated manuscript, GWU Archives.
*"toned-down"* Letter from Freeman to Walter Lewin, December 20, 1967, GWU Archives.
*Freeman drew a laugh from the packed* Freeman, *Autobiography*, p. 18-11.
*A letter writer to the* American Journal Letter from Joseph Winn to Editor, *American Journal of Psychiatry*, December 7, 1967, GWU Archives.
*Subsequent studies of suicide* Rich, 1980, pp. 261, 263.
*Freeman also studied the frequency* Freeman, "Suicide After Frontal Lobotomy" (undated manuscript), GWU Archives.
*"We'd visit her, and nothing"* Paul Freeman, Interview, 2002.

295 *One morning in late 1967* Freeman, *Autobiography*, p. 19-11.
*By December 1967 he had to admit* Letter from Freeman to Francis Braceland, December 19, 1967, GWU Archives.
*Randy invited his father to accompany* Freeman, *Autobiography*, p. 20-4.
*During the summer of 1967* Ibid., p. 2-13.

296 *"This was a very trying time"* Ibid.
*Soon, however, he was back to hiking* Letter from Freeman to Francis Braceland, December 19, 1967, GWU Archives.
*"The one thing that kept me from an act"* Freeman, *Autobiography*, p. 2-14.
*The idea of a book focusing on the lives* Ibid., p. 22-1.
*"I had high hopes"* Ibid., p. 22-5.

297 *The* Journal of the American Psychiatric Association Ibid., p. 18-12.
*"but punctuationally obnoxious"* Letter from patient to Freeman, January 27, 1968, GWU Archives.
*"Psychiatrists, I think, take themselves"* Freeman, *Autobiography*, p. 22-6.
*At age ninety-one* Letter from Freeman to Vesta Amiden, December 6, 1963, GWU Archives.
*equipped with a file cabinet* Freeman, Dictation, 1968.
*"This was an ambitious project"* Freeman, *Autobiography*, p. 18-8.

298 *"I do need money, Mr. Freeman"* Letter from patient to Freeman, May 20, 1968, GWU Archives.
*To another former patient* Letter from Freeman to patient, June 5, 1968, GWU Archives.

298 *While on the road* Freeman, Dictation, 1968.
*"the Great Man-Hunt of 1968"* Letter from Freeman to James Watts, October 1968, GWU Archives.
*Surviving on meals made up of* Freeman, Dictation, 1968.
*once deviating two hundred miles* Shutts, 1982, p. 241.
*"The neurosurgeons in Seattle"* Paul Freeman, Interview, 2002.
299 *"Deep down inside he suffered"* Shutts, 1982, p. 245.
*After making a weekend stop at McLean Hospital* Beam, 2001, p. 91.
*"On the whole I enjoyed having them"* Freeman, *Autobiography*, p. 18-15.
*Soon after returning to California* Letter from Freeman to James Watts, October 1968, GWU Archives.
*"the foot still itches"* Ibid.
300 *"She is a pleasant faced, soft spoken"* Freeman, Follow-up report, April 19, 1969, GWU Archives.
*"Lura has had her ups and downs"* Letter from Freeman to James Watts, April 26, 1969, GWU Archives.
*"In his records," Watts wrote* James Watts, Untitled and undated manuscript, GWU Archives.
*"I removed the file case from the Cortez"* Freeman, *Autobiography*, p. 18-19.
301 *"I was almost outraged"* Ibid., pp. 16-29–16-30.
*"I suppose I'd like to be remembered"* "Salute to Teaching," p. 3.
*but was bothered by the depiction of his left hand* Freeman, *Autobiography*, p. 10-18.
*"You may be interested to note"* "Salute to Teaching," pp. 2–3.
302 *In just two years a group* Shorter, 1997, p. 300.
*"Most of it was over my head"* Freeman, *Autobiography*, p. 16-31.
303 *"I refused amputation because it was obvious"* Ibid., p. 19-11.
*"We gathered for a graveside service"* Ibid., p. 19-12.
*"was pleased to be relieved of the responsibility"* Ibid., p. 21-7.
*he had trouble reading street signs* Letter from Freeman to Sadao Hirose, December 1971, Paul Freeman collection.
*Wielding his cane in a dapper manner* Lebensohn, 1972, p. 357.
*"I have put it away in my bureau"* Freeman, *Autobiography*, p. 5-22.
*"It is with distress that I write"* Letter from Freeman to Sadao Hirose, February 24, 1972, Paul Freeman collection.
304 *He finished one last paper* Valenstein, 1986, p. 283.
*"REGRET TO INFORM"* Telegram from Lorne Freeman Canter to Sadao Hirose, June 1, 1972, Paul Freeman collection.
*"Working with Walter Freeman"* James Watts, Untitled and undated manuscript, p. 20, GWU Archives.

## 14: GHOST

305 *"The movement's basic argument"* Shorter, 1997, p. 273.
*They had once engaged in an acrimonious* Breggin, 1991, p. 32.
306 *Organic brain disorders, these researchers* Shutts, 1982, pp. 251–252.
*The year after Freeman's death* American Psychiatric Assocation, 1984, p. 269.
*Bans or restrictions on lobotomy* Shutts, 1982, p. 254.
*while the National Commission* American Psychiatric Association, 1984, p. 270.
*The latter group* Shutts, 1982, p. 255.
*In 1978 the U.S. Department of Health* Ibid., p. 258.

307 *In 1975 James Watts was called out of retirement* Letter from Joseph T. Bodell Jr. to James Watts, December 8, 1975, GWU Archives; "Jury Exonerates Brain Surgeon."
*in the case of* Kaimowitz v. Department of Mental Health Shutts, 1982, pp. 252–253.

308 *"It could not be simultaneously true"* Shorter, 1997, p. 146.
*As early as 1948* Ibid., p. 311.
*In one telling study at Camarillo* Ibid., p. 313.
*Twenty-five years later* Ibid., pp. 309–310.

309 *Starting in the 1970s* Ibid., p. 245.
*Prozac, an antidepressant* Ibid., p. 324.
*as well as others in London* Binder, 2000.

310 *"Surgery is then offered as a therapy"* Feldman, "Psychosurgery: A Historical Overview," p. 25.
*Although fewer than three hundred* Feldman, "Contemporary Psychosurgery and a Look to the Future," p. 952.
*"In OCD, for example"* Carey, 2003.
*"He seemed like such an honest"* Isay, 2004.

311 *"In some ways he was a pioneer"* El-Hai, 2001, p. 31.
*Others involved in functional neurosurgery* Carey, 2003.
*"Despite advances in our understanding"* Feldman, "Psychosurgery: A Historical Overview," p. 25.
*In at least one nation* Goldbeck-Wood, 1996.
*"I used to go back"* Paul Freeman, Interview, 2002.

312 *"Put simply"* Pressman, 1998, p. 438.

# Bibliography

"Activity a Safeguard." February 17, 1944. *Kansas City Times.*

American Psychiatric Association. 1984. *The Psychiatric Therapies.* Washington: American Psychiatric Association.

Arnold, William. 1978. *Shadowland.* New York: McGraw Hill.

Aronson, Arnold E. 2000. *Aronson's Neurosciences Pocket Lectures.* San Diego: Singular Publishing Group.

Barton, James W. May 1943. "Surgery Helps Relieve Strain of Mental Pain." *Spokane Chronicle.*

Beam, Alex. 2001. *Gracefully Insane: The Rise and Fall of America's Premier Mental Hospital.* New York: Public Affairs.

Bernstein, Leon L. 1952. Book review, "Psychosurgery." *Bulletin of the Menninger Clinic.*

Berrios, G. E. 1997. "The Origins of Psychosurgery: Shaw, Burckhardt and Moniz." *History of Psychiatry* 8:61–81.

Berryman, Florence S. Janury 13, 1946. "Exhibiton at National Museum Introduces Alfred Jonniaux, Belgian Portrait Painter." *Washington Star.*

Binder, Devin K., and Bermans J. Iskandar. 2000. "Modern Neurosurgery for Psychiatric Disorders." *Neurosurgery* 47 (1):9–23.

Breggin, Peter R. 1991. *Toxic Psychiatry.* New York: St. Martin's.

Brinton, D. G. 1879. *Biographical Dictionary of Physicians.* Philadelphia.

"Body of D.C. Youth Sought Under Falls." July 10, 1946. *Washington News.*

Book review, "Psychosurgery." 1951. *Journal of the International College of Surgeons.*

Book review, "Psychosurgery." 1951. *Medical Annals of the District of Columbia.*

"Brain Operations on 13 Mental Patients." August 20, 1947. *Seattle Post Intelligencer.*

Brody, Eugene B., and Frederick C. Redlich. 1953. "The Response of Schizophrenic Patients to Comic Cartoons Before and After Prefrontal Lobotomy." *Folia Psychiatrica, Neurologica et Neurochirugica Neerlandica* 56 (5):623–635.

Burton, Hal. April 11, 1953. "How to Prevent a Murder (Sometimes)." *Newsday.*

Carey, Benedict. August 4, 2003. "New Surgery to Control Behavior." *Los Angeles Times.*

"Carnegie Hero Fund Commission Case Minute." 1947. Carnegie Hero Fund Commission.

Cheng, Sylvia, Sinclair Tait, and Walter Freeman. 1956. "Transorbital Leucotomy Versus Electroconvulsive Therapy in the Treatment of Mentally Ill Tuberculous Patients." *American Journal of Psychiatry* 113 (1):32–33.

"Cites a Sliver Lining." February 16, 1944. *Kansas City Star.*

Cleveland, David. Book review, "Psychosurgery." *Surgery, Gynecology and Obstetrics.*

"Clumsy with Hand? Blame Front Lobe." January 16, 1948. *Hartford Times.*

Cohen, Charles J. 1922. *Rittenhouse Square: Past and Present.* Privately printed.

Cohen, Lucille. October 16, 1948. "New Surgery Used on 5 Patients." *Seattle Post Intelligencer.*

———. October 16, 1948. "Five Undergo Operation for Mental Cure." *Seattle Post Intelligencer.*

"'Confessions' to Reds Are Laid to Surgery." September 2, 1950. *New York Times*, p. 13.

"Cultural Side of D.C. Doctors Revealed in Artistic Exhibit." November 3, 1948. *Washington Post.*

Dannecker, Harry A. October 1942. "Psychosurgery Cured Me." *Coronet.*

Davey, Lycurgus M. 1998. "John Farquhar Fulton." *Neurosurgery* 43 (1):185–187.

"Dayton Navy Veteran Dies in Futile Attempt to Rescue Boy." July 10, 1946. *Dayton Herald.*

"D.C. Boy, 11, Swept to Death Over 325-Foot Yosemite Falls." July 10, 1946. *Washington Times Herald.*

"D.C. Doctor Helpless as Son Is Swept Over Yosemite Falls." July 10, 1946. *Washington Post.*

"D.C. Residents Cut 15 Cords of Wood in Park Experiment." November 22, 1943. *Washington Star.*

"D.C. Surgeon Pioneered Lobotomy." April 7, 1980. *Washington Post.*

"Death at Yosemite Waterfall." July 10, 1946. *San Francisco Chronicle.*

Desmond, James. December 12, 1947. "Surgeon Claims 'Cure' for Reds." *New York News.*

*Dictionary of American Medical Biography*. 1984. Edited by M. G. Kaufman, Stuart Galishoff, and Todd L. Savitt. Vol. 1. Westport, Conn.: Greenwood Press.

"Died. Antonio Caetano de Abreu Freire Egas Moniz." December 26, 1955. *Time.*

"Do Their Minds Clear?" August 4, 1958. *Newsweek.*

"Dr. Freeman Lectures on Lobotomy." February 27, 1947. *Hartford Daily Courant.*

"Dr. Freeman Lectures on Neurology." January 28, 1944. *Hartford Daily Courant.*

El-Hai, Jack. 1999. "Minnesota in the Age of Lobotomy." *Minnesota Medicine* 82:20–26.

———. February 4, 2001. "The Lobotomist." *Washington Post Magazine*, pp. 16–20, 30–31.

Elithorn, Alick. 1952. Book review, "Psychosurgery." *British Medical Journal.*

Elliot, Edith Farmer. 1978. *Look Back in Love*. Portland, Ore.: Gemaia Press.

Ensz, Gus. October 24, 1948. "Mental Patients Benefit from Revolutionary Operation." *Lincoln Sunday Journal and Star.*

Ewald, Florence, Walter Freeman, and James Watts. 1947. "Psychosurgery: The Nursing Problem." *American Journal of Nursing* 47 (4):210–212.

"Expert Talks to Surgery Candidate." October 17, 1948. *Seattle Post Intelligencer.*

Feinstein, Anthony. 1997. "Psychosurgery and the Child Prodigy: The Mental Illness of Violin Virtuoso Josef Hassid." *History of Psychiatry* viii:55–60.

Feldman, Robert P., Ronald L. Alterman, and James T. Goodrich. 2001. "Contemporary Psychosurgery and a Look to the Future." *Journal of Neurosurgery* 95:944–956.

Feldman, Robert P., and James T. Goodrich. 2001. "Psychosurgery: A Historical Overview." *Neurosurgery* 48 (3):647–659.

Frayer, W. C. 2001. "Doctor Rush's Eye Water and the Opening of the American West." *Survey of Ophthalmology* 46 (2):185–189.

"Freeman Boy to Rest Near Scene of Death." July 18, 1946. *Washington Post.*

Freeman, Franklin. December 2000. Interview.

———. October 7, 2002. Interview.

Freeman, Paul. October 9, 2002. Interview. San Francisco.

Freeman, Paul, and Walter Freeman Jr. December 2000. Interview. San Francisco.

Freeman, Walter. Undated. Autobiography. In *Freeman Watts Collection, George Washington University Archives*. Washington, D.C.

———. Letters, Clippings, Photographs, 1939–1948 (scrapbook). In *Personal Collection of Franklin Freeman.*

———. Undated. Letters. In *Personal Collection of Paul Freeman.*

———. 1928. "Biometrical Studies in Psychiatry: III. The Chances of Death." *American Journal of Psychiatry* 8:425–441.

———. 1931. "Torula Infection of the Central Nervous System." *Journal of Psychology and Neurology* 43:2243–2345.

———. 1933. "Psychological Plagues." In *Freeman Watts Collection, George Washington University Archives*. Washington, D.C.

———. 1941. Letters and Journals. In *Personal Collection of Paul Freeman*.

———. 1948. "Transorbital Lobotomy: Preliminary Report of Ten Cases." *Medical Annals of the District of Columbia* XVII (5):257–261.

———. 1952. "Transorbital Lobotomy: The Problem of the Thick Orbital Plate." *American Journal of Psychiatry* 108 (11):825–827.

———. 1954. "Transorbital Lobotomy in State Mental Hospitals." *Journal of the Medical Society of New Jersey* 51:148–149.

———. 1956. "Egas Moniz (1874–1955): His Life and Work." *American Journal of Psychiatry* 112 (10):769–772.

———. 1957. "The Contributions of Egas Moniz." *Revista Medicina Contemporanea* 147–150.

———. 1958. "Psychosurgery: Present Indications and Future Prospects." *California Medicine* 88:429–434.

———. 1960. "Bedside Neurology." *Northwest Medicine* 491–498.

———. 1961. "Adolescents in Distress: Therapeutic Possibilities of Lobotomy." *Diseases of the Nervous System* xxii (10):555–558.

———. 1962. "West Virginia Lobotomy Project: A Sequel." *Journal of the American Medical Association* 181:1134–1135.

———. 1967. "Multiple Lobotomies." *American Journal of Psychiatry* 123 (11):1450–1452.

———. 1968. *Recorded dictation*.

Freeman, Walter, Hiram W. Davis, Isaac C. East, H. Sinclair Tait, Simon O. Johnson, and Weaver Rogers. 1954. "West Virginia Lobotomy Project." *Journal of the American Medical Association* 156:939–943.

Freeman, Walter, and James Watts. 1936. "Prefrontal Lobotomy in Agitated Depression: Report of a Case." *Medical Annals of the District of Columbia* 5:326–328.

———. 1942. *Psychosurgery: Intelligence, Emotion and Social Behavior Following Prefrontal Lobotomy for Mental Disorders*. Springfield, Ill.: Charles C. Thomas.

———. 1945. "Intelligence Following Prefrontal Lobotomy in Obsessive Tension States." *Archives of Neurology and Psychiatry* 53:244–245.

Freeman, Walter, and Others. 1949. "Proceedings of the First Postgraduate Course in Psychosurgery." *Digest of Neurology and Psychiatry*.

Freeman, Walter J. Photograph Albums of Walter J. Freeman. Yale University Archives.

———. In *Walter J. Freeman and James W. Watts collection*. George Washington University Archives, Washington, D.C.

———. 1933. *Neuropathology: The Anatomic Foundation of Nervous Diseases*. Philadelphia: W. B. Saunders Co.

———. 1948. "Transorbital Leucotomy." *The Lancet*.

———. 1958. Head-and-Shoulder Hunting in the Americas. *Medical Annals of the District of Columbia* xxvii (7):336–345.

———. 1968. *The Psychiatrist: Personalities and Patterns*. New York: Grune & Stratton.

Freeman, Walter Jr. 2000. Letter from Walter Freeman Jr. to Jack El-Hai. In *Jack El-Hai collection*.

———. October 8, 2002. Interview. Berkeley, Calif.

Freeman, Walter J., and James W. Watts. 1950. *Psychosurgery in the Treatment of Mental Disorders and Intractable Pain*. Oxford: Blackwell Scientific Publications.

"Frontal Lobotomy." 1941. *Journal of the American Medical Association.*

Fulton, John Farquhar. Papers of John Farquhar Fulton. Yale University Archives.

———. Diary of John Farquhar Fulton. Minnesota Historical Society.

Gibson, Barbara and Ted Schwarz. 1995. *Rose Kennedy and Her Family: The Best and Worst of Their Lives and Times.* New York: Birch Lane Press.

Goldbeck-Wood, Sandra. 1996. Norway Compensates Lobotomy Victims. *British Medical Journal* 313 (7059): 708.

Goodwin, Doris Kearns. March 23, 1987. "The First Tragedy: Rosemary Kennedy's Ill-Fated Lobotomy." *Washington Post.*

Grob, Gerald N. 1994. *The Mad Among Us: A History of the Care of America's Mentally Ill.* New York: The Free Press.

Gropper, Charles. 1997. "Heart-Sinks of the Rich and Famous." *The Lancet* 349 (9061):1331.

Haseltine, N. S. November 26, 1947. "Brain Surgery Is Successful in Easing Pain." *Washington Post.*

Hebb, Donald. 1951. Book review, "Psychosurgery." *Journal of Abnormal and Social Psychology.*

Henry, Thomas R. November 20, 1936. "Brain Operation by D.C. Doctors Aids Mental Ills." *Washington Evening Star.*

———. November 30, 1936. "Southern Doctors." *Time.*

———. April 19, 1942. "Psychosurgery." *Washington Star.*

Herner, Torsten. 1961. *Treatment of Mental Disorders with Frontal Stereotaxic Thermo-Lesions: A Follow-Up Study of 116 Cases.* Copenhagen: Ejnar Munksgaard.

Hirose, Sadao. 1963. "Orbito-Ventromedial Undercutting, 1957–1963: Follow-Up Study of 77 Cases." *American Journal of Psychiatry* 121 (12).

Horwitz, Norman H. 1998. "John F. Fulton." *Neurosurgery* 43 (1):178–184.

"Ill Men's Faces, Filmed, Unravel Brain Maladies." June 8, 1931. *New York Herald Tribune.*

Isay, David. 2003. Interview with Larry, former Herrick Memorial Hospital staff member.

———. 2004. Interview with Ellen Ionesco and Angelene Forester.

James, W. W. Keen. 2002. *Keen of Philadelphia: The Collected Memoirs.* Dublin, N.H.: William L. Bauhan.

Joanette, Yves, Brigitte Stemmer, Gil Assal, and Harry Whitaker. 1993. "From Theory to Practice: The Unconventional Contribution of Gottlieb Burckhardt to Psychosurgery." *Brain and Language* 45:572–587.

Jones, Charles H., and James G. Shanklin. 1948. "Transorbital Lobotomy: Preliminary Report of Forty-One Cases." *Northwest Medicine* 47 (6).

Kaempffert, Waldemar. May 24, 1941. "Turning the Mind Inside Out." *Saturday Evening Post.*

———. January 11, 1942. Psychosurgery. *New York Times.*

Kandela, Peter. 1999. Antisepsis. *The Lancet* 353 (9156):937.

Keen, William Williams. 1924. *Everlasting Life: A Creed and a Speculation.* Philadelphia: J. B. Lippincott Co.

Kelly, Gerald. 1957. *Medico-Moral Problems.* St. Louis, Mo.: The Catholic Hospital Association.

Lal, G. B. 1947. "Reformed by Brain Surgery." *American Weekly,* March 9, 1947.

Lamb, Peter. 1956. "Suppressed Lobotomy: Commies' Secret for World Domination." *Suppressed* 3 (5).

Laughlin, Henry P. 1952. "Some Areas of Psychiatric Interest." Psychological Strategy Board, Central Intelligence Agency.

Laurence, William L. June 7, 1937. "Surgery Used on the Soul-Sick: Relief of Obsessions Is Reported." *New York Times,* pp. 1, 10.

Leamer, Laurence. 2001. *The Kennedy Men, 1901–1963: The Laws of the Father*. New York: William Morrow.

Lebensohn, Zigmond M. 1972. "In Memoriam: Walter Freeman 1895–1972." *American Journal of Psychiatry* 129 (3):356–357.

Lerch, Walter. November 29, 1948. "'Ice-Pick' Pain Cure Described." *Cleveland Plain Dealer*.

Lichtenstein, Robert. 2001. Interview, December 2001.

Ligon, Lee B. 1998. "The Mystery of Angiography and the 'Unawarded' Nobel Prize: Egas Moniz and Hans Christian Jacobaeus." *Neurology* 43 (3):602–611.

Macmillan, Malcolm. 2000. *An Odd Kind of Fame: Stories of Phineas Gage*. Cambridge, Mass.: MIT Press.

McDonough, Stephen J. June 6, 1941. "Brain Surgery Is Credited with Cure of 50 'Hopelessly' Insane Persons." *Houston Post*.

———. 1942. "Patient Even Converses." *Concord Record*.

McKenzie, K. G., and G. Kaczanowski. 1961. "Prefrontal Leukotomy: A Five-Year Controlled Study." *Canadian Medical Association Journal* 91:1192–1196.

"Medicine." June 22, 1931. *Time*.

Menninger, Karl A. 1988. *Selected Correspondence of Karl A. Menninger, 1919–1945*, edited by H. J. Faulkner and V. D. Pruitt. New Haven: Yale University Press.

"Mentally Sick Patients Aided by Brain Surgery." February 3, 1947. *Boston Traveller*.

Meyer, Frank S. 1961. *The Moulding of Communists: The Training of the Communist Cadre*. New York: Harcourt, Brace and Co.

Moniz, Egas. 1948. "The Discovery of Lobotomy." In *The Age of Madness*, edited by T. Szasz. New York: Anchor.

"Mrs. T. D. Hammatt, Former Topekan, Dies in Washington." Unknown.

"Mystery in Hammatts Dual Death." Unknown.

"New and Delicate Lobotomy Performed for First Time." November 2, 1948. Unknown.

"New $100,000 Surgery Unit for Institute." February 27, 1947. *Hartford Courant*.

"New Technique in Delicate Brain Operation Shown Here." November 18, 1948. *Paterson Evening News*.

Nuttin, Bart, Loes Gabriels, Paul Cosyns, and Jan Gybels. 2000. "Electrical Stimulation of the Brain for Psychiatric Disorders." *CNS Spectrums* 5 (11):35–39.

"One Hundred and Fourth Annual Meeting of the American Psychiatric Association." Paper read at One Hundred and Fourth Annual Meeting of the American Psychiatric Association, May 17–20, 1948.

"Oxygen Treatment Found Insanity Aid." June 12, 1931. *New York Times*.

"Parole Is Denied." January 16, 1952. *New York Times*, p. 12.

Partridge, Maurice. 1950. *Pre-Frontal Leucotomy: A Survey of 300 Cases Personally Followed Over 1½–3 Years*. Oxford: Blackwell Scientific Publications.

Patrick, William C. September 29, 1951. "Brain Surgery Progress Noted at Meet." *Salt Lake Tribune*.

Patterson, George A. November 23, 1948. "Psycho-Surgery Is Effective Here: 8 Leave Hospitals, Majority Free of Fear Complex." *Paterson Evening News*.

Peery, Thomas M. 1979. Interview with James Winston Watts.

"Personality Shift Is Laid to Surgery." December 14, 1947. *New York Times*.

Porteus, D. March 9, 1944. "Neutralizing the Worry Center." *Honolulu Star-Bulletin*.

Pressman, Jack D. 1998. *Last Resort: Psychosurgery and the Limits of Medicine*. Cambridge, Mass.: Cambridge University Press.

"Psychosurgery." 1942. *Journal of the American Medical Association*.

"Rare Operation Used on 2 Berkeley Women by Noted Brain Surgeon." November 2, 1948. Unknown.

Rasmussen, Steven, Benjamin Greenberg, Per Mindus, Gerhard Friehs, and George Noren. 2000. "Neurosurgical Approaches to Intractable Obsessive-Compulsive Disorder." *CNS Spectrums* 5 (11):23–26.

Ratey, John J. 2000. *A User's Guide to the Brain*. New York: Vintage Books.

Rensberger, Royce. May 9, 1967. "Psychiatrists Top the List as Potential Suicide Cases." *Detroit Free Press*.

Rezai, Ali R. 2000. "Surgery for Psychiatric Disorders: Part Two." *CNS Spectrums* 5 (11):20.

Rich, Charles L., and Ferris N. Pitts Jr. 1980. "Suicide by Psychiatrists: A Study of Medical Specialists Among 18,730 Consecutive Physician Deaths During a Five-Year Period, 1967–72." *Journal of Clinical Psychiatry* 41:261–263.

Robinson, Mary Frances. 1950. "Personality Changes After Psychosurgery." In *Psychosurgery in the Treatment of Mental Disorders and Intractable Pain*, edited by Walter Freeman and James W. Watts. Oxford: Blackwell Scientific Publications.

Robinson, Mary Frances, and Walter Freeman. 1954. *Psychosurgery and the Self*. New York: Grune & Stratton.

Rosenbloom, Michael. 2002. "Chlorpromazine and the Psychopharmocologic Revolution." *Journal of the American Medical Association* 287:1860–1861.

Rovit, Richard L., and William T. Couldwell. 2002. "A Man for All Seasons: W. W. Keen." *Neurosurgery* 50 (1):181–190.

Rylander, Gösta. 1973. "The Renaissance of Psychosurgery." In *Surgical Approaches in Psychiatry*, edited by L. V. Laitinen and K. E. Livingston. Lancaster, Penn.: Medical and Technical Publishing Co. Ltd. pp. 3–7.

"Salute to Teaching." Spring 1970. *GW Medicine*.

Sargant, William. 1967. *The Unquiet Mind*. Boston: Little, Brown & Co.

Scheflin, Alan W., and Edward M. Opton Jr. 1978. *The Mind Manipulators*. New York: Paddington Press.

"Scholar Is Seized as Yale Murderer." October 10, 1950. *New York Times*, p. 26.

Sessler, Betty. 1952. "Doors Leading Back." *Richmond Times Dispatch*, January 27, 1952.

Shorter, Edward. 1997. *A History of Psychiatry: From the Era of the Asylum to the Age of Prozac*. New York: John Wiley & Sons.

Shutts, David. 1982. *Lobotomy: Resort to the Knife*. New York: Van Nostrand Reinhold Co.

Slaughter, Frank G. 1958. *Daybreak*. Garden City, N.Y.: Doubleday & Co.

"Surgery Advocated for Schizophrenia." January 27, 1944. *Hartford Times*.

"Surgery for Insanity?" April 14, 1952. *Newsweek*.

"Trent-Lyon Committed." November 17, 1950. *New York Times*, p. 41.

Untitled advertisement. November 16, 1960. *Stockton Record*.

Valenstein, Elliot S. 1986. *Great and Desperate Cures: The Rise and Decline of Psychosurgery and Other Radical Treatments for Mental Illness*. New York: Basic Books.

Verhandlungen des X. *Internationalen Medicinischen Congresses*. 1891. Berlin: August Hirschwald.

Wallace, Irving. October 20, 1951. "The Operation of Last Resort." *Saturday Evening Post*.

———. 1965. *The Sunday Gentlemen*. New York: Bantam Books.

"Walter Jackson Freeman." In *Federal Bureau of Investigation file*.

"Waterfalls Kill 2 in 325-Foot Drop." July 10, 1946. *Philadelphia Inquirer*.

Watts, James. 1965. "Neurology and Neurological Surgery: Its Story at George Washington University." *Medical Annals of the District of Columbia* 34 (5):225–228.

# INDEX